THE GREAT MORTALITY

An Intimate History
of the Black Death,
the Most Devastating Plague of All Time

JOHN KELLY

NEW YORK • LONDON • TORONTO • SYDNEY

HARPER PERENNIAL

A hardcover edition of this book was published in 2005 by HarperCollins Publishers.

P.S.™ is a trademark of HarperCollins Publishers.

FIRST HARPER PERENNIAL EDITION PUBLISHED 2006.

Designed by Joy O'Meara
Map by Springer Cartographics LLC

The Library of Congress has catalogued the hardcover edition as follows:

Kelly, John.
The great mortality : an intimate history of the Black Death, the most devastating plague of all time / John Kelly—1st ed.
p. cm.
Includes bibliographical references and index.
ISBN 0-06-000692-7
1. Black Death—History. I. Title.

RC172.K445 2005
614.5'732—dc22 2004054213

ISBN-10: 0-06-000693-5 (pbk.)
ISBN-13: 978-0-06-000693-8 (pbk.)

21 ❖/LSC 23

Praise for *The Great Mortality*

Editors' recommendation, *San Francisco Chronicle*

"John Kelly's *Great Mortality* is timely and welcome. . . . It conveys in excruciating but necessary detail a powerful sense of just how terribly Europe suffered."
—Jonathan Yardley, *Washington Post*

"The images could have come from one of Hieronymus Bosch's nightmarish paintings of hell: dusty roads with frightened refugees . . . dogs and rats running wild on deserted streets; fields littered with dead bodies. . . . John Kelly gives the reader a ferocious, pictorial account of the horrific ravages of [the] plague."
—Michiko Kakutani, *New York Times*

"There is a seething vitality here. . . . [The book] leaves an optimistic impression of the human race as a kind of hardy bacteria—at once tiny and fragile but also determined to thrive and very, very hard to erase." —*Minneapolis Star Tribune*

"Splendidly written." —*Philadelphia Inquirer*

"While Kelly combines distinguished scholarship in science, medicine, and European history, other capabilities earn *Mortality* a place in plague literature with seventeenth and eighteenth century eyewitness accounts, respectively, in Samuel Pepys's diary and Daniel Defoe's *A Journal of the Plague Year*. Kelly meets the need for verisimilitude faced by all serious historians by confronting some of the world's darkest days as if he were a forensic sleuth who must re-create the ambience of the victims' world before tracking down their deaths."
—*Houston Chronicle*

"A compellingly vivid account. Kelly's prose moves with brushfire speed—often as exciting as a first-class TV drama-documentary." —*Guardian* (London)

"Kelly is a reliable guide, with a gift for providing racy and vivid background for those who know nothing of the Middle Ages."
—*Independent on Sunday* (London)

"Kelly, the author of numerous nonfiction works . . . skillfully draws imaginative word pictures, for instance . . . Dorset is full of 'brooding Heathcliff vistas.' The selling of Papal indulgences turns the medieval church into a 'spiritual Pez dispenser.'" —*JAMA: The Journal of the American Medical Association*

"Kelly's narrative offers us an intimate exploration of a world falling apart. It's a world that, sadly, feels a lot like our own in its fears of an impending apocalypse from mysterious forces." —*San Francisco Chronicle*

"John Kelly's *Great Mortality* is one of those works that proves history can be a wonderful read and not merely a dry recounting of events and dates."
—*Cleveland Plain Dealer*

"*The Great Mortality* makes the pestilence personal, from the royals to the rowdies."
—Associated Press

"*The Great Mortality* skillfully draws on eyewitness accounts to construct a journal of the plague years."
—*New York Times Book Review*

"A fascinating account of the plague. A frightening reminder of what could happen today."
—Nelson DeMille, *Birmingham News*

"The Black Death is history's best-known pandemic, but until now its full history has not been written. In *The Great Mortality*, John Kelly gives a human face to the fourteenth-century disaster that claimed seventy-five million lives—a third of the world's population."
—*Oakland Tribune*

"A compelling and bone-chilling account."
—*Tampa Tribune*

"Compelling. Explores both the human and scientific dimensions of the tragedy with skill and aplomb."
—*Dallas Morning News*

"This sweeping, viscerally exciting book contributes to a literature of perpetual fascination."
—*Booklist* (starred review)

"A ground-level illustration of how the plague ravaged Europe . . . putting a vivid, human face on an unimaginable nightmare."
—*Kirkus Reviews*

"Accessible and engrossing."
—*Publishers Weekly*

"A compelling and eminently readable portrait."
—*Library Journal*

"*The Great Mortality* is a chilling account of a global siege: public pits, death carts, silent villages, and empty streets."
—*Charleston Post and Courier*

"Written with a keen eye for the details of the past, it might also be a warning about our future."
—Jack Weatherford, DeWitt Wallace Professor of Anthropology at Macalester College and author of *Genghis Khan and the Making of the Modern World*

"Rich and evocative . . . written in the tradition of Barbara Tuchman."
—Richard Preston, author of *The Hot Zone*

For Suzanne, Jonathan, and Sofiya—

to a future without plague

CONTENTS

Probable Routes of Transmission
Possible Routes of Transmission
0 250 Miles
0 250 Km
Bergen
Oslo
Sweden
Norway
Stockhol
North Sea
Scotland
Goteborg
Edinburgh
Copenhagen
Ireland
Dublin
York
Chester
England
Hamburg
Norwich
Bristol
London
Antwerp
Poznan
Bruges
Ghent
Holy
Weymouth
Calais
Brussels
Cologne
Leipzig
Atlantic
Ocean
Rouen
Frankfurt
Prague
Caen
Mainz
Nuremberg
Paris
Roman
Brno
Strasbourg
Nantes
France
Vienna
Munich
Bay
of
Biscay
Basel
Zurich
Innsbrück
Empire
Geneva
Lyon
Bordeaux
Milano
Venice
Aquitaine
Genoa
Papal
Avignon
Porto
Valladolid
Marseille
Toulon
Pisa
Florence
Sienna
Split
States
Perugia
Portugal
Madrid
Barcelona
Corsica
Rome
Adriatic Sea
Toledo
Aragon
Kingdom
Lisbon
Castile
of
Valencia
Naples
Sardinia
Naples
Cordoba
Majorca
Tyrrhenian
Sea
Granada
Gibraltar
Almeria
Mediterranean Sea
Messina
Ceuta
Algiers
Kingdom
Tunis
of
Sicily
Africa
Tripoli

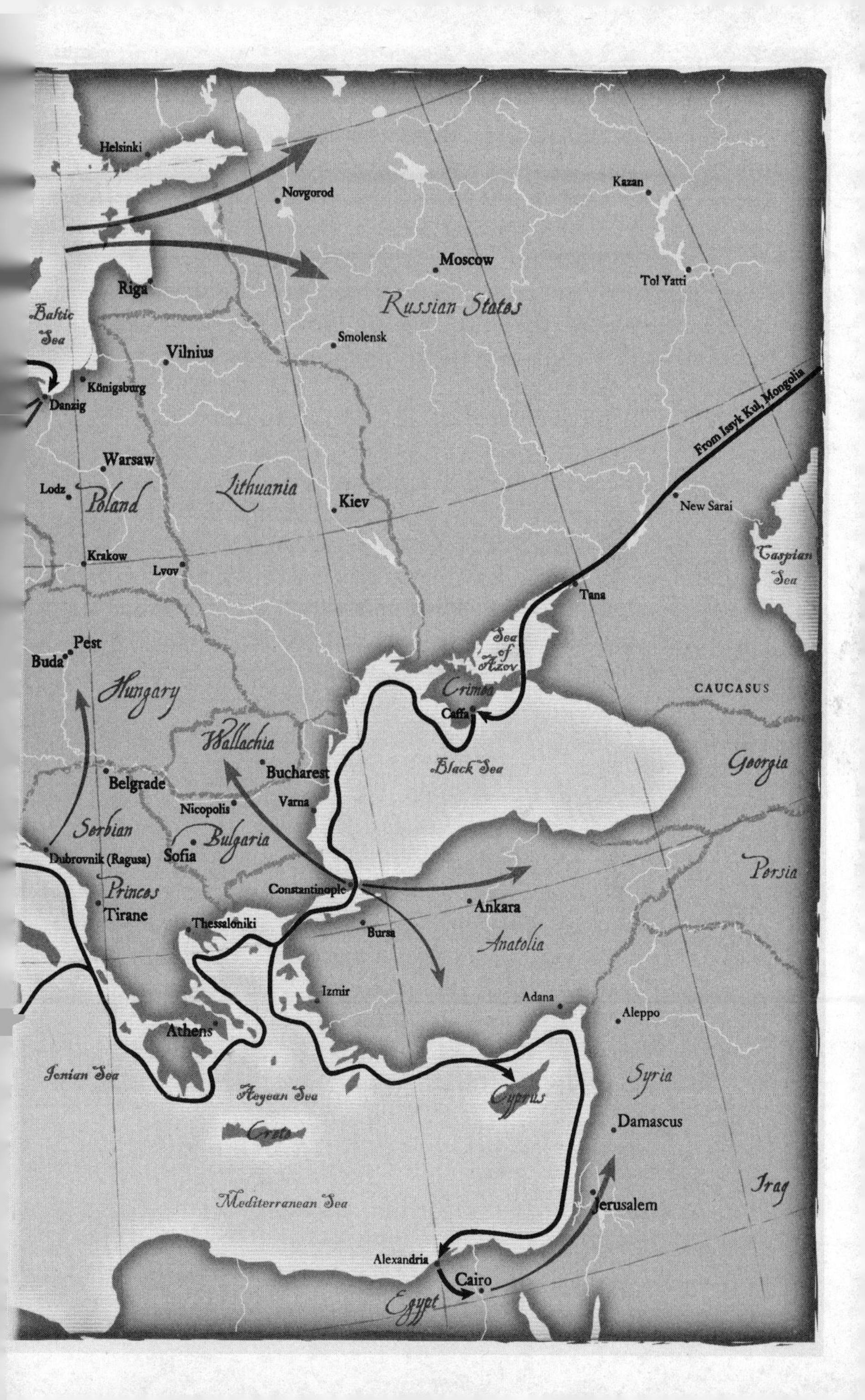

Helsinki
Novgorod
Kazan
Moscow
Tol Yatti
Riga
Russian States
Baltic Sea
Smolensk
Vilnius
Königsburg
Danzig
From Issyk Kul, Mongolia
Warsaw
Lodz
Poland
Lithuania
Kiev
New Sarai
Krakow
Lvov
Caspian Sea
Tana
Pest
Buda
Sea of Azov
Hungary
Crimea
CAUCASUS
Caffa
Wallachia
Bucharest
Black Sea
Georgia
Belgrade
Varna
Nicopolis
Serbian
Bulgaria
Sofia
Dubrovnik (Ragusa)
Persia
Princes
Constantinople
Tirane
Ankara
Thessaloniki
Bursa
Anatolia
Izmir
Adana
Aleppo
Athens
Ionian Sea
Syria
Aegean Sea
Cyprus
Crete
Damascus
Iraq
Mediterranean Sea
Jerusalem
Alexandria
Cairo
Egypt

INTRODUCTION

THIS BOOK BEGAN AS AN INQUIRY INTO THE FUTURE AND ENDED as an investigation of the past.

Five years ago, when I first began thinking about writing a book on the plague, I had in mind plague in the generic sense of a major outbreak of epidemic disease, and I was looking ahead to the twenty-first century, not backward toward the fourteenth. From an earlier book on experimental medicine, *Three on the Edge: The Story of Ordinary Families in Search of a Medical Miracle,* I had gotten a glimpse of the power of unfettered pandemic disease. One of the people I wrote about was an AIDS patient. During the two years I followed him in the early 1990s—a time when effective HIV treatments were rare—the man lost a former lover, three friends, and a colleague at work. A hallmark of pandemic disease is its ability to destroy worlds, not just individuals, but it was one thing to know that, quite another to witness it.

In 1995, the year I finished the book, Jonathan Mann, a professor at the Harvard University School of Public Health, warned that AIDS was the beginning of a frightening new era. "The history of our time will be marked by recurrent eruptions of newly discovered diseases," he predicted. Two years later, 1997, the *New England Journal of Medicine* issued a similar warning when a virulent new strain of bubonic plague was identified. "The finding of multi-drug-resistant [strain of plague] reinforces the concern . . . that the threat from emerging infectious diseases is not to be taken lightly," declared the *Journal*. Scarcely twenty years after the eradication of smallpox in 1979, an accomplishment widely described in the media as humanity's ultimate triumph over infectious disease, we seemed to be slipping back into the world of our ancestors, a world of sudden, swift, and uncontrollable outbreaks of epidemic illness. The book I was thinking about would explore the nature of this threat, and in particular the danger posed by newly emerging illnesses such as Ebola fever, Marburg disease, hantavirus pulmonary syndrome, SARS, and avian flu.

The book I ended up writing is quite different, though, indirectly, it rehearses many of the same themes. Its subject is an outbreak of a particular infectious disease in a particular time and place. Seven hundred years after the fact, what we call the Black Death—and what medieval Europeans called the Great Mortality, and medieval Muslims, the Year of Annihilation—remains the greatest natural disaster in human history.

Apocalyptic in scale, the Black Death affected every part of Eurasia, from the bustling ports along the China Sea to the sleepy fishing villages of coastal Portugal, and it produced suffering and death on a scale that, even after two world wars and twenty-seven million AIDS deaths worldwide, remains astonishing. In Europe, where the most complete figures are available, in many places the plague claimed a third of the population; in others, half the population; and in a few regions, 60 percent. The affliction was not limited to humans. For a brief moment in the middle of the fourteenth century, the words of

Genesis 7—"All flesh died that moved upon the earth"—seemed about to be realized. There are accounts of dogs, cats, birds, camels, even of lions being afflicted by the "boil," the telltale bubo of bubonic plague. By the time the pestilence ended, vast stretches of the inhabited world had fallen silent, except for the sound of the wind rustling through empty, overgrown fields.

What led me from the book I planned to the book I wrote was an encounter with the literature of the Black Death. Before plunging into a book about the future of epidemic disease, I wanted to acquaint myself with its past, and the medieval plague, as the most famous example of the phenomenon, seemed the appropriate place to start. Thus, in the autumn of 2000 I began shuttling between the main reading room of the New York Public Library and Butler Library at Columbia University. I read a number of excellent academic histories, but it was the original source material, the literature of the Great Mortality—the chronicles, letters, and reminiscences written by contemporaries—that turned my gaze from the future to the past. I had approached this material with some trepidation. If, as an English writer once observed, the past is a foreign country, no part of the past seems more foreign, more "otherly," to a modern sensibility than the Middle Ages.

My wariness proved unwarranted. Much has changed since the 1340s, the decade the Black Death arrived in Europe, but not human nature. The plague generation wrote about their experiences with a directness and urgency that, seven hundred years after the fact, retains the power to move, astonish, and haunt. After watching packs of wild dogs paw at the newly dug graves of the plague dead, a part-time tax collector in Siena wrote, "This is the end of the world." His contemporaries provided vivid descriptions of what the end of the world looked like, circa 1348 and 1349. It was corpses packed like "lasagna" in municipal plague pits, collection carts winding through early-morning streets to pick up the previous day's dead, husbands abandoning dying wives and parents abandoning dying children—for fear of contagion—and knots of people crouched over latrines and sewers inhaling the noxious fumes in hopes of inoculating themselves against the plague. It was dusty roads packed with panicked refugees, ghost ships crewed by corpses, and a feral child running wild in a deserted mountain village. For a moment in the middle of the fourteenth century, millions of people across Eurasia began to contemplate the end of civilization, and with it perhaps the end of the human race.

The medieval plague was one of the seminal events of the past millennium. It cast a deep shadow across the centuries that followed, and it remains part of the collective memory of the West. In discussions of AIDS and other emerging infectious diseases, the Black Death is constantly evoked as a warning from the past, something humanity must avoid at all costs. Yet, as a historical event, the medieval plague remains little known. It originated in inner Asia, somewhere in the still-remote region between Mongolia and Kirgizia, and might well have remained there, had not the Mongols unified much of Eurasia in the thirteenth century, thereby facilitating the growth of three activities that continue to play an important role in the spread of infectious disease: trade, travel, and larger and more efficient communications. In the Mongols' case, the improvement in communication was the Yam, the Tartar version of the pony express. From inner Asia, one prong of the plague swept eastward into China, another westward across the steppe into Russia. Sometime in the mid-1340s the western wing reached Italian trading outposts in Crimea; from there, fleeing sailors carried the disease to Europe and the Middle East.

Despite the enormous size of Eurasia and the slowness of medieval travel—in 1345 it took eight to twelve months to travel from the Crimea to China overland—the plague spread to almost every corner of the continent within a matter of decades. It seems to have erupted in inner Asia sometime in the first quarter of the fourteenth century, and by the autumn of 1347 it was in Europe. In late September Sicily was infected by a fleet of Genoese galleys. One chronicler reported that the Genoese tumbled off the infected ships with "sickness clinging to their very bones." From Sicily the pestilence moved swiftly northward to continental Europe. By March of 1348 much of central and northern Italy was or soon would be infected, including Genoa, Florence, and Venice, where gondolas slipped through wintery canals to collect the dead; by spring the plague was in Spain, southern

France, and the Balkans, where one contemporary reports that packs of wolves came down from the hills to attack the living and feed on the dead. By summer the disease had reached northern France, England, and Ireland, where it took a great toll among the English in port towns like Dublin, but left the native Irish in the hills relatively unscathed. By the late autumn of 1348 the plague was in Austria and menacing Germany, and by 1349 and 1350 it had infected the regions along the European periphery: Scotland, Scandinavia, Poland, and Portugal. In a century when nothing moved faster than the fastest horse, the Black Death had circumnavigated Europe in a little less than four years.

To many Europeans, the pestilence seemed to be the punishment of a wrathful Creator. In September 1349, as the disease raced toward an anxious London, the English king Edward III declared that "a just God now visits the sons of men and lashes the world." To many others, the only credible explanation for death on so vast a scale was human malfeasance. Evildoers were using poisons to spread the plague, warned Alfonso of Cordova. To many of Alfonso's contemporaries, that could only mean one thing. The Great Mortality occasioned the most violent outburst of anti-Semitism in the Middle Ages, a period already marked by violent anti-Semitic outbursts.

Few events in history have evoked such extremes of human behavior as the medieval plague. There was the horrendous brutality of the Flagellants, who trampled the roads of Europe, flailing their half-naked bodies and murdering Jews, and the sweet selflessness of the sisters of Hôtel-Dieu, who sacrificed their own lives to care for the plague victims of Paris. There was the fearfulness of Pope Clement VI, who fled Avignon, seat of the medieval Church, a few months after the pestilence arrived, and the fearlessness of his chief physician, Gui de Chauliac, who stayed in the afflicted city until the bitter end, so as "to avoid infamy." The Great Mortality produced many examples of great cunning and compassion, charity and greed, and, in a testimonial to the stubbornness—some would say the irredeemable wickedness—of human character, it also provided a backdrop for both the most notorious royal murder trial and greatest comic opera of the Middle Ages, the former featuring, as defendant, the beautiful Queen Joanna of Naples; the latter, the Roman tribune Cola di Rienzo, possibly the silliest man in Europe. Throughout the worst months of the plague, somewhere in Europe men were always waging war on one another.

The Great Mortality attempts to bring alive the world described in the letters, chronicles, and reminiscences of contemporaries. It is a narrative of a supreme moment in human history told through the voices, personalities, and experiences of the men and women who lived through it. However, since it is impossible to understand the pestilence without understanding its historical context, it is also a book about a time as well as an event. As the British historian Bruce Campbell has observed, the decades preceding the plague were "exceptionally hazardous and unhealthy for both humans and domesticated animals." Almost everywhere in Europe there was war, overpopulation (relative to resources), economic stagnation and decline, filth, overcrowding, epidemic (nonplague) disease, and famine, as well as climatic and ecological instability.

In retrospect, contemporaries interpreted the reign of woe as a portent of the coming apocalypse, and in a sense it was. The economic and social conditions of the early fourteenth century, and the environmental instability of the period, made Europe an unhealthy place to live.

The Great Mortality also looks at new theories about the nature of the medieval plague. For well over a century, it has been considered settled fact that the pestilence was a catastrophic outbreak of bubonic plague and a variant called pneumonic plague, which attacks the lungs. However, since in modern settings neither form of plague looks or acts much like the illness described in the Black Death literature, a number of historians and scientists have recently begun to argue that the mortality was caused by a different infectious illness, perhaps anthrax, perhaps an Ebola-like ailment.

While history may never repeat itself, "man," as Voltaire once observed, "always does." The factors that allowed the Black Death to escape the remoteness of inner Asia and to savage the cities of medieval Europe, China, and the Middle East still operate. Except, of course, today they operate on a vastly larger scale. Trade and human expansion, the twin anthems of modern globalization, have opened up ever more remote areas of the globe, while transportation has enhanced the mobility of both men and microbes incalculably. A journey that took the plague bacillus decades to complete in the fourteenth century today takes barely a day. And for all the triumphs of modern science, infectious disease retains the ability to render us as impotent as our medieval ancestors.

In the spring of 2001 English journalist Felicity Spector was reminded of that fact when an epidemic of hoof-and-mouth disease spun out of control and the British government, helpless to contain the outbreak, was forced to resort to methods that made Britain seem "suddenly, shockingly medieval." We thought, wrote Spector in the *New York Times,* that "modern medicine had brought us beyond the days when the only solution to an infectious disease was to burn entire herds of livestock, to close vast swaths of countryside, to soak rags and spread them on the roads. . . . [M]odernity, it seems, is a very fragile thing."

The Great Mortality

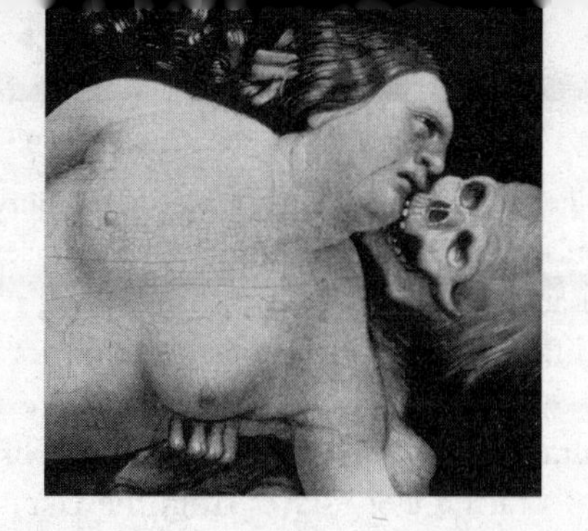

CHAPTER ONE

Oimmeddam

FEODOSIYA SITS ON THE EASTERN COAST OF THE CRIMEA, A REC-tangular spit of land where the Eurasian steppe stops to dip its toe into the Black Sea. Today the city is a rusty wasteland of post-Soviet decay. But in the Middle Ages, when Feodosiya was called Caffa and a Genoese proconsul sat in a white palace above the harbor, the city was one of the fastest-growing ports in the medieval world. In 1266, when the Genoese first arrived in southern Russia, Caffa was a primitive fishing village tucked away far from the eyes of God and man on the dark side of the

Crimea—a collection of windswept lean-tos set between an empty sea and a ring of low-rising hills. Eighty years later, seventy thousand to eighty thousand people coursed through Caffa's narrow streets, and a dozen different tongues echoed through its noisy markets. Thrusting church spires and towers crowded the busy skyline, while across the bustling town docks flowed Merdacaxi silks from Central Asia, sturgeon from the Don, slaves from the Ukraine, and timber and furs from the great Russian forests to the north. Surveying Caffa in 1340, a Muslim visitor declared it a handsome town of "beautiful markets with a worthy port in which I saw two hundred ships big and small."

It would be an exaggeration to say that the Genoese willed Caffa into existence, but not a large exaggeration. No city-state bestrode the age of city-states with a more operatic sense of destiny—none possessed a more fervent desire to cut a *bella figura* in the world—than Genoa. The city's galleys could be found in every port from London to the Black Sea, its merchants in every trading center from Aleppo (Syria) to Peking. The invincible courage and extraordinary seamanship of the Genoese mariner was legendary. Long before Christopher Columbus, there were the Vivaldi brothers, Ugolino and Vadino, who fell off the face of the earth laughing at death as they searched for a sea route to India. Venice, Genoa's great rival, might carp that she was "a city of sea without fish, . . . men without faith, and women without shame," but Genoese grandeur was impervious to such insults. In Caffa, Genoa built a monument to itself. The port's sunlit piazzas and fine stone houses, the lovely women who walked along its quays with the brocades of Persia on their backs and the perfumes of Arabia gracing their skin, were monuments to Genoese wealth, virtue, piety, and imperial glory.

As an Italian poet of the time noted,

And so many are the Genoese
And so spread . . . throughout the world
That wherever one goes and stays
He makes another Genoa there.

Caffa's meteoric rise to international prominence also owed something to geography and economics. Between 1250 and 1350 the medieval world experienced an early burst of globalization, and Caffa, located at the southeastern edge of European Russia, was perfectly situated to exploit the new global economy. To the north, through a belt of dense forest, lay the most magnificent land route in the medieval world, the Eurasian steppe, a great green ribbon of rolling prairie, swaying high grass, and big sky that could deliver a traveler from the Crimea to China in eight to twelve months. To the west lay the teeming port of Constantinople, wealthiest city in Christendom, and beyond Constantinople, the slave markets of the Levant, where big-boned, blond Ukrainians fetched a handsome price at auction. Farther west lay Europe, where the tangy spices of Ceylon and Java and the sparkling diamonds of Golconda were in great demand. And between these great poles of the medieval world lay Caffa, with its "worthy port" and phalanx of mighty Russian rivers: the Volga and Don immediately to the east, the Dnieper to the west. In the first eight decades of Genoese rule the former fishing village doubled, tripled, and quadrupled in size. Then the population quadrupled a second, third, and fourth time; new neighborhoods and churches sprang up; six thousand new houses rose inside the city, and then an additional eleven thousand in the muddy flats beyond the town walls. Every year more ships arrived, and more fish and slaves and timber flowed across Caffa's wharves. On a fine spring evening in 1340, one can imagine the Genoese proconsul standing on his balcony, surveying the tall-masted ships bobbing on a twilight tide in the harbor, and thinking that Caffa would go on growing forever, that nothing would ever change, except that the city would grow ever bigger and wealthier. That dream, of course, was as fantastic a fairy tale in the fourteenth century as it is today. Explosive growth—and human hubris—always come with a price.

Before the arrival of the Genoese, Caffa's vulnerability to ecological disaster extended no farther than the few thousand meters of the Black Sea its fishermen fished and the half moon of sullen, windswept hills behind the city. By 1340 trade routes linked the port to places half a world away—places even the Genoese knew little about—and in some of the places strange and terrible things were happening. In the 1330s there were reports of tremendous environmental upheaval in China. Canton and Houkouang were said to have been lashed by cycles of torrential rain and parching drought, and in Honan mile-long swarms of locusts were reported to have blacked out the sun. Legend also has it that in this period, the earth under China gave way and whole villages disappeared into fissures and cracks in the ground. An earthquake is reported to have swallowed part of a city, Kingsai, then a mountain, Tsincheou, and in the mountains of Ki-ming-chan, to have torn open a hole large enough to create a new "lake a hundred leagues long." In Tche, it was said that 5 million people were killed in the upheavals. On the coast of the South China Sea, the ominous rumble of "subterranean thunder" was heard. As word of the disasters spread, the Chinese began to whisper that the emperor had lost the Mandate of Heaven.

In the West, news of the catastrophes evoked horror and dread. Gabriel de' Mussis, a notary from Piacenza, wrote that "in the Orient at Cathay, where the world's head is . . . , dreadful signs and portents have appeared." A musician named Louis Heyligen, who lived in Avignon, passed on an even more alarming tale to friends in Flanders. "Hard by greater India, in a certain province, horrors and unheard of tempests overwhelmed the whole province for the space of three days," Heyligen wrote. "On the first day, there was a rain of frogs, serpents, lizards, scorpions, and many venomous beasts of that sort. On the second, thunder was heard, and lightning and sheets of fire fell upon the earth, mingling with hail stones of marvelous size. . . . On the third day, there fell fire from heaven and stinking smoke which slew all that were left of man and beasts and burned up all the cities and towns in those parts."

The Genoese, who were much closer to Asia than de' Mussis and Heyligen, undoubtedly heard rumors about the disasters, but in the 1330s and early 1340s they faced so many immediate dangers in Caffa, they could not have had much time to worry about events in faraway India or China. The port of Caffa was held under a grant from the Mongols, rulers of the greatest empire in the medieval world—indeed, in the fourteenth century, rulers of the greatest empire the world had ever seen. For the Tartars, Caffa was only a small part of a vast domain that stretched from the Yellow River to the Danube, from Siberia to the Persian Gulf, but, like a pebble in a boot, it was an annoying part—or, rather, its colonial power was. To the Mongols, the Genoese seemed vainglorious, supercilious, and deeply duplicitous, the kind of people who would name their children after you—as the Dorias of Genoa had named three sons after three Mongol notables; Huegu, Abaka, and Ghazan—while picking your pocket. When the founder of the Mongol empire, Genghis Khan, railed against "eaters of sweet greasy food [who wear] garments of gold . . . [and] hold in their arms the loveliest of women," he might have had the Genoese in mind. In 1343 decades of economic and religious tension between the two powers finally erupted in a major confrontation at Tana, a trading station at the mouth of the Don, famous as the starting point of the land route to China. "The road you take from Tana to Peking," begins *La Practica della Mercatura*, Francesco Balducci di Pegolotti's fourteenth-century travel guide for eastern-bound merchants.

According to notary de' Mussis,* the brawl grew out of a confrontation between Italian merchants and local Muslims on a Tana street. Apparently insults were exchanged, fists waved, punches thrown. Market stalls tumbled, pigs squealed, a knife flashed, and a Muslim fell to the ground, dead. Shortly thereafter, a Mongol khan named Janibeg, a self-proclaimed defender of Islam, appeared outside Tana, and with him, a large Tartar force. An ultimatum was sent into the besieged town and, according to a Russian historian named A. A. Vasiliev, a response, insolent even by Genoese standards, was sent back. Enraged, Janibeg flung his Mongols into Tana. Amid plumes of black smoke and the thundering cries of sword-slashing Tartar horsemen, the Italians, outnumbered but stout, made a fighting retreat to the harbor; from there, a race westward to Caffa commenced, with

* For centuries de' Mussis was thought to have been an eyewitness to the events he describes. But in the nineteenth century a curious editor discovered that the notary was in Piacenza when Caffa was besieged. The source of de' Mussis's information is unclear, but his account is probably based on talks with merchants and/or mariners recently returned from the Crimea.

the pursued Italians traveling by ship, the pursuing Mongols by horse.

"Oh God," writes de' Mussis of the Mongols' arrival on the hills above Caffa. "See how the heathen Tartar races, pouring together from all sides, suddenly invest . . . Caffa [attacking] the trapped Christians . . . [who] hemmed in by an immense army, could hardly breathe." To the Genoese caught inside the city, the siege seemed like the end of the world, but they were wrong. In 1343 the end of the world was still several thousand miles away, on the eastern steppe.

Medieval Europeans like de' Mussis and musician Louis Heyligen were aware that plague as well as ecological upheaval raged in Asia. The new global economy had made the world a little smaller. In his account of the siege of Caffa, de' Mussis writes: "In 1346, in the countries of the East, countless numbers . . . were struck down by a mysterious illness." Heyligen, too, mentions the plague in his account of "unheard of calamities . . . hard by Greater India." The musician says that "the terrible events" in India culminated in an outbreak of the pestilence that infected "all neighboring countries . . . by means of the stinking breath." However, the best medieval guide to the Black Death's early history in Asia is Ibn al-Wardi, an Arab scholar who lived in the Syrian town of Aleppo, an important international trading center and listening post in the Middle Ages.

Al-Wardi, who like de' Mussis also got his information from merchants, says that the pestilence raged in the East for fifteen years before arriving in the West. This timeline fits the plague's pace of dissemination, which is relatively slow for an epidemic disease. A 1330s starting date would also explain the references to a mysterious illness that begin to appear in Asian documents around the same time. Among them are the Chronicles of the Great Mongol Khanate of Mongolia and Northern China, which state that, in 1332, the twenty-eight-year-old Mongol Great Khan Jijaghatu Toq-Temur and his sons died suddenly of a mysterious illness. In 1331, the year before the Great Khan's death, the Chinese records also make reference to a mysterious illness; this one, a treacherous epidemic, swept through Hopei province in the northeast region of the country and killed nine-tenths of the population.

Most modern historians believe that what we call the Black Death originated somewhere in inner Asia, then spread westward to the Middle East and Europe and eastward to China along the international trade routes. One frequently mentioned origin point is the Mongolian Plateau, in the region of the Gobi desert where Marco Polo says the night wind makes "a thousand fantasies throng to mind." In an account of the pestilence, the medieval Arab historian al-Maqrizi seems to speak of Mongolia when he says that before the Black Death arrived in Egypt, it had raged "a six month ride from Tabriz [in Iran, where] . . . three hundred tribes perished without apparent reason in their summer and winter encampments . . . [and] sixteen princes died [along with] the Grand Khan and six of his children. Subsequently, China was depopulated while India was damaged to a lesser extent."

Another often mentioned origin point is Lake Issyk Kul, where medieval travelers would come to pick up the fast road into China. Surrounded by dense forest and snow-capped mountains, in Kirgizia, near the northwest border of China, the lake region is located close to several major plague foci. (Foci are regions were plague occurs naturally.) More to the point, something terrible happened around the lake a few years before the pestilence arrived in Caffa. In the late nineteenth century a Russian archaeologist named D. A. Chwolson found that an unusually large number of headstones in local cemeteries bore the dates 1338 and 1339, and several of the stones contained a specific reference to plague. One reads:

In the year . . . of the hare [1339]
This is the grave of Kutluk.
*He died of the plague with his wife Magnu-Kelka.**

* Recently, Chwolson has been accused of misreading the inscriptions on the tombstones. Allegedly, he mistranslated "pestilence" as "plague." If true, the accusation would not materially alter the case for or against the Black Death's visit to Issyk Kul. In the Middle Ages, both plague, a biblical term used to describe an affliction associated with divine displeasure, and pestilence were applied to all kinds of epidemic disease. The appearance of either word on the Issyk Kul tombstones suggests but does not prove that the Black Death visited the lake region.

After Issyk Kul, the Black Death remains a shadowy presence for the next several years; there is no reliable information on its movements, except that it always seems to be glimpsed moving westward through the steppe high grass. The year after Kutluk and his wife, Magnu-Kelka, died, one account puts the disease in Belasagun, a rest stop to the west of Issyk Kul where riders of the Yam, the Mongol pony express, changed mounts and Marco Polo's father, Niccolo, and uncle Maffeo stopped on their way into China. A year or so later, the plague is spotted in Talas, to the west of Belasagun, and then to the west of Talas in Samarkand, a major Central Asian market town and crossroads where medieval travelers could pick up the road south to India or continue onward toward the Crimea. But only in 1346 do the first reliable accounts become available. That year one Russian chronicle speaks of the plague arriving on the western shore of the Caspian Sea and attacking several nearby cities and towns, including Sarai, capital of the Mongol Principality of the Golden Horde and home to the busiest slave market on the steppe. A year later, while Sarai buried its dead, the pestilence lurched the final few hundred miles westward across the Don and Volga to the Crimea, came up behind the Tartar army in the hills above Caffa, and bit it in the back of the neck.

The Genoese, who imagined that God was born in Genoa, greeted the plague's arrival with prayers of thanksgiving. The Almighty had dispatched a heavenly host of warrior angels to slay the infidel Mongols with golden arrows, they told one another. However, in de' Mussis's account of events, it is Khan Janibeg who commands the heavenly host at Caffa. "Stunned and stupefied" by the arrival of the plague, the notary says that the Tartars "ordered corpses to be placed in catapults and lobbed into the city in hopes that the intolerable stench would kill everyone inside. . . . Soon rotting corpses tainted the air . . . , poisoned the water supply, and the stench was so overwhelming that hardly one man in several thousand was in a position to flee the remains of the Tartar army."

On the basis of de' Mussis's account, Janibeg has been proclaimed the father of biological warfare by several generations of historians, but the notary may have invented some of the more lurid details of his story to resolve an inconvenient theological dilemma. Self-evidently—to Christians, at least—the plague attacked the Tartars because they were pagans, but why did the disease then turn on the Italian defenders? Historian Ole Benedictow thinks de' Mussis may have fabricated the catapults and flying Mongols to explain this more theologically sensitive part of the story—God did not abandon the gallant Genoese, they were smitten by a skyful of infected Tartar corpses, which, not coincidentally, was just the kind of devious trick good Christians would expect of a heathen people. Like most historians, Professor Benedictow believes the plague moved into the port the way the disease usually moves into human populations—through infected rats.* "What the besieged would not notice and could not prevent was that plague-infected rodents found their way through the crevices in the walls or between the gates and the gateways," says the professor.

The siege of Caffa ended with both sides exhausted and decimated by war and disease. In April or May of 1347, as the hills above Caffa turned green under a soft spring sun, the dying Tartar army faded away, while inside the pestilential city, many of the Genoese defenders prepared to flee westward. There are no accounts of life in the besieged port that fateful spring, but we do have images of Berlin in 1945 and Saigon in 1975, enough information to suggest what Caffa's final days may have looked like. As the death toll mounted, the streets would have filled with feral animals feeding on human remains, drunken soldiers looting and raping, old women dragging corpses through rubble, and burning buildings spewing jets of flame and smoke into the Crimean sky. There would have been swarms of rodents with staggering gaits

*Khan Janibeg does have one stout modern defender, Mark Wheelis, a professor of microbiology at the University of California. The professor notes that in a recent series of 284 plague cases, 20 percent of the infections came from direct contact—that is, the victim touched an object contaminated with the plague bacillus, *Y. pestis*. "Such transmissions," he says, "would have been especially likely at Caffa, where cadavers would have been badly mangled by being hurled, and many of the defenders probably had cut or abraded hands from coping with the bombardment." Professor Wheelis also thinks the rat scenario favored by many historians ignores a crucial feature of medieval siege warfare. To stay out of arrow and artillery range, besiegers often camped a kilometer (six-tenths of a mile) away from an enemy stronghold—normally beyond the range of the sedentary rat, who rarely ventures more than thirty or forty meters from its nest. (Mark Wheelis, "Biological Warfare at the 1346 Siege of Caffa," *Emerging Infectious Diseases* 8, No. 9 [2002]: 971–75.)

and a strange bloody froth around their snouts, piles of bodies stacked like cordwood in public squares, and in every eye, a look of wild panic or dull resignation. The scenes in the harbor, the only means of escape in besieged Caffa, would have been especially horrific: surging crowds and sword-wielding guards, children wailing for lost or dead parents, shouting and cursing, everyone pushing toward teeming ships, and beyond the melee, on the departing galleys, prayerful passengers hugging one another under great white sheets of unfurling sail, ignorant that below deck, in dark, sultry holds, hundreds of plague-bearing rats were scratching themselves and sniffing at the cool sea air.

Caffa was almost certainly not the only eastern port the plague passed through en route to Europe, but for the generation who lived through the Black Death, it would forever be the place where the pestilence originated, and the Genoese, the people who brought the disease to Europe. The chronicler of Este spoke for his contemporaries when he wrote that Genoa's "accursed galleys [spread the plague] to Constantinople, Messina, Sardinia, Genoa, Marseilles and many other places. . . . The Genoese wrought far more slaughter and cruelty . . . than even the Saracens."

Plague is the most famous example of what the Pima Indians of the American Southwest call *oimmeddam,* wandering sickness; and an ancient Indian legend evokes the profound dread *oimmeddam* produced in premodern peoples.

"Where do you come from?" an Indian asks a tall, black-hatted stranger.

"I come from far way," the stranger replies, "from . . . across the Eastern Ocean."

"What do you bring?" the Indian asks.

"I bring death," the stranger answers. "My breath causes children to wither and die like young plants in the spring snow. I bring destruction. No matter how beautiful a woman, once she has looked at me she becomes as ugly as death. And to men, I bring not death alone, but the destruction of their children and the blighting of their wives. . . . No people who looks upon me is ever the same."

Plague is the most successful example of *oimmeddam* in recorded history. Worldwide, the disease has killed an estimated 200 million people, and no outbreak of plague has claimed as many victims or caused as much anguish and sorrow as the Black Death. According to the Foster scale, a kind of Richter scale of human disaster, the medieval plague is the second greatest catastrophe in the human record. Only World War II produced more death, physical destruction, and emotional suffering, says Canadian geographer Harold D. Foster, the scale's inventor. Harvard historian David Herbert Donald also ranks the Black Death high on a list of history's worst catastrophes. However, the greatest—if most backhanded—tribute to the plague's destructiveness comes from the United States Atomic Energy Commission, which used the medieval pestilence to model the consequences of all-out global nuclear war. According to the commission's *Disaster and Recovery,* a Cold War–era study of thermonuclear conflict, of all recorded human events, the Black Death comes closest to mimicking "nuclear war in its geographical extent, abruptness of onset and scale of casualties."

The sheer scope of the medieval plague was extraordinary. In a handful of decades in the early and mid-fourteenth century, the plague bacillus, *Yersinia pestis,* swallowed Eurasia the way a snake swallows a rabbit—whole, virtually in a single sitting. From China in the east to Greenland in the west, from Siberia in the north to India in the south, the plague blighted lives everywhere, including in the ancient societies of the Middle East: Syria, Egypt, Iran, and Iraq. How many people perished in the Black Death is unknown; for Europe, the most widely accepted mortality figure is 33 percent.* In raw numbers that means that between 1347, when the plague arrived in Sicily, and 1352, when it appeared in the plains in front of Moscow, the continent lost

* Estimates of the Black Death mortality rate fluctuate almost as often as the stock market. Recently, one historian argued that 60 percent of Europe perished in the Black Death. However, 33 percent is the most frequently cited and enduring figure. Interestingly, it is also not far from the estimate contemporaries arrived at. In the wake of the plague, a Church commission put the death toll at almost 24 million, remarkably close to a mortality of one-third in a Europe of 75 million. (William Naphy and Andrew Spicer, *The Black Death: A History of Plagues* [Stroud, Gloucestershire: Tempus Publishing, 2000], p. 34. See also Ole J. Benedictow, *The Black Death: The Complete History* [Woodbridge, Suffolk: Boydell Press, 2004], p. 383.)

twenty-five million of its seventy-five million inhabitants. But in parts of urban Italy, eastern England, and rural France, the loss of human life was far greater, ranging from 40 to 60 percent. The Black Death was particularly cruel to children and to women, who died in greater numbers than men, probably because they spent more time indoors, where the risk of infection was greater, and cruelest of all to pregnant women, who invariably gave birth before dying.

Contemporaries were stunned by the scale of death; almost overnight, it seemed, one out of every three faces vanished from the human community, and on the broad shires of England, the little villages along the Seine, and the cypress-lined roads of Italy—where the afternoon light resembles "time thinking about itself"—one out of two faces may have disappeared. "Where are our dear friends now?" wrote the poet Francesco Petrarch. "What lightning bolt devoured them? What earthquake toppled them? What tempest drowned them. . . . There was a crowd of us, now we are almost alone."

In the Islamic Middle East and North Africa, mortality rates also were in the one-third range. To the Muslim historian Ibn Khaldun, it seemed "as if the voice of existence in the world had called out for oblivion." In China the presence of chronic war makes it difficult to assess plague mortalities, but between 1200 and 1393 the population of the country fell 50 percent, from about 123 million to 65 million. Today a demographic disaster on the scale of the Black Death would claim 1.9 billion lives.

The Black Death would be an extraordinary accomplishment for any wandering sickness, but it is an especially extraordinary one for a sickness not even native to humans. Plague is a disease of rodents. People are simply collateral damage, wastage in a titanic global struggle between the plague bacillus *Yersinia pestis* and the world's rodent population.* *Y. pestis*'s natural prey are turbots, marmots, rats, squirrels, gerbils, prairie dogs, and roughly two hundred other rodent

* At the moment, plague seems to have the upper hand in the battle. Recent studies suggest that the disease has become so virulent in rodents, it may be disrupting the natural selection process in several species. (Dean E. Biggins and Michael Kosvol, "Influences of Introduced Plague on North American Mammals," *Journal of Mammalogy* [November 2001]: 906–16.)

species. For the pathogen to ignite a major outbreak of human disease on the scale of the Black Death, a number of extraordinary things had to have happened. And while we will never know what all of them were, from about 1250 onward, social, economic, and perhaps ecological changes were making large parts of Eurasia an increasingly unhealthy place to live.

One new risk factor was increased mobility. Along with facilitating international trade, the Mongol unification of the steppe brought merchants, Tartar officials, and armies into proximity with some of the most virulent, and heretofore isolated, plague foci in the world. Rodents (and more to the point, their fleas) that once would have died a lonely, harmless death on a Gobi sand dune or Siberian prairie now could be transported to faraway places by caravans, marching soldiers, and riders of the Mongol express, who could travel up to a hundred miles a day on the featureless, windswept prairies of the northern steppe.

Environmental upheaval may also have played a role in the origin of the plague. Like a vain old matinee idol, *Y. pestis* is fond of ecological drum rolls. In the mid-sixth century, during the pestilence's first (documented) visit to Europe, the Plague of Justinian, there were reports of blood-colored rain in Gaul, of a yellow substance "running across the ground like a shower of rain" in Wales, and of a dimming of the sun everywhere throughout Europe and the Middle East. "We marvel to see no shadow on our bodies at noon, to feel the mighty vigor of the sun's heat wasted into feebleness," wrote the Roman historian Flavius Cassiodorus.

Similar, if less flamboyant, accounts of environmental instability appeared in the decades prior to the Black Death. In the West as well as the East, there were reports of volcanic eruptions (Italy), earthquakes (Italy and Austria), major floods (Germany and France), a tidal wave (Cyprus), and swarms of locusts "three German miles" long (Poland). However, since the medieval world viewed natural disasters as portents and expressions of Divine Wrath, these accounts have to be read with caution. Undoubtedly, many of the calamities described by European—and Chinese—chroniclers were invented or exaggerated beyond recognition after the fact to provide the Black Death with a suitably apocalyptic overture.

That said, tree ring data indicates that the early fourteenth century was one of the most severe periods of environmental stress in the last two thousand years—perhaps due to unusual seismic activity in the world's oceans. And modern experience shows that ecological upheaval in the form of droughts, floods, and earthquakes can play a role in igniting plague, usually because such events dislodge remote wild rodent communities, the natural home of *Y. pestis,* from their habitats and drive them toward human settlements in search of food and shelter.

Social and demographic conditions are also risk factors in plague. Like other infectious illnesses, the disease requires a minimum population base of four hundred thousand people to sustain itself. When human numbers fall below that base—or people are dispersed too widely—the chain of infection begins to break down. Sanitary conditions are important, too. A principal vector in human plague, the black rat—*Rattus rattus*—feeds on human refuse and garbage, so the filthier a society's streets and homes and farms, the larger its plague risk. Since the flea is an even more critical disease vector, personal sanitation matters as well; people who wash rarely are more attractive to an infected flea than those who wash regularly. Humans who live with farm animals are also at greater risk because they are exposed to more rats and fleas; and if a population lives in homes with permeable roofs and walls, the risk is even greater.

The role of malnutrition in human plague is controversial, though perhaps unjustifiably so. It is true that bacteria, which require many of the same nutrients as humans, have more difficulty reproducing in malnourished hosts. But experience with plague in early-twentieth-century China and India suggests that nutritional status, like sanitation, is a risk factor in the disease, and emerging research suggests that nutrition may also affect susceptibility in another, more subtle way. Recent studies have found that exposure to malnutrition in utero damages the developing immune system, creating a lifelong vulnerability to illness in general.

* * *

From Caffa to the jungles of Vietnam,* war has also been an important predisposing factor in human plague. War creates human remains and refuse, which attract rats; filthy bodies, which attract fleas; and stresses, which can lower immune system function. Marching soldiers and cavalry also help to make a pestilence more mobile.

The historical evidence suggests that the existence of only a few of these conditions is not enough to ignite a pandemic, or major outbreak of plague. The Victorian West, for example, was far more densely interconnected and populated than medieval Europe, but when a major wave of plague swept through China and India a century ago, relatively healthy populations, relatively good sanitation and public health standards, and a sturdy physical plant—wood and brick houses—prevented the plague from gaining a toehold in either America or Europe. The disease reached the West, but after causing a few hundred deaths in Oakland, San Francisco, Glasgow, Hamburg, and several other cities, it died out.

The era that was once called the Dark Ages and is now referred to (less judgmentally, if no more accurately) as the Early Middle Ages also had several conditions associated with plague, including widespread violence, disorder, malnutrition, and filth—if early medieval Europeans washed or changed their clothes more than once or twice a year, it was the best-kept secret in Christendom. However, foreign trade had virtually disappeared, and the rise of hostile new Muslim states in the Middle and Near East put the plague foci of Central Asia and Africa at a further remove from Europe. Also, the early Middle Ages was a period of profound depopulation. In the sixth and seventh centuries a crumbling Roman Europe lost as much as one-half to two-thirds of its population. From Scotland to Poland, the inheritors of a once great civilization lived huddled together like fugitives in forest clearings. Even if, by some fluke, *Y. pestis* had managed to travel to the early medieval West, it would have failed as miserably as it did in the streets of Victorian San Francisco.

* During the Vietnam War, roughly twenty-five thousand cases of plague were reported; almost all the victims were Vietnamese. (*Plague Manual: Epidemiology, Distribution, Surveillance and Control* [Geneva: World Health Organization, 1999], pp. 23–24.)

* * *

By contrast, the environment of the fourteenth century was well suited to *Y. pestis*. By modern standards, the population of medieval Europe was relatively low: about 75 million, compared to today's nearly 400 million. However, compared to the resources available to the population, the continent had become dangerously overcrowded. The years between 1000 and 1250 were a period of great economic and demographic growth in the medieval West, but when the economy began to stall out after 1250, Europe found itself trapped in what historian David Herlihy has called "Malthusian deadlock." Medieval Europeans were still able to feed, clothe, and house themselves, but because the balance between people and resources had become very tight, just barely. A worsening climate made the margin between life and death even narrower for tens of millions of Europeans. Between 1315 and 1322 the continent was lashed by waves of torrential rain, and by the time the sun came out again in some places 10 to 15 percent of the population had died of starvation. In Italy especially, malnutrition remained widespread and chronic, right until the eve of the plague.

In the fourteenth century war was almost as much a constant as hunger. Italy, where the papacy and Holy Roman Empire were fighting for ascendancy, had descended into a Hobbesian state of all against all. There were large, small, and in-between-sized wars raging in the papal states around Orvieto, Naples, and Rome. At sea, Italy's "two torches," as Petrarch called Genoa and Venice, were locked in an interminable maritime conflict. And almost everywhere up and down the peninsula, roving bands of *condottieri* (mercenaries) were waging fierce little freelance wars. To the north and west, conflict was raging in Scotland, Brittany, Burgundy, Spain, and Germany, and in the ports and plains and cities of northern France the English and French were fighting the first battles of the Hundred Years' War.

"The city makes men free," medieval Germans told one another, but a combination of people, rats, flies, waste, and garbage concentrated inside a few square miles of town wall also made the medieval city a human cesspool. By the early fourteenth century so much filth had collected inside urban Europe that French and Italian cities were naming streets after human waste. In medieval Paris, several street names were inspired by *merde,* the French word for "shit." There were rue Merdeux, rue Merdelet, rue Merdusson, rue des Merdons, and rue Merdiere—as well as a rue du Pipi. Other Parisian streets took their names from the animals slaughtered on them. There was a Champs-Dolet—roughly translated as "field of suffering and cries"—and l'Echorcheire: "place of flailing." Every town of any size in Europe had its equivalent of l'Echorcheire: an outdoor slaughterhouse, where butchers in bloodstained clothing cut and chopped and sawed amid discarded body parts and offal and the agonizing moans of dying animals. One irate Londoner complained that the runoff from the local slaughterhouse had made his garden "stinking and putrid," while another charged that the blood from slain animals flooded nearby streets and lanes, "making a foul corruption and abominable sight to all dwelling near." In much of medieval Europe, sanitation legislation consisted of an ordinance requiring homeowners to shout, "Look out below!" three times before dumping a full chamber pot into the street.

The medieval countryside, where 90 percent of the population lived, was an even more dangerous place than the medieval city. Thinly walled, peasant homes were highly permeable, and the rat-to-person ratio tended to be very high in rural areas. Urban rat colonies usually divided their attention among several homes on a street, but in the country, not uncommonly, a single peasant family would find itself the target of an entire rodent colony.

The medieval body was in as shocking a state as the medieval street. Edward III scandalized London when he bathed three times in as many months. Friar Albert, a monk in Boccaccio's *Decameron,* displays a more typical medieval attitude toward personal hygiene. "I shall do something today that I have not done for a very long time," the friar announces cheerfully. "I shall undress myself." When the assassinated Thomas à Becket was stripped naked, an English chronicler reports that vermin "boiled over like water in a simmering cauldron" from his body.

Arguably, medieval Europe's reigning religious and medical orthodoxies also created a vulnerability to plague by promoting a public health program based on clothing the wicked, inhaling fragrances, and prayer. The educated classes, influenced by the theories of a former sports physician and Roman celebrity doctor named Galen, believed that the pestilence arose from miasmas—dense clouds of infected air. "Corrupted air, when breathed in, necessarily penetrates to the heart and corrupts the substance of the spirit there," warned the Paris medical faculty, the most eminent medical body of the day. The "prince of physicians," the Italian Gentile da Foligno, recommended the inhalation of herbs as an antidote for "corrupt" air.

To the Church and common folk, the plague was seen as a form of divine retribution for human wickedness. Henry Knighton, an English monk who hovers over the Black Death like a cackling Shakespearean witch, exemplifies this school of thinking. Knighton believed that God smote a third or more of Europe because medieval England's most glamorous young women were becoming tournament groupies. "Whenever and wherever tournaments were held," Knighton wrote a few decades after the Black Death, "a troupe of ladies would turn up dressed in a variety of male clothing . . . and mounted on chargers. There were sometimes as many as forty or fifty of them, representing the showiest and most beautiful, though not most virtuous, women of the realm. . . . [They] wore thick belts studded with gold and silver slung across their hips, below the navel . . . and were deaf to the demands of modesty. But God, present in these things, as in everything, supplied a marvelous remedy"—plague.

Plague is among the slowest moving of wandering sicknesses. New strains of influenza can leap around the world in a year or two, but *Y. pestis,* like the AIDS virus, is tied to a complicated chain of infection that can take decades to unfold. The principal vector in the disease is not the rodent, the animal most often associated with it, but the rodent flea. When an infected host dies, the flea leaps to a new host, transferring the plague bacillus, *Y. pestis,* to the host by way of a skin bite. Sometimes humans are infected directly by one of the many flea species that prey on wild rodents, such as squirrels, prairie dogs, and marmots; however, most often the agent of infection in human plague is the more familiar black rat flea, *Xenopsylla cheopis.*

In human plague, the chain of infection can take several forms. For example, an ecological disaster, which destroys the food supply, or a dramatic spike in the rodent population, which puts tremendous pressure on it, may drive a colony of infected animals toward a human settlement, where members of the colony exchange fleas with domestic rats. Another possible scenario that may have relevance to the Black Death is that a group of travelers stumble into a wild rodent community in the midst of a plague outbreak; infected rodents—or their fleas—infiltrate the travelers' saddlebags and carts, and when the group arrives in the next town or village, the hitchhikers leap from their hiding place and spread the disease to the domestic rodent population. The next-to-last stage in the sequence is the involvement of *X. cheopis,* the rat flea, which becomes a disease vector because, in one way or another, its rat hosts have become infected by wild rodent fleas.

The jump of plague into man is driven by *X. cheopis*'s desperation. It does not particularly like human blood, but as plague kills off the local domestic rat community, the flea's only alternatives are starvation or *Homo sapiens*. Once embedded in a human population, the rat flea becomes a very efficient disease vector. *X. cheopis* can survive up to six weeks without a host—long enough to travel hundreds of miles in grain or cloth shipments. It is also an extremely aggressive insect. It has been known to stick its mouth parts into the skin of a living caterpillar and suck out the caterpillar's bodily fluids and innards. However, *X. cheopis*'s greatest attribute as a disease vector lies in the vagaries of its digestive system.

In an uninfected rat flea, blood from a skin bite flows directly to the stomach, satiating hunger. In an infected flea, plague bacilli build up in the foregut, producing a blockage; this enhances the insect's ability to spread infection in two ways. First, because no nutrients are reaching the stomach, *X. cheopis,* chronically hungry, bites constantly; and second, as undigested blood builds up in the foregut, the flea becomes a living hypodermic needle. Every time it bites, it gags on the undigested blood, now tainted with plague bacilli, and vomits it into the new bite.

The way to get from the digestive problems of a minor insect to 25 million to 30 million dead in Europe, a third of the Middle East wiped out, and China "depopulated," is by multiplying. Normally a rodent only carries a half dozen or so fleas, but in the midst of an epizootic, when hosts become rare, surviving rodents often become the equivalent of flea towns, carrying a hundred to two hundred insects—and sometimes flea cities. Researchers counted nine hundred infected fleas on one unfortunate ground squirrel in Colorado.

Three forms of plague prey on humans. *Bubonic* plague, the most common form, is transmitted by a flea bite and has a two- to six-day incubation period. "Behold, the swelling, the warning signs sent by the Lord," wrote a contemporary, of the Black Death's most characteristic symptom, the egg-shaped bubo. Medieval chroniclers frequently described the bubo as tumorlike, and the analogy is an apt one. Like malignant cancer cells, once inside the body, plague bacilli multiply with an aggressive wildness. Typically, the site of the flea bite determines the site of the *gavocciolo,* as contemporaries called the bubo. Bacilli from leg and ankle bites produce buboes in the abdominal region or thigh; upper-body bites, buboes under the arms or on the neck. Exquisitely sensitive to pressure, *gavocciolo* often create odd deformations in their victims. Thus, a neck bubo may produce a head permanently cocked in the opposite direction, a thigh bubo a hopping limp, an underarm bubo an outstretched or raised arm. Buboes are also oddly noisy creatures. Human plague speaks to its victims in the strange gurgling tongue of the bubo.

According to the chroniclers, three other symptoms were also quite common in the bubonic plague of the Black Death. One were *petechiae*. These bruiselike purplish splotches often appeared on the chest, back, or neck, and were also known as "God's tokens" because their appearance meant the victim had a fatal case of plague. Legend has it that the "tokens" were the inspiration for a Black Death–era nursery rhyme still sung today:

Ring around the rosie, pocket full of posies
*Ashes, ashes [the hemorrhages], we all fall down.**

Malodorousness was another frequent symptom of historic bubonic plague. Victims not only looked as if they were about to die, according to many Black Death chroniclers, they often smelled as if they were. After visiting a plague-stricken friend, one man wrote, "The stench [of] sweat, excrement, spittle, [and] breath [was] overpowering." A number of contemporary accounts also suggest that the medieval plague disrupted the nervous system. There are reports of delirious, agitated victims shouting madly from open windows or walking around half-naked or falling into a stupor.

Oddly, these last three symptoms are uncommon in modern bubonic plague. Dr. Kenneth Gage, chief of the Plague Division at the U.S. Centers for Disease Control, has encountered "God's tokens" in his fieldwork, but so infrequently that, when asked if he had ever seen a plague victim with hemorrhagic bruises, he had to stop and think for a moment. While the CDC official has encountered many cases of malodorousness, he describes the foul odor as a by-product of poor nursing care—the plague victim was not being changed or bathed regularly, or lived in a hovel. The smells described by Black Death chroniclers—or at least some of them—seemed to emanate from inside the victim, as if his insides were gangrenous. Dr. Gage, who has fought plague in Asia and North and South America, cannot recall ever encountering a case of central nervous system involvement in a plague victim.

Bubonic plague is the most survivable of the three forms of the disease. Untreated, it has a mortality rate of about 60 percent.

Pneumonic is the second type of plague, and—uniquely—it can spread directly from person to person. However, like other forms of the infection, it is borne in the rodent/insect connection. In some cases of bubonic plague, bacilli escape the lymph system and infect the

*Recently, some scholars have challenged the Black Death connection, arguing that the poem originated in the early nineteenth century.

lungs, causing secondary pneumonic plague. As the victim begins to cough and spit up blood—the principal symptoms of the "coughing" plague—the disease breaks free of the flea connection and spreads into the population like a cold or flu—through the air. Though summer outbreaks occur, pneumonic plague is more frequent in winter, when the colder temperatures favor the transmission of pulverized and frozen sputum and cough droplets.

As with bubonic plague, there are also some notable differences between the modern variant of the "coughing plague" and its Black Death counterpart. One is incidence. Relatively uncommon today, during the first year of the Black Death pneumonic plague seemed to be everywhere in Italy and southern France. The other notable difference involves contagiousness. Modern pneumonic plague is not a particularly "catchy" disease, nor should it be. Plague bacteria are larger than viruses and thus harder to transmit directly from person to person—the bigger bacilli require bigger air droplets, and if they do reach another person, tend to get "stuck" in the upper respiratory system before they can reach the lungs.

Even making allowance for the medieval propensity to exaggerate, one gets the impression that the pneumonic plague of the Black Death was not merely highly contagious but explosive in the manner of a nuclear chain reaction. "Breath," wrote one horrified Sicilian chronicler, "spread the infection among those speaking together . . . and it seemed as if the victim[s] were struck all at once by the affliction and [were] shattered by it. . . . Victims violently coughed up blood, and after three days of incessant vomiting for which there was no remedy, they died, and with them died not only everyone who talked with them but also anyone who had acquired or touched or laid hands on their belongings."

The "coughing plague" is extremely lethal. If it goes untreated, the mortality rate in its victims is between 95 and 100 percent.

No one survives untreated *septicemic plague,* the third form of the disease. The shocklike movement of massive amounts of plague bacilli directly into the blood system creates such enormous toxicity that even insects normally incapable of transmitting *Y. pestis,* such as body lice, can become disease vectors. During one outbreak of septicemic plague in the early twentieth century, the average survival time from onset of symptoms to death was 14.5 hours.

There have been suggestions that the horrible disfigurement caused by septicemic plague—the extremities become as black and hard as coal—inspired the term Black Death, but septicemic disease is uncommon, and, in any event, the application of the terms "Black Death" to the medieval plague grew out of an old historical error. In 1631, a historian named Johannes Isaacus Pontanus, perhaps thinking of Seneca's use of the Latin term for Black Death—*Arta mors*—to describe an outbreak of epidemic disease in Rome, claimed that the phrase had been current during the fourteenth-century mortality. The Swedes, who began using the expression around 1555 (*swarta döden*), the Danes, who adopted it fifty years later (*den sorte Død*), and the rest of Europe, which began using the expression "Black Death" in the eighteenth century, may have been laboring under the same misapprehension. The generation who lived through the medieval pestilence called it *la moria grandissima, la mortalega grande, très grande mortalité, grosze Pestilentz, peligro grande,* and *huge mortalyte*: names that translate roughly as the "Great Mortality," or, more colloquially, the "Big Death."

One of the great mysteries of the medieval plague is how the fleeing Genoese survived the sixteen-hundred-mile sea journey from Caffa to Sicily, where the disease enters European history. Even if the escaping galleys stopped first in Constantinople and other ports en route to Italy as seems likely, getting caught on an open sea with *Y. pestis* would have been akin to getting caught in a revolving door with a rattlesnake. The only current explanation for the riddle is lucky genes. Recent research suggests that an allele,* CCR5-Δ32, may confer protection against plague. Possibly some of the crew members had the requisite lucky allele.

Clearer is what happened once Caffa slipped below the horizon.

* An allele is one of at least two forms of how a particular gene might be expressed. For example, the gene for human eye color might be expressed in several forms: brown, blue, green, and so forth.

On the second or third day at sea a mariner awakes feeling feverish; after he falls asleep again, a shipmate steals his flea-infested jacket; a few days later, the thief is ill. As word of the men's illness spreads through the ship, panicky crew members gather in the horse stalls on the lower deck to share rumors and conspire. That night there is a splash off the aft side of the ship, then a second; no one raises an alarm as the bodies sink below the surface in a cone of rippling moonlight.

As the days lengthen and the disease takes hold, men begin to turn on one another, as they will later in Europe, when the plague arrives. There are beatings, murders, summary executions, mutinies; only the progress of the pestilence prevents complete anarchy. Men become too ill to kill, then too ill to work. A helmsman with a neck bubo is strapped to the helm; a ship's carpenter with a bloody cough, to his bench. A rigger shaking with fever is lashed to the mast.

Gradually each escaping vessel becomes a menagerie of grotesques. Everywhere there are delirious men who talk to the wind and stain their pants with bloody anal leakages; and weeping men who cry out for absent mothers and wives and children; and cursing men who blaspheme God, wave their fists at an indifferent sky, and burble blood when they cough. There are men who ooze pus from facial and body sores and stink to high heaven; lethargic men who stare listlessly into the cruel, gray sea; mad men who laugh hysterically and dig filthy fingernails into purple, mottled flesh; and dead men, whose bloated bodies roll back and forth across pitching decks until they hit a rail or mast and burst open like piñatas.

In the thousand days between the autumn of 1347—when the Genoese arrived in Sicily, so diseased "that if anyone so much as spoke with one of them, he was infected"—and the winter of 1351–52, when the plague crossed the icy Baltic back into Russia, *Y. pestis* drew a hangman's noose around Europe.

From Sicily, where it raged unceasingly for a dozen months, one strain of the pestilence swept westward along the Mediterranean coast to Marseille, where half the city may have perished in the bitter winter of 1347–48. Sweeping up the Rhône to Avignon, in April 1348 the plague ended one of the mythic love stories in Western literature, exposed the moral weakness of a pope, and inspired nightly marches to the local cemeteries by Avignon's hungry pigs. Farther to the east, in the Adriatic port city of Ragusa, authorities celebrated spring's arrival by ordering all citizens to make out a will. In June the plague visited Paris, where the municipal cemetery ran out of burial space and the renowned Paris medical faculty pronounced the cause of the disease to be "an unusual conjunction of Saturn, Mars, and Jupiter at one on the afternoon on March 20th, 1345." Later that bleak summer the plague forked like a serpent's tongue. One strain swept northward toward Tournai on the Flemish border, where church bells rang unceasingly for two days to announce its arrival; while a second strain, enticed by the scent of war and death around recently besieged Calais, rolled up the coast and peered westward across the channel toward England. On the other side, from Dover to Land's End, anxious Englishmen scanned the summer seas as they would not scan them again until the Battle of Britain in the summer of 1940.

In July *Y. pestis* slipped through the cordon of watchers and entered the little port of Melcombe; a month later the town was still, except for the pounding of rain on village rooftops and the crash of surf against the chalky Dorset cliffs to the south. In the terrible month of September, the pestilence pivoted and wheeled eastward through an incessant late summer downpour toward London, where a grieving king mourned a beloved child. "No fellow human being could be surprised if we were inwardly desolated by the sting of this bitter grief, for we are human, too," wrote Edward III of his plague-dead daughter, fifteen-year-old Princess Joan. In the spring of 1349, as the green hills of Wales echoed with birdsong, a local poet wrote, "Death comes into our midst like black smoke." In the merry month of May *Y. pestis* arrived in Derbyshire, where in three short months it killed peasant William de Wakebridge's wife, father, sister, a sister-in-law, and an aunt. Across the Irish sea, in Dublin, where the living had surrendered the streets to the dead, Franciscan John Glynn wrote, "I . . . am waiting among the dead for death to come."

Another strain of the plague entered Europe through Genoa,

where several galleys laid anchor on the last day of 1347. As a raw winter wind whipped through the city's nighttime streets, a candle glowed in the window of local notary Antonio de Benitio, who remained in the infected city to make out wills for clients unable to flee. Swinging inland across the narrow plains of central Italy, the plague swept into Florence on a cold March day and killed so many of its citizens, church bells were stilled to preserve public morale; "the sick hated to hear [them] and it discourages the heathy as well," wrote a survivor. In June, when the plague arrived in Siena, a tax collector and former shoemaker named Agnolo di Tura declared, "This is the end of the world." Nearby Pistoia greeted the pestilence more pragmatically. "Henceforth . . . ," declared the town fathers, "each grave shall be dug two and a half arms' lengths deep." In August the pestilence reached Perugia, where Gentile da Foligno, one of Italy's most celebrated physicians, cast his lot with the poor. As the wealthy and well-born of Perugia fled, wealthy, well-born da Foligno remained at his post, visiting the stinking hovels of the needy until, at last, the plague claimed him.

Descending from the Alpine passes into Austria in the fall of 1348, *Y. pestis* killed with such fetid abundance, one observer reports that the wolves who preyed on local sheep "turned and fled back into the wilderness . . . as if alarmed by some infallible warning." Arriving in Central Europe, the pestilence ignited an unparalleled burst of anti-Semitism. In September 1348, in Chillon, a town near Lake Geneva, a Jewish surgeon and a Jewish mother were accused of fomenting plague, forcing the surgeon to choose between himself and his community, and the mother between herself and her son.

In January 1349 Basel burned its Jews on an island in the Rhine, while hygiene-conscious Speyer, fearing pollution, put its dead Jews in wine barrels and rolled them into the river. In February, as a prophylactic measure, Strasbourg marched its Jews to a local cemetery and burned them. Entering the cemetery, several beautiful young Jewesses refused salvation at Christian hands and insisted on going to the stake. The plague struck Strasbourg anyway. In Worms the local Jewish community, faced with death at the hands of Christian neighbors, locked themselves in their homes and set themselves ablaze. In Constance, under a gray March sky, a group of Jews marched into a fire, singing and laughing.

As the plague made its way through the primeval forests of Germany, another demon bubbled up from the medieval Teutonic psyche: the Flagellants, who believed the curse of the mortality could be lifted through self-abuse of the flesh and slaying Jews. Twenty years later one spectator could still recall the hysteria the Flagellants aroused. The men, he wrote, "lashed themselves viciously on their naked bodies until the blood flowed, while crowds, now weeping now singing, shouted, 'Save us!'"

In May 1349 an English wool ship brought the plague to Bergen, in Norway. Within days of arriving the passengers and crew were all dead. By the end of the short Scandinavian summer, the pestilence was moving in an easterly arc toward Sweden, where King Magnus II, believing the mortality to be the work of an angry God, ordered foodless Fridays and shoeless Sundays to appease His divine wrath. Approaching the east coast of Greenland, *Y. pestis* encountered towering ice cliffs rising out of a frigid, white-capped sea like the parapets of an Arctic Xanadu; undaunted, it persevered. Later an observer would write that, from that moment on, "no mortal has ever seen that [eastern] shore or its inhabitants."

In the three and a half years it took *Y. pestis* to complete its circle of death, plague touched the life of every individual European: killing a third of them, leaving the other two-thirds grieving and weeping.

Here is the story of that epic tragedy.

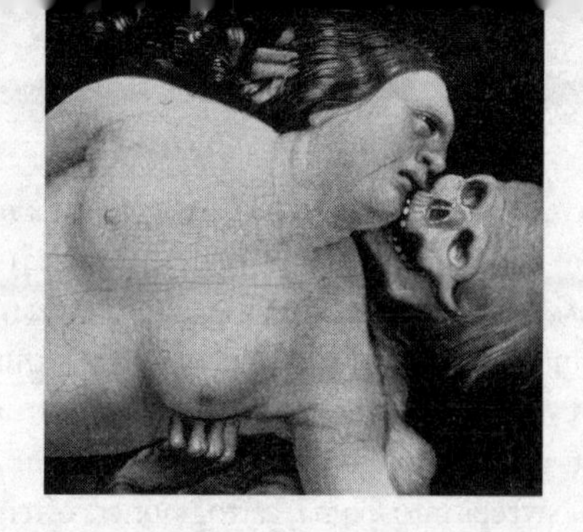

CHAPTER TWO

"They Are Monsters, Not Men"

ON A MAP, THE EURASIAN STEPPE LOOKS LIKE A TRAVELER'S PARADISE, but the steppe of the cartographers, the broad green crayon slash that sweeps effortlessly across the belly of the continent from Byelorussia to China, is a polite semifiction available only in spring, when the air is warm, the grass not yet knee-high, and the wind still fragrant with the smell of wildflowers. As Napoleon learned and Hitler after him, in winter waist-high snowfields transform the western steppe into an immense featureless sea that billows and swirls when the Arctic wind whips down from the Siberian tundra. The cartographer's map also ignores the summer sun, which hangs so low over the treeless August plains, a traveler can almost reach up and touch it, and the incessant buzz of the mosquitoes, which grow almost as big as a man's thumb on parts of the steppe and can leave a bite the size of a small tumor.

Farther east, on the Mongolian Plateau, where the steppe skirts the Gobi desert as it sweeps into China, cartographers often take note of the change in terrain with another polite semifiction—a splash of sandy color. Parts of the eastern steppe resemble a seabed baked dry by a billion years of sun. Between canyons of rust-colored cliff and sandy elevations more hill than mountain, low undulations of rocky ground flow to a limitless horizon like ocean swells, while overhead, above noisy flocks of circling carrion, the enormous sky oppresses with an infiniteness that crushes the soul. Even in high spring, the only two crops that grow in this part of the steppe are tufts of hard, spiky grass and the bones of the men and animals who failed to survive the winter snows.

In *La Practica della Mercata,* Francesco Balducci Pegolotti attempted to ease the burdens of the medieval steppe traveler with reassurances—"the road you take from Tana to Peking is perfectly safe"; with sex tips—"the merchant who wishes to take a woman with him from Tana can do so"—and dos and don'ts—"do not try to save money on [a translator] . . . by taking a bad one."* But *La Practica* was a polite fiction, too. Upon leaving Caffa, a traveler could expect to spend eight to twelve months on the back of a Mongol pony or a jostling cart, to see nothing but horizon and prairie in every direction, and to feel no warmth at night except the body warmth of a traveling companion. As alien as the terrain were the fearsome Mongols who inhabited the Asian plains. "They [are] like beasts," wrote one Westerner. "They live on wild roots and on meat pounded tender under the saddle . . . are ignorant of the use of the plow and of fixed habitation. . . . If you inquire . . . whence they come and where they were born, they cannot tell you." William of Rubruck, a Flemish cleric who

* Pegolotti's credibility as a travel writer is not enhanced by the fact that he never got any farther east than the east side of his native Florence. His information was based on interviews with Italian merchant-travelers.

visited thirteenth-century Mongolia, described the Tartar women as "astonishingly fat," with "hideously painted faces," and the men as grotesques, with short, stocky bodies and "monstrously oversized heads." Both sexes were also incredibly filthy; the Mongols refused to wash, believing it made God angry.

The French historian René Grousset has called the "discovery of Asia . . . as important to men of the Middle Ages as the discovery of America was to men of the Renaissance." But it might be more accurate to describe the medieval discovery of Asia as a rediscovery. During Antiquity, news from the Orient would occasionally drift westward along the Silk Road, which wound through the necklace of desert between China and Arabia, or across the snowy passes of the Pamir Mountains in Central Asia, where representatives of Rome and China would meet to exchange goods. But from the seventh century onward, Europe became isolated on the western edge of Eurasia, a prisoner of its own disorder and collapse. To the extent that the reawakening West of the eleventh and twelfth centuries had a knowledge of the East, it was of the Middle East, and, more particularly, a thin strip of the coastal Middle East where Genoese and Venetian merchants were allowed to buy Asian goods from Arab middlemen at exorbitant markups. Beyond Arabia, everything faded into a myth wrapped in a fable. There were stories about strange Asian races like the Dog Men, who were said to have human bodies and canine faces, and the Headless Men, who were thought to have no heads at all; about Gog and Magog, who were believed to be related to the lost tribes of Israel; Prester John, a mysterious Christian King of the Orient; and the Garden of Eden, which was thought to be somewhere in India. But until the mid-thirteenth century, when the Mongols unified the steppe from Kiev to China, no one in the West was able to investigate any of these Eastern wonders firsthand.

The first Europeans to travel to Asia were clerics like John de Marignolli, a papal emissary, who pronounced the Tartar Great Khan "delighted, yea exceedingly delighted," by the pope's gifts, and John of Monte Corvino, who translated the New Testament into Mongol script, and whose long years in China aged him beyond his years. "I myself am grown old and grey, more with toil and trouble than with years," John wrote after eleven years in the East. The group also included the rambunctious William, a Franciscan friar who endured every hardship of steppe travel, including the hardest of all, an alcoholic translator. From William, medieval Europe received its first description of Chinese script, of a potent Mongol liquor called *koumiss,* and of the Tebets, a Tibetan tribe whose members formerly ate their parents when they died but had given up the practice. William also was the first European to correctly identify the Caspian as a land-locked sea, not an ocean inlet, and—his proudest achievement—to participate in what may have been the first theological Super Bowl. On a May evening in 1254, in the Mongol capital of Karakorum on the edge of the Gobi desert, William strode into a crowded tent, and, in the presence of the Grand Khan himself, defended the Western concept of monotheism against a *tunis,* a Buddhist priest.

"It is fools who say that there is only one God," declared the wily Buddhist. "Are there not many great rulers on earth? . . . The same is true with God. . . . [T]here are ten Gods in Heaven and none is all-powerful."

"So then," replied William, "not one of your Gods is capable of rescuing you, inasmuch as [if you encounter a predicament] . . . the God has no power over, he will be unable to help you."

On the basis of that exchange alone, William felt he had won the day, but, alas, the three Mongol judges who scored the debate disagreed and declared the Buddhist priest victor.

The second wave of European visitors were merchants, most Genoese or Venetian, enticed to the East by the opportunity to buy Asian goods at the source. No one knows for sure how many of these trader-travelers followed in the footsteps of Marco Polo, the daring young merchant's son from Venice who crossed the steppe in the early 1270s; but by the early fourteenth century there were bustling Italian colonies in several Chinese cities, including Peking, and the two east-west trade routes open to Europeans buzzed with activity. Asia by sea could take up to two years—but oh! what sights the traveler saw along the way. The sea route could be picked up in Trebizond, a Greek colony in the

Black Sea, or Tabriz, an Iranian city of thrusting minarets, so fabulously wealthy, a European visitor declared it "worth more to the Great Khan than his whole kingdom to the King of France." From the Crimea and Iran, the route led down to the Port of Ormuz on the tip of the Persian Gulf, thence across the Indian Ocean to Quilon, an Indian kingdom where all the wonders of the seven seas seemed to have gathered under swaying palm trees. Quilon had lumbering elephants and chattering monkeys, local markets that smelled of pepper and cinnamon in the sultry heat, and a port crowded with huge, oceangoing Chinese ships whose sailors sang "la la la" as they rowed. The final stop on the journey was Hangchow, Venice of the East and one of the great wonders of the medieval world. A hundred miles around and guarded by twelve great gates, the city had blue-water canals, fire brigades, hospitals, and fine broad streets lined with houses upon whose doors were listed the names of every occupant. Along Hangchow's canals, spanned by twelve thousand bridges and sailed by gaily colored boats, strolled the greatest wonder of all. This city has "the most beautiful women in the world," declared a breathless Western visitor. In a nearby palace a Tartar khan was served his daily meals by five singing virgins.

However, because sea travel took so long, many Western merchants preferred the quicker overland route. In the Middle Ages there were several variations available, including the fabled Silk Road. But around 1300 a new route across the northern steppe began to gain favor. Travelers found the broad, flat terrain in the north easier on men, animals, and carts, but the new route had a significant disadvantage, though none of the newcomers realized it. It skirted the tarabagan colonies of Siberia, Mongolia, and northwest China.

Prized for its fur, the limpid-eyed, squirrellike tarabagan was—and still is—greatly feared on the steppe for its powers of contagion. In *Memories of a Hunter in Siberia,* A. K. Tasherkasoff, a nineteenth-century Russian writer, described how generations of nomad steppe hunters were weaned on stories of a mysterious tarabagan illness that could jump to humans foolish enough to trap sick animals (identifiable by a wobbly, staggering gait). According to steppe legend, the mysterious, highly contagious tarabagan illness was supposed to be caused by "small worms, invisible to the naked eye," but in 1905, when the first infected animals were autopsied, the invisible worms turned out to be the plague bacillus *Yersina pestis,* leading one scientist to compare the "tarabagan gardens" of the Asian plains to "a heap of embers where plague smolders continuously and from which sparks of infection may dart out . . . to set up conflagrations."

More recent research on the tarabagan also has relevance for the Black Death. The tarabagan is a member of the marmot family, and, according to the Russian scientists who have studied the animals, the strain of *Y. pestis* that circulates among marmots is the most virulent in the world. Besides extreme lethality, marmot plague, as the Russians call it, has another Black Death–like feature. It is the only form of the plague in rodents that is pneumotropic—that is, in tarabagan and other marmots, and only in them, plague has a tendency to spread to the lungs and become pneumonic. On the steppe, dead tarabagans are often found with a bloody froth around the nose and mouth—telltale signs of pneumonic infection.

American microbiologists tend to be skeptical about the Russian claims for marmot plague, but the Russians are so convinced of its virulence and lightning contagion that during the Cold War they bet their national defense on it. According to Wendy Orent, who has worked closely with many Russian scientists, whenever the Soviet Union drew up plans for a new plague weapon, Major General Nikolai Urakov, a leader of the USSR's biological weapons program, would shout to his staff, "I only want one strain"—marmot plague.

Reconstructing any pathogen's genetic history is necessarily an exercise in guesswork, but like the Black Death, *Y. pestis* seems to have originated on the steppe of Central Asia. Microbiologist Robert Brubaker thinks the big bang in *Y. pestis*'s life may have been the end of the last Ice Age. As the ice sheet retreated, the rodent population on the freshly thawed steppe would have exploded, creating an urgent need for a Malthusian pruning mechanism. The development of agriculture, another demographic landmark in rodent history, would have further heightened the need for such an agent.

Y. pestis is only fifteen hundred to twenty thousand years old, young enough to fit Dr. Brubaker's scenario, and its ragged genetic structure certainly suggests an agent slapped together in a hurry to meet an evolutionary emergency. *Y. pestis*'s genome has a great many nonfunctioning genes and three ungainly plasmids. However, in pathogens as in people, appearances can be deceiving. *Y. pestis* has all the properties an infectious agent needs to be a world beater, including biological oomph. One reason many infectious agents fail to rise to the level of lethality is that their bacilli cluster together at an infection site (like a flea bite), instead of spreading to vital body organs. Consequently, nothing more serious than some local swelling and redness develops. *Y. pestis* has solved the cluster problem by evolving special enzymes that deliver plague bacilli to the liver and spleen, from whence the bacilli can be quickly recycled to the rest of the body. Equally important, the bacillus has also learned how to elude almost everything sent to kill it, including flea antigens and human antigens. In the case of flea antigens, the elusiveness gives *Y. pestis* time to multiply in the flea gut, which is a key step in the transmission of plague. In the case of the human antigens, the elusiveness buys the pathogen time to jump from the lymph nodes to the liver and spleen. Like HIV, *Y. pestis* is extremely adept at confusing the human immune system. Often by the time the body can mount a defense, the pathogen has become uncontainable.

Y. pestis can also kill nearly anything put in front of it, including humans, rats, tarabagans, gerbils, squirrels, prairie dogs, camels, chickens, pigs, dogs, cats, and, according to one chronicler, lions. Like other major pathogens, *Y. pestis* has become a successful killer by learning how to be an adaptable killer. It can be transmitted by thirty-one different flea species, including *X. cheopis,* the most efficient vector in human plague, and *Pulex irritans,* the most controversial. Some researchers believe the bite of the ubiquitous *P. irritans,* the human—and pig—flea, contains too few bacilli to transmit plague effectively; but other investigators* suspect that the human

* France is a hotbed of *P. irritans* support. French scholars, including the leading modern historian of the plague, Jean-Noël Biraben, believe that British and American medievalists have seriously underestimated the role of the human flea as a plague vector.

flea plays an important, if unappreciated, role in the spread of the disease. The pro–*P. irritans* school is supported by the work of General Shiro Ishii, commander of the Japanese army's biological warfare unit during World War II.*

Assessing the Japanese deployment of a plague weapon against the Chinese city of Changteh early in the war, an admiring U.S. Army report notes that "one of Ishii's greatest achievements . . . was his use of the human flea, *P. irritans*. . . . This flea is resistant to air drag, naturally targets humans, and could also infect the local rat population to prolong the epidemic. . . . Within two weeks [of the attack] individuals in Changteh started dying of plague."

Y. pestis does have limitations. It cannot survive very long on surfaces like chairs, tables, and floors, and it operates optimally only within a fairly narrow climatic range—air temperatures between 50 and 80 degrees Fahrenheit and humidity above 60 percent and, ideally, 80 percent. A number of animals are also resistant to the plague bacillus, including the Siberian polecat, black bear, skunk, and coyote. Man may also enjoy a degree of immunity to *Y. pestis,* though that is another question fraught with controversy. Despite a recent finding suggesting that CCR5-Δ32, the allele that protects against HIV, may also protect against plague, many scientists remain skeptical of humanity's ability to resist *Y. pestis,* except for a temporary resistance acquired after exposure to the disease.

Among the species that do develop at least a partial immunity to *Y. pestis* is its host population, rodents. Indeed, the plague bacillus's exquisite attunement to the rodent community is one of the great marvels of nature. Most of the time *Y. pestis* and the rodent kingdom live in a state of unhappy but workable coexistence. The scientific term for the modus vivendi is enzootic. Animals continue to get sick and die, but usually there are enough partially resistant rodents in any given

* *P. irritans* and General Ishii may help provide an answer to one of the great mysteries of the Black Death: why do so few medieval sources mention rat die-offs? In modern outbreaks of plague, dead rats usually litter the streets a few weeks before *Y. pestis* jumps into humans. Many authors have tried to explain away the discrepancy by saying that dead rats were such a common sight on the medieval street that no one thought them worth writing about. But there may be another explanation: medieval man's poor hygiene may have made *P. irritans* a significant vector in the Black Death.

community to keep the smoldering embers of infection in check. There are a number of theories about why, from time to time, this biological firewall suddenly collapses and the colony is consumed by the flames of an epizootic—a full-scale outbreak of plague. These include genetic change in the plague bacillus, which makes it more virulent, and demographic changes in the wild rodent community, which make its members more vulnerable to plague. A third theory, not incompatible with the first two, is that epizootics are activated by surge years—sudden, dramatic spurts in the rodent population. No one is sure what causes surges, but a number of scientists believe that they may be related to sun spots, which follow roughly the same cyclical pattern as surge years in many (though not all) rodent species, approximately ten to twelve years.

The connection is not as odd as it sounds. Sun spot cycles—which influence rainfall, tropical cyclones, and tree growth—may affect the wild rodent food supply. Climate changes may make vegetation more abundant, encouraging a burst of overbreeding, perhaps by affecting rodent fertility. Certain species of hare are known to experience cyclical bursts of fecundity. Clearer is what happens during a surge year; rodent populations breed themselves into a classic Malthusian dilemma: too many animals, too little food. And, as Malthus noted, when that happens nature usually prunes the population back to a sustainable level by way of a major demographic catastrophe, such as a famine or infectious disease. In the case of rodents, one component of the pruning mechanism may be an alteration in the community's genetic composition. As rodent numbers spike during the surge, the pool of older, partially immune animals—the community firewall—is diluted by a rapidly expanding group of younger animals who have not yet acquired resistance to *Y. pestis*. This pool of unprotected young may constitute the biological equivalent of an oil slick; throw a match on it and it bursts into flame.

Aside from providing an insight into nature's secret harmonies, the dynamics of rodent populations would not matter much to humans but for the fact that during surge years towns, villages, and campsites are more likely to be invaded by hungry rodents. In one thirty-four-year period on the eastern steppe, four of five plague outbreaks occurred in tarabagan surge years, and the victims were local hunters, men schooled in the dangers of trapping sick animals. If experienced steppe veterans proved vulnerable to *Y. pestis,* it does not take a large leap to imagine what would happen to a group of unwary outsiders unlucky enough to brush up against a tarabagan community in Mongolia, Manchuria, or Siberia in the midst of a surge year, particularly if the tarabagans' food supply was under threat not only from the demographic pressures of the surge but also from long-term ecological changes.

Indeed, it is not even necessary to imagine what might happen; a historical precedent is available. Between 1907 and 1910, the world price of tarabagan skins quadrupled from 0.3 rubles to 1.2 rubles, producing a corresponding increase in the number of tarabagan hunters. Many of the newcomers were unskilled Chinese, looking to turn a fast ruble and ignorant of steppe lore about hunting staggering tarabagan. In April 1910 pneumonic plague broke out among a colony of hunters in Manchuria; within a year sixty thousand people were dead.

In the case of the Black Death, the first group of outsiders to be infected may have been Mongol herdsmen looking for new pastureland. During the fourteenth century, the wind patterns of Eurasia changed, altering the climate of the continent; Europe became wetter, while Asia became drier. The arrival of desertlike weather in the East may have forced Tartar herdsmen out of their traditional pasturelands and into the "tarabagan gardens" of the northern steppe, a region whose dangers they were as ignorant of as were the Chinese hunters in early-twentieth-century Manchuria. From the herdsmen, the plague would have spread to other outsiders: Arab, Persian, Italian, or Central Asian merchants; Tartar horsemen and soldiers; Chinese or Ukrainian laborers—or some combination of all or most of the above. Also easy to imagine is how political and economic changes like the rise of the Mongol Empire and the development of a nascent global economy would have allowed *Y. pestis* to overcome the vast distances, thin populations, and other firewalls that had kept it trapped in the remoteness of inner Asia for centuries. After unifying the fractious steppe under a *pax Mongolica,* the Tartars threw several overlapping communication networks across the vast, open prairies of Asia and Russia, including the Yam, a pony express service, new trade routes, and a network of caravansaries.

William McNeill, author of *Plagues and Peoples,* thinks the caravan rest stops may have played a key role in the early spread of plague. "Assuredly the far-flung network of caravanserais extending throughout Central Asia . . . made a ready made pathway for the propagation of [plague] across thinly inhabited regions. Each resting spot for caravans must have supported a complement of fleas and rats attracted there by the relative massive amount of foodstuffs necessary to keep scores even hundreds of travelling men and beasts going."

If the Black Death originated in or near the Gobi desert, *Y. pestis* would have visited a half dozen such rest stops before climbing a mile in the sky to Lake Issyk Kul, the hot lake where it first burst into history. Warmed by thermal springs that can produce water temperatures of 85 to 95 degrees Fahrenheit, Issyk Kul sits five thousand feet above sea level in a bowl of snow-capped mountain and thick forest. Today the lake region is full of ghosts—Soviet, czarist, even a few Tartar, including the ripply outlines of a submerged village a few hundred feet from the shoreline. But in the mid-fourteenth century, Issyk Kul was a bustling trading center astride the northern steppe route. From the region, eastbound travelers could pick up the fast road into China; westbound travelers, the road home to Caffa, Tabriz, or Baghdad. Medieval Issyk Kul also had a substantial local population. Many of its inhabitants were Nestorians, a Christian sect of Syrian origin that spread across Asia in the early Middle Ages. On arriving in China, evangelical pioneers like John of Monte Corvino were astonished to see church spires rising above cities like Hangchow. "We have found many Christians scattered all over the east and many fine churches, lofty, ancient and of good architecture," declared one Western visitor.

The Nestorians were both an accomplished and a flowery people. The inscriptions on the headstones found in two local Issyk Kul cemeteries speak in the florid language of the funeral oration. One informs the passerby: "This is the grave of Shliha, the celebrated commentator and teacher, who illuminated all the monasteries with light. . . . His voice rang as high as a trumpet." Another: "This is the grave of Pesoha, the renowned evangelist and preacher who enlightened all. . . . Extolled for wisdom and may our Lord unite his spirit with the saints."

In comparison, the inscription on a third Nestorian headstone, that of a husband and wife, Kutluk and Magnu-Kelka, has an almost ominous starkness. No accomplishments are mentioned, no holiness praised. The headstone tells us only enough to suggest the following scenario: one morning in 1339, perhaps a fragrant early-summer morning when the air temperature almost matched the water temperature on the lake, Kutluk awoke with the early symptoms of plague. On that first day he felt lightheaded and nauseous, symptoms so unobtrusive Magnu-Kelka did not even realize her husband was ill until dinner, when Kutluk suddenly vomited into his meal. On the second day of his illness, Kutluk awoke with a terrible pain in his groin; overnight, a hard, apple-sized lump had formed between his navel and his penis. That afternoon, when Magnu-Kelka probed the tumor with a finger, the pain was so terrible, Kutluk rolled over on his side and vomited again.

Toward evening, Kutluk developed a new symptom; he began to cough up thick knots of bloody mucus. The coughing continued for several hours. As night gathered around the lake, a sweaty, feverish Kutluk fell into a delirium; he imagined he saw people hanging by their tongues from trees of fire, burning in furnaces, smothering in foul-smelling smoke, being swallowed by monstrous fish, gnawed by demons, and bitten by serpents. The next morning, while Kutluk was reliving the terrible dream, the cough returned—this time even more fiercely. By early afternoon, Kutluk's lips and chin had become caked with blood, and the inside of his chest felt as if it had been seared by a hot iron. That night, while Magnu-Kelka was sponging Kutluk, the tumor on his groin gurgled. For a moment Magnu-Kelka wondered if the swelling were alive; quickly, she made the sign of the cross. On the fourth day of his illness, Kutluk stained his straw bed with a bloody anal leakage, but Magnu-Kelka failed to notice. After vomiting twice in the morning, she slept until dark. When she awoke again, it was to the sound of crickets chirping in the evening darkness; she listened for a moment, then vomited on herself. On the fifth day of his illness, Kutluk was near death. All day Magnu-Kelka lay on a straw mat on the other side of the cottage, listening to her husband's hacking cough and breathing in the fetid air. Toward evening Kutluk made a strange rattling sound in his throat and the cottage fell silent. As Magnu-Kelka gazed at her husband's still body, she felt an odd sensation—like the fluttering of butterfly wings against the inside of her chest. A moment later, she began to cough.

Kutluk and Magnu-Kelka were almost certainly not the first victims of the Black Death, but their remote little lakeside cottage is where the most terrible natural disaster in history begins to enter the human record.

Two other notable names in the history of plague are Justinian, the sixth-century Byzantine emperor, and Alexandre Yersin, a dreamy young Swiss scientist who became the Yersin in *Yersinia pestis* during the Third Pandemic of a century ago. The human equivalent of an epizootic, a pandemic is a catastrophic outbreak of infectious disease; including the Black Death, three times in recorded human history plague has risen to the level of pandemic. The first time, the Plague of Justinian, is where the story of man and *Y. pestis* begins, and the last time, the Third Pandemic, is where the mysteries of the plague bacillus were finally unraveled. The reemergence of large-scale pestilence in late-nineteenth-century China horrified the self-confident Victorians, occasioning an early example of what today is called Big Science. In the 1890s, as *Y. pestis* swept through China and India, researchers from dozens of countries focused their energies on a single, urgent question: "What causes plague?"

In the end, the worldwide race narrowed down to one city, Hong Kong, and two young men, each a surrogate for the two great microbiologists of the Victorian age, the Frenchman Louis Pasteur and the German Robert Koch. Koch's surrogate, a former student named Shibasaburo Kitasato, was a heavyset, ambitious young man who wore a starched wing-tip collar even in the sultry Hong Kong heat and enjoyed the seemingly unbeatable advantages of modern equipment, a large staff, and devious mind. Pasteur's surrogate was the moony Yersin, a Somerset Maugham-ish figure, who gave up a life of privilege in the West to search for Higher Truth in the East. In a film about the race to identify *Y. pestis,* Leslie Howard would have played Yersin.

Despite having fewer resources and lacking the inspired duplicity of Kitasato ("The Japanese . . . have bribed the staff of the hospital so that they will not provide me with [any bodies for] autopsy!" Yersin complained to his mother), the young Swiss investigator became, in 1894, the first person to accurately describe the plague pathogen. "The pulp of the buboes always contains short, stubby bacilli," he noted in one of the most important papers ever written about human disease. A few years later a Frenchman named Paul-Louis Simond identified the rat and rat flea, *X. cheopis,* as plague vectors. In 1901 Kitasato's mentor, Robert Koch, summarized the new research this way. Plague, said Koch, is "a disease of rats in which men participate." A few decades later the first effective antiplague medications began to become available.

If the Third Pandemic is where the story of man and plague ends—at least for now—then the sixth-century Plague of Justinian is where it begins. There are several references to what sounds like plague in the Bible, but *Y. pestis* did not officially enter human history until A.D. 542, when it strode off a ship in the Egyptian port of Pelusium. From a modern perspective, the most striking thing about the Plague of Justinian is how closely it resembles the Black Death. There is, first of all, the crucial role of commerce in disseminating disease. Until a trade route from Egypt made it more accessible, Ethiopia, the probable home of the First Pandemic, was as remote from the major population centers of late Antiquity as the other plague foci of the premodern world: the Eurasian steppe (including Siberia and the Gobi desert), Yunnan in China, and perhaps Kurdistan in Iran. Like the Black Death, the Plague of Justinian also occurred during a period of extreme ecological change. A recent study of two thousand years of tree ring data reveals that four years in the last two millennia were periods of extraordinarily severe weather, and two of those four years were situated in or around a plague pandemic. One was 1325, roughly the time *Y. pestis* may have been at work on the rodent population of the Gobi or some other region of inner Asia; the other was 540, two years before *Y. pestis* arrived in Pelusium, and roughly around the time that Chinese scribes were describing yellow dust falling like snow and Europeans were complaining about the bitter cold that ushered in the Dark Ages. "The winters [are] grievous and more severe than usual," wrote the sixth-century monk Gregory of Tours. "The streams are held in chains of frost and furnish . . . a path for people like dry ground. Birds too [are] affected by the cold and hunger and [are] caught in the hand without any snare when the snow [is] deep."

In the Plague of Justinian, one also hears for the first time a sound that becomes overpowering during the medieval pestilence: the sound of human beings drowning in death. "At the outset of this great misfortune," wrote a lawyer named Evagrius, "I lost many of my children, my wife and other relatives and numerous estate dwellers and servants. . . . As I write this in the 58th year of my life . . . I [have recently] lost another daughter and the son she has produced quite apart from other losses."

No one knows how many people died in the Plague of Justinian, but in Constantinople, where the daily mortality rate is said to have reached ten thousand at the peak of the epidemic, people put on name badges so that they could be identified if they fell on the streets. The mortality was also very severe in the Middle East. "In every field from Syria to Thrace the harvest lacked a harvester," wrote John of Ephesus, who went to bed each night expecting to die and awoke each morning surprised to find himself still alive. Untold thousands also perished in Italy and in France, where Gregory of Tours reported that "soon no coffins or bearers were left."

The Plague of Justinian marked an important turning point in Europe's relationship with infectious disease. The centuries preceding the First Pandemic were a period of chronic, devastating epidemics. In the second and third centuries, smallpox and measles outbreaks may have killed a quarter to a third of the population in parts of the Roman Empire. The centuries after Justinian were, if not disease-free, then close to it. During the early Middle Ages, all forms of infectious illness became uncommon and plague (as far as is known), nonexistent. For this disease-free interim, the collapse of civilization deserves some credit. A disease is more than just a pathogen plus a transportation system. To ignite an epidemic, a friendly environment is also necessary, and after Rome fell, the environment in Europe, particularly northwest Europe, became increasingly unfriendly to epidemic disease. In the early Middle Ages, the population plunged precipitously, shrinking the pool of potential host-victims. In the second and third centuries, Roman Europe had 50 to 70 million people; by 700, Europe had 25 to 26 million. The disappearance of urban life removed two other necessities: concentrations of people, and filthy, rodent-infested streets. At its height, the city of Rome had a population variously estimated at half a million to ten million; by 800, no city in Europe contained more than twenty thousand residents. "In the middle of the debris of great cities," wrote one Dark Ages scribe, "only scattered groups of wretched peoples survive."

The resurgent forest provided a further barrier against infectious illness. By A.D. 800, dense woodland had reclaimed 80 percent of a depopulated Europe's surface, severely restricting trade and travel and providing a firewall, which helped to keep local epidemics local. The international situation added a final layer of insulation. By the ninth century, the principal east-west trade routes were all in unfriendly Muslim hands.

Around the year 1000, this process began to reverse itself and Europe started to re-create the environmental conditions associated with demographic collapse in premodern societies. And indeed, four hundred years later the West would suffer a second great demographic catastrophe, but that gets us ahead of our story, which begins where stories of pandemics often do: in a burst of human progress.

Sometime between 750 and 800, Europe entered the Little Optimum,* a period of global warming. Across the continent, temperatures in-

* The Big Optimum lasted from the end of the Ice Age to roughly 1300 B.C. Perhaps significantly, this may also have been the period when *Y. pestis* evolved. ("Climatology," *Dictionary of the Middle Ages,* ed. Joseph Strayer [New York: Charles Scribner, 1982], p. 456.)

creased by an average of more than 1 degree Celsius, but, rather than producing catastrophe, as many current theorists of global warming predict, the warm weather produced abundance.* England and Poland became wine-growing countries, and even the inhabitants of Greenland began experimenting with vineyards. More important, the warm weather turned marginal farmland into decent farmland, and decent farmland into good farmland. In the final centuries of Roman rule, crop yields had fallen to two and three to one—a yield represents the amount of seed harvested to the amount planted: a return so meager, the Roman agricultural writer Columella feared that the land had grown old. In the eleventh and twelfth centuries, as winters became milder and summers warmer and drier, European farms began to produce yields of five and six to one, unprecedented by medieval standards.

A burst of technological innovation added to agricultural productivity. Someone figured out that one easy (and cheap) way to get a horse to pull more was to redistribute weight away from its windpipe, so when it moved forward it wouldn't choke. Thus was born the horse collar, which increased horsepower by a factor of four. Another simple innovation, the horseshoe, increased it even more, by improving the horse's endurance. The new *carruca* plow, with its large, sharp rectangular blade, also represented an important improvement, particularly in northern Europe, where the soil was heavier and harder to turn. However, the true technological marvels of the age were the watermill and the windmill; for the first time in human history, a society had harnessed a natural source of power. "Behold," wrote an admiring monk, in a soliloquy to his abbey's watermill, "the river . . . throws itself first impetuously into the mill . . . to grind the wheat . . . separating the flour from the bran. [Then] . . . [it] fills the cauldron . . . to prepare drinks for the monks. . . . Yet, the river does not consider itself discharged. The fullers [wool workers] near the

* Says Dr. Phillip Stott, professor emeritus of bio-geography at the University of London, "What has been forgotten in all the discussion of global warming is a proper sense of history. . . . During the medieval warm period, the world was warmer than even today and history shows that it was a wonderful period of plenty for everyone." (Phillip Stott, interview, *Daily Telegraph*, 4/6/2003.)

mill call [it] to them. . . . Merciful God! What consolations you grant to your poor servants." Another important innovation was a new crop rotation system, which kept more of the land under cultivation during the year.

As agricultural productivity improved, living standards rose, producing a baby boom of historic proportions. The demographic surge of the Central Middle Ages was as dramatic as the decline of five hundred years earlier. Between 1000 and 1250 the population doubled, tripled, and may even have quadrupled. Around 1300 Europe held at least seventy-five million—and some scholars think as many as a hundred million—people, up from twenty-six million during the Dark Ages.* In France the population jumped from five million to about sixteen to twenty-four million; in England, from a million and a half to five to six to seven million; in Germany, from three million to twelve million; and in Italy, from five million to ten million. In 1300 parts of Europe were more populous than they would be again until the eighteenth and nineteenth centuries. Britain, for example, would not see six million people again until the American Revolution, and France would not reach seventeen million again until Napoleon's time, while Tuscany would not have two million inhabitants again until 1850.

As the population soared, urban life reawakened. Pre–Black Death Paris had about 210,000 residents; Bruges, site of the rapidly growing cloth industry, 50,000; London, 60,000 to 100,000; and Ghent, Liege, and Ypres, 40,000 each. In Florence, banker Giovanni Villani boasted that "five to six thousand babies are born in the city each year." But Florence, with a population of 120,000, took a backseat to Milan, with its population of 180,000. Siena, Padua, Pisa, and Naples were the little brothers of the Italian peninsula, with populations of 30,000-plus, but even they would have been major cities in the year 1000.

The medieval countryside also filled up. In Germany's Moselle Valley, the 340 villages of the year 800 quadrupled to 1,380 villages by 1300. Many parts of rural France experienced equally spectacular growth; Beaumont-le-Roger county in Normandy would not have

* Medieval demographic data are very rough approximations and should be read that way.

thirty thousand residents again until the twentieth century; and San Gimignano, in Tuscany, is still smaller than it was in 1300.

In the village of Broughton in England, the population reached a medieval high of 292 souls around 1290.

As the population went up, the forests came down. During the Great Clearances of the twelfth and thirteenth centuries, Europeans burst out of the enormous woodlands that had held them prisoner since the Dark Ages and began to reassert human dominance over the environment. From Scotland to Poland, the great dark woods echoed with the song of human progress: sawing and banging, and the boom and thud and thump of falling timber. Swamps were drained, pastures cleared, fields laid out, crops planted, homes built, and villages erected. Sunlight fell on land that had not felt its warmth since before Justinian's time. Under the press of an expanding population, the continent also thrust outward. In the south, land-hungry Christian kings and colonists—their hearts full of God and avarice—pushed the once invincible Muslims down the Spanish peninsula until, by 1212, only Grenada on the southern tip of Iberia failed to fly the flag of the "Reconquista." In the East, German and Flemish colonists pushed across the Elbe to settle the still-dense forests of eastern Germany and Prussia; in the Danube Valley, streams of carts and horses were flowing into what would become Austria and Hungary.

As the population grew, trade revived. In the year 1000 an Italian merchant had virtually no chance of doing business in England. By 1280 a trader—or a pathogen—could travel though a reinvigorated, reconnected Europe with relative ease. The Atlantic and Mediterranean regions were linked by a land route that wound through the meadows of the high Alpine passes and a sea route that looped around the Pillars of Hercules—Gibraltar—and ended in the busy port of London, where dinosaur-necked wooden cranes were used to unload arriving ships. There were also dozens of new regional trade networks: some originated in Flanders, home of a wealthy new bourgeoisie mad for jewels and spices; some in Germany, seat of the Hanseatic League, an association of Baltic merchants. Another important commercial network sprang up around Champagne, site of the Champagne (Trade) Fairs where, once a year, local servant girls, laundresses, and tradeswomen would become prostitutes-for-a-day to entertain merchants from as far away as Iceland, and where crafty Sienese and Florentine bankers offered loans with so many strings attached that a borrower could be excommunicated from the Church and face eternal damnation should he default.

From the bustling ports of Venice and Genoa, another set of trade routes led southeastward across the Mediterranean to the great trading cities of the Middle East, where the air smelled of mango and palm and the call to prayers echoed through alabaster streets five times a day. In Alexandria (Egypt), in Aleppo (Syria), in Acre or Tyre (the Kingdom of Jerusalem), a shopper could buy sugar from Syria, wax from Morocco and Tunis, and camphor, alum, ivory tusks, muslin, ambergris, musk, carpets, and ebony from Quinlon, Baghdad, and Ceylon—but, alas, were he a Christian shopper, only at unconscionably high prices. In Alexandria, local tolls added 300 percent to the price of Indian goods, and that 300 percent was in addition to the enormous markup that the Arab middlemen took off the top.

Early in the thirteenth century the Venetians, who described themselves as "rulers of half and a quarter of the Roman Empire," devised an ingenious bit of mischief to circumvent the greedy Arabs and deal directly with the East. Venetian authorities offered a group of French Crusaders free passage to the Holy Land, then rerouted the Crusaders east to capture Constantinople. While the plan succeeded brilliantly—the Venetians even managed to steal four great gilded horses for St. Mark's Cathedral—Constantinople, where the rival Genoese soon had a base, was still a long way from the timber, fur, and slaves of the Crimea and southern Russia, farther away from the great market towns of Central Asia like Samarkand and Merv, and light-years away from the emerald city of Hangchow.

For the "two torches of Italy," as Petrarch called Venice and Genoa, the Bosporus nights were full of frustration and longing—but relief was nigh.

According to legend, on a cold morning in 1237, three anonymous riders emerged out of a lightly falling snow in front of Ryazan, a town near the eastern border of medieval Russia. The small party halted for a moment; then one rider broke free and dashed across the snowy ground toward Ryazan, shouting. Attracted by the noise, a crowd gathered at the town gate. "A witch," said one townsman, pointing at the rider, who had turned out to be a woman of astonishing ugliness. "No," said a second townsman, "a sorceress." As the two argued, the rider continued to dash back and forth in front of Ryazan, shouting, "One-tenth of everything! Of horses, of men, of everything! One tenth!"

In a second version of the Sorceress of Ryazan story, the female rider, apparently selected for her knowledge of the local dialect, demanded "one tenth of everything" from a group of Russian princes gathered in Ryazan. But in both versions, the end is the same. The Tartar demand for tribute is spurned, the mysterious sorceress vanishes, and her visit is forgotten.

Then, on a winter morning a few months later, a thunderous rumble awakens the town. Doors fly open, heads appear, half-dressed men rush into the street. Someone shouts; fingers point. To the east, a black band of horsemen is hurtling across the horizon toward Ryazan under a dawn sky. Hurriedly, children are slipped under floorboards or hidden under blankets and quilts; doors are bolted; swords unsheathed, prayers whispered. As the short-legged Mongol ponies clear the earthworks in front of Ryazan, the morning streets fill with slashing, cutting horsemen. People scream, body parts fly, pools of blood form in the fresh snow. Plumes of black smoke rise into a vermilion sky. All morning and into the early afternoon, under a dim winter sun, Ryazan is methodically, systematically exterminated. Children are killed along with parents, girls along with boys, old along with young, princes along with peasants. Later a Russian chronicler will write that the citizenry was slaughtered "without distinction to age or rank."

Ryazan was not the Mongols' first appearance on the western steppe. Twenty years earlier, the Tartars had made another brief foray into medieval Russia, but that raid had been more in the nature of an evil rumor. Afterward the chronicler of Novgorod wrote, "For our sins, unknown tribes came among us. . . . God alone knows who they are or where they came from." By contrast, Ryazan was part of a grand design of world conquest. Genghis Khan means "Emperor of Mankind," and though the founder of the Mongol Empire was ten years dead when Ryazan fell, his universalist ambitions lived on in his sons and grandsons. After subduing most of northern China in the 1210s and Central Asia in the 1220s, the Mongol leadership held a *kuriltai* (grand assembly) in 1235, where it was decided to move against the West.

Europe knew nothing about the *kuriltai,* but during the 1230s enough rumors had drifted westward across the steppe to create a profound sense of unease. There were stories of terrible massacres in Central Asia and, after Ryazan and other Russian towns fell, almost daily rumors of a Tartar invasion. In 1238 the fisheries of Yarmouth shut down because their German customers had become too frightened to travel. In the late 1230s the immediacy of the danger was underscored when one of Christendom's most implacable foes, the "Old Man of the Mountain," leader of the fanatical Muslim "Assassins of Iran," reportedly sent an envoy to Europe to propose a joint alliance against the Tartars. Whether true or only a rumor, the story was as shocking to contemporaries as the Hitler-Stalin pact was in its time.

On April 9, 1241, with a conquered Russia in ruins, the cream of European arms gathered on a Polish field to meet the Mongol onslaught. After the battle, the Tartars sent home nine sacks of ears. Two days later a large Hungarian army was crushed at Mohi; shortly thereafter, a small Tartar force appeared in the vicinity of Vienna. As eastern Europe filled with refugees, panic gripped western Europe. In Germany rumors that the Tartars were Gog and Magog, the two lost tribes of Israel, ignited pogroms against the Jews. In France a knight warned Louis IX that the Mongols would soon be on the Somme. In England the monk Matthew Paris predicted a bloodbath of unimaginable proportions. The Mongols, he said, are "monsters rather than men, . . . inhuman and beastly, thirsting for and drinking blood and devouring the flesh of dogs and men, and striking everyone with terror and incomparable horror."

In Rome the pope received a letter from Grand Khan Ogedi. It read: "You personally, as the head of all kings, you shall come, one and all, to pay homage to me and serve me. Then we shall take note of your submission. If . . . you do not accept God's order, we shall know that you are our enemies."

However, Europe's string of good fortune had not quite run its course. Just as the apocalypse seemed about to burst upon Christendom, dissension erupted in the Mongol royal family and offensive operations were halted in the West. This provided a breathing space for clerical visitors like William to improve relations between East and West. Thus, in the 1250s, when the Mongols next went on the offensive, it was not against Christendom, but against an older enemy. In the 1220s the Mongol hordes had crushed Islamic power in Central Asia; now they would crush it again, this time in the Muslim heartland, the Near East and Middle East.

On hearing of Baghdad's fall in 1259, a Christian chronicler exulted: "Now, after five hundred years, the measure of the city's inequity [is] fulfilled and she [is] punished for all the blood she [has] shed."

A decade later, the Genoese were in Caffa and the Venetians in Tana at the mouth of the Don, and a few years after that young Marco Polo was crossing the Gobi Desert, observing the local wildlife. The ubiquitousness of one species in particular struck him. "There are great numbers of Pharaoh's rats in burrows on [these] plains," he noted.

"Pharaoh's rat" was a medieval term for the tarabagan.

CHAPTER THREE

The Day Before the Day of the Dead

HIDDEN IN A DEEP AND SECLUDED VALLEY, THE VILLAGE OF Broughton has two brooks, two streets, and not enough acreage to warrant the attention of the larger world. On most maps the village lies in the terra incognita of gray-green space between Huntingdon and Peterborough. Indeed, except for the local church spire, which rises above the valley wall like the hand of a drowning man, Broughton would be a rural Atlantis, secreted away on a few thousand acres of Oxford clay in the green and pleasant English countryside.

Like many medieval villages, Broughton began life as a forest clearing. Three hundred years before villager John Gylbert was born, the tree line came right up to the front door, but by 1314—the year John turned nineteen—the enveloping forest had been cut down, replaced by neat checkerboard squares of gold and green farmland and pasture. Coming up the road from Huntingdon on a summer morning, Broughton would rise up before the medieval traveler like a thatched-roof island adrift on a sunlit sea of swaying oats and barley. In John's time Broughton had some 268 residents, down slightly from its medieval high of 292, but not significantly down. The size of the local animal population is unrecorded, but cows, chickens, pigs, and horses, just beginning to replace oxen at the plow, were ubiquitous in Broughton. Animals roamed the village lanes and gardens like curious sightseers, peering into doors, sunning themselves in rosebeds, eyeing the old men in front of the alewife's house. In the evening, while two-footed Broughton drank, cooked, argued, and made love in one room, four-footed Broughton slept, ate, and defecated in another room—or sometimes the same room.

As far as medieval Broughton can be said to have left behind a collective biography of itself, it resides in the annual round of births, deaths, marriages, misdemeanors, bills of sale, and suits noted in the local court records. These show that while John Gylbert was growing up in the first decade of the fourteenth century, Broughton was anglicizing itself. In 1306 William Piscator became William Fisser or Fisher (the English equivalent of Piscator); a few years later Richard Bercarius became Richard Sheppared (the English equivalent of Bercarius), and Thomas Cocus, Thomas Coke. John was probably born Johannes, and his friend Robert Crane, Robertuses. Only the eponymously named John de Broughton resisted the anglicizing trend, perhaps because de Broughton, a humbly born man who had come up a bit in the world, could not resist that fancy-sounding French "de."

Local court records also show that Broughton, like many small villages, had its share of scandals. Between 1288 and 1299 John's great-aunt Alota was arrested four times for brewing substandard ale. The records do not give a reason for the arrests, but it was not uncommon for an alewife to spike her product with hen excrement to hasten fermentation. Alota's husband, Reginald, also makes an appearance in the court records; in 1291 Reginald was charged with committing adultery with "a woman from Walton." As far as is known, Alota made no public comment about the case, but it is perhaps significant that after her next arrest, Alota appeared in court on the arm of another village man, John Clericus, who lived a few doors down from the Gylberts.

John Gylbert's name also appears in village court documents. In early February 1314 John was fined for drinking ale and playing alpenypricke—a kind of hurling game—with Robert Crane and Thomas Coke in a wood near Broughton when he should have been at Ramsey Abbey, working. Broughton was part of the abbey manor, which in the ethos of feudalism meant that its villagers owed the monks a portion of their labor.

As an abbey villein, or serf, John was required to spend two days each week toiling in the monks' demesne, or personal farmland. In return for his labor, on work days John would receive an *alebedrep,* a lunch served with ale, or, if the monks were in an ungenerous mood, an *waterbedrep,* a lunch served with water. But even the *alebedrep,* which came with thick slices of warm bread and the smiles of the servant girls, was meager compensation for the bite of the February wind on the abbey fields and the kick of a heavy iron plow against an aching shoulder. At harvest time, when John's abbey obligations doubled, he would spend ten hours in the monks' fields under a blazing August sun, walk back to Broughton in the gathering twilight, work into the night on the Gylberts' farm, then fall asleep on a straw mat listening to the heavy breathing of the oxen in the next room.

In Broughton, John's future surrounded him like a death foretold. It was there in his father's lame leg and in his uncle's deformed back (spinal deformations, arthritis, and osteoarthritis were rife among the medieval peasantry), and it was there, too, in the worn faces of the village's thirty-year-olds. John would work hard, die young—probably before forty—and, as sure as the sun rose into the cozy English sky above Broughton each morning, the day after his death an abbey official would be at the door to claim his best horse or cow from his widow as a heriot, or death tax.

Thus it had always been. But at least in the boom years of the thirteenth century, a peasant had a reasonable chance of being rewarded for his hard labor. Good weather—and good soil—made it relatively easy to grow surplus crops, and the booming towns provided an eager market, not only for the extra wheat and barley, but also for peasant handicrafts. If a man owned a little land, as peasants increasingly did in the thirteenth century, he could also count on its value rising. By John Gylbert's time, all these compensations were vanishing.

Between 1250 and 1270 the long medieval boom sputtered to an end. One of the great ironies of the Black Death is that it occurred just as the medieval global economy, the vehicle of *Y. pestis*'s liberation, was nearing collapse. However, in Europe, it was the implosion of the vastly larger domestic economy, particularly the agricultural economy, that people felt most keenly. The implosion was continentwide, but in England, a nation of meticulous record keepers, it was documented with great diligence. Around 1300 the acreage under plow decreased, while the land still in use either declined in productivity or stagnated. After centuries of heady advancement, the medieval peasant's mistakes had caught up with him. Some of the good land brought into service during the Great Clearances of the twelfth century had been overfarmed, and some of the more marginal land, which never should have been cleared in the first place, was giving out entirely.

Paradoxically, the decline in productivity was accompanied by a long-term decline in the price of staples like wheat and barley. As the economy faltered, living standards fell and large pockets of grinding poverty began to appear. Many despairing peasants simply gave up. First individual farms were abandoned, then whole villages. In 1322 officials in west Derbyshire reported that six thousand acres and 167 cottages and houses lay empty. Urban trade and commerce also declined. In the early fourteenth century, rents in central London were cheaper than they had been in decades, and the serpentine London lanes were full of gaunt-faced beggars and panhandlers. In the postboom collapse, even imports of claret, that staple of the English well-to-do, fell. In the villages and towns of France, Flanders, and Italy, the story was much the same. By 1314 millions of people were living in abject poverty, and millions more were only a step away from it.

Europe's abrupt descent into semidestitution invites a Malthusian interpretation of the Black Death. During the twelfth and thirteenth centuries, population expanded faster than resources, and as sure as night follows day, in the fourteenth century the continent paid for its heedless growth with economic ruin and demographic disaster. However, the facts tell a more nuanced story. In a traditional Malthusian scenario—say, a tarabagan community in a surge year—population continues to grow recklessly until disaster slips up on it like a mugger in the night. In Europe that did not happen; the baby boom and economic boom both ended around the same time—somewhere between 1250 and 1270. After the stall, living standards fell in many regions and stagnated in others, indicating that the balance between resources and people had become very tight, but since demographic disaster was averted for nearly a century before the plague, a Malthusian reckoning may not have been inevitable. "Many . . . went hungry and many were undoubtedly malnourished," says historian David Herlihy, "but somehow people managed to survive. . . . Circa 1300, the community was successfully holding its numbers."

Rather than a reckoning, the image of postboom Europe that comes to mind is that of a man standing up to his neck in water. Drowning may not be inevitable, but the man's position is so fraught, even a very slight rise in the next tide could kill him. As Dr. Herlihy asserts, a crowded Europe may well have been able to hang on for "the indefinite future," but, like the man in the water, after the land gave out and the economy collapsed, the continent had no margin for error. Just to continue keeping its head above water, everything else had to go right, and in the early fourteenth century, a great many things began to go terribly wrong, beginning with the climate.

The Swiss farmers in the Saaser Visp Valley may have been the first people in Europe to notice that the weather was changing. Sometime around 1250 the resurgent Allalin glacier began to reclaim the farmers' traditional pasturelands. Or the Greenlanders may have been the first to notice the change, alerted by the sudden chill in the August nights and the appearance of ice in places it had never been seen before. "The ice now comes . . . so close to the reefs none can sail the old route without risking his life," wrote the Norwegian priest Ivar Baardson. Or the first Europeans to realize that the Little Optimum was over may have been the fishermen on the Caspian Sea, where torrential rains produced a rise in the water level at the end of the thirteenth century.

In the European heartland, the Little Optimum gave way to the Little Ice Age around 1300.* People noticed that the winters were growing colder, but it was the summers, suddenly cool and very wet, that alarmed them. By 1314 a string of poor and mediocre harvests had sent food prices skyrocketing. That fall, every peasant in every sodden field knew: one more cold, wet summer, and people would be reduced to eating dogs, cats, refuse—anything they could get their hands on. As the summer of 1315 approached, prayers were offered up for the return of the sun, but, like a truculent child, the cold and wet persisted. March was so chilly, some wondered if spring would ever return to the meadows of Europe. Then, in April, the gray skies turned a wicked black, and the rain came down in a manner no one had ever seen before: it was cold, hard, and pelting; it stung the skin, hurt the eyes, reddened the face, and tore at the soft, wet ground with the force of a plow blade. In parts of southern Yorkshire, torrential downpours washed away the topsoil, exposing underlying rock. In other areas, fields turned into raging rivers. Everywhere in Europe in the bitter spring of 1315, men and animals stood shivering under trees, their heads and backs turned against the fierce wind and rain. "There was such an inundation of waters, it seemed as though it was the Flood," wrote the chronicler of Salzburg.

Flanders experienced some of the worst downpours. Day after day, the crackle and boom of thunder echoed above Antwerp and Bruges like a rolling artillery barrage. Occasionally a bolt of lightning would strike, illuminating the network of cascading urban rivers

* The starting date of the Little Ice Age is a source of controversy. Most authorities date it from 1300, when the Alpine glaciers began to advance again, but some experts insist that the true Little Ice Age did not begin until the 1600s, when temperatures turned bitterly cold.

below. Along the riverbanks, rows of soot-stained rectangular houses leaned into the narrow Flemish streets like drunks in blackface. Everywhere, ceilings and floors leaked, fires refused to light, bread molded, children shivered, and adults prayed. Occasionally the rain would stop and people would point to a golden tear in the gray sky and say, "Thank God, it's over!" Then the next day, or the day after that, the sky would mend itself and the rain would begin all over again.

All through the terrible summer of 1315, angry walls of rain swept off the turbulent Atlantic: bursting dikes, washing away villages, and igniting flash floods that killed thousands. In Yorkshire and Nottingham, great inland seas developed over the lowlands. Near the English village of Milton, a torrential rain inundated the royal manor. In some areas, farmland was ruined for years to come; in other places, it was ruined forever.

Poorer peasants, who had been pushed onto the most marginal farmland during the Great Clearances of the twelfth century, suffered the greatest devastation. In three English counties alone, sixteen thousand acres of plow land were abandoned. "Six tenants are begging," wrote a resident of one Shopshire village. By the end of summer, the six would become hundreds of thousands. Everywhere in Europe in the early autumn of 1315, the poor huddled under trees and bowers, listening to the rain beat a tattoo against leaf and mud. They walked the fields, "grazing like cattle"; stood along the roads, begging; searched behind alehouses and taverns for moldy pieces of food. Visiting a friend, a French notary encountered "a large number of both sexes . . . barefooted, and many, even excepting the women, in a nude condition." To the north in Flanders, one man wrote that "the cries that were heard from the poor would move a stone."

The harvest of 1315 was the worst in living memory. The wheat and rye crops were stunted and waterlogged; some oat, barley, and spelt was redeemable, but not very much. The surviving corn was laden with moisture and unripened at the ears. In the lower Rhine "there began a dearness of wheat [and] from day to day prices rose." The French chronicles also mention the *"chierté"* (dearness) of food prices *"especiaument à Paris."* In Louvain the cost of wheat increased

320 percent in seven months; in England, wheat that sold for five shillings a quarter in 1313 was priced at forty shillings just two years later. Across the English countryside that autumn, the poor did their sums; a year's worth of barley, the cheapest grain, cost a family sixty shillings, the average laborer's annual wage was half that amount. The price of beans, oats, peas, malt, and salt rose comparably. Even when food was available, washed-out bridges and roads often prevented it from being transported.

The early winter months of 1316 brought more suffering. As food grew costlier, people ate bird dung, family pets, mildewed wheat, corn, and finally, in desperation, they ate one another. In Ireland, where the thud of shovels and the tearing of flesh from bone echoed through the dark, wet nights, the starving "extracted the bodies of the dead from the cemeteries and dug out the flesh from their skulls and ate it." In England, where they consider the Irish indecorous, only prisoners ate one another. "Incarcerated thieves," wrote the monk John de Trokelowe, ". . . devoured each other when they were half alive." As the hunger intensified, the unspeakable became spoken about. "Certain people . . . because of excessive hunger devoured their own children," wrote a German monk; another contemporary reported, "In many places, parents, after slaying their children, and children their parents, devoured the remains."

Many historians think the accounts of cannibalism are overblown, but no one doubts that human flesh was eaten.

In the spring of 1316, public order began to break down. In Broughton, Agnyes Walmot, Reginald Roger, Beatrice Basse, and William Horseman were exiled for stealing food. In Wakefield, Adam Bray had his son John arrested for removing a bushel of oats from the family farm. In dozens of other English villages, there were violent disputes over gleaning. Traditionally, corn discarded by harvesters became the property of the very poor, but with destitution everywhere, even wealthy peasants were on their knees in the sodden fields. That summer, more than one man had his throat slit over the leavings of a failed crop. As the violence mounted, men began to take up arms; the knife, the sword, the club, and the pike became the new tools of the peasant. Food or anything redeemable for food was stolen, and the stealing went on at sea as well as on land. With incidents of piracy mounting daily, in April 1316 an alarmed Edward II, the English king, instructed his sailors to "repulse certain malefactors who have committed manslaughter and other enormities on the sea upon men of this realm and upon men from foreign parts coming to this realm with victuals."

All through May and June 1316, the rain continued. In Canterbury desperate crowds gathered under a brooding channel sky to pray for "a suitable serenity of the air," but to no avail. In Broughton torrential downpours pressed the wheat and barley against the sodden earth with such force, the stalks looked as if they had been ironed. In Yorkshire the waterlogged fields of Bolton Abbey, tormented by eighteen months of unceasing rain, gave out entirely. The abbey's 1316 rye crop was 85.7 percent below normal. The second failed harvest in succession broke human resistance. There was the "most savage, atrocious death," "the most tearful death," "the most inexpressible death." Emaciated bodies winked out from half-ruined cottages and forest clearings, floated facedown in flooded fields, coursed through urban rivers, protruded from mud slides, and lay half hidden under washed-out bridges. In Antwerp burly stevedores serenaded the waking city with cries of "Bring out your dead!" In Erfurt, Germany, rain-slicked corpses were tossed into a muddy ditch in front of the town wall. In Louvain the collection carts "carried pitiable little bodies to the new cemetery outside the town . . . twice or thrice a day." In Tournai Gilles li Muisis, a local abbot, complained that "poor beggars were dying one after the other."

As if in sympathy with the human suffering, Europe's animals began to die in great numbers; some sheep and cattle succumbed to liverfluke; some, possibly, to anthrax. But rinderpest—a disease that produces discharges from the nose, mouth, and eyes, chronic diarrhea, and an overpowering urge to defecate—may have been the most common killer. In the watery June and July 1316, the music of summer included the agonizing bleats of dying animals vainly trying to relieve themselves in muddy pastures.

Strange diets, putrid food, and a generally lower resistance to disease also produced a great many hard human deaths. Of ergotism, which seems to have been especially common, one English monk wrote, "It is a dysentery-type illness, contracted on account of spoiled food . . . from which follow[s] a throat ailment or acute fever." However, this description does not do justice to the full horrors of ergotism, which was called St. Anthony's fire in the Middle Ages. First, the ergot fungus, a by-product of moldy wheat, attacks the muscular system, inducing painful spasms, then the circulatory system, interrupting blood flow and causing gangrene. Eventually the victim's arms and legs blacken, decay, and fall off; LSD-like hallucinations are also common. If the Irish Famine of 1847 is a reliable indicator, vitamin deficiencies were also rife. Between 1315 and 1322, when the rain finally stopped, many people must have become demented from pellagra (a niacin deficiency) or been blinded by xerophthalmia (a vitamin A deficiency). Typhus epidemics may have killed many thousands more.

The fortunate died of starvation, a condition whose end stage symptoms include brown and brittle skin, the abundant growth of facial and genital hair, and ebbing away of the desire for life.

John Gylbert, whose name vanishes from Broughton's records after 1314, may have died such a death. After months of wandering through mist and rain with a patriarch's beard and dead man's eyes, one day John may have sat down in a field, looked up into the sky, and, like thousands of other Europeans of his generation, concluded that there was no point in ever getting up again.

The Great Famine, the collective name for the crop failures, was a tremendous human tragedy. A half-million people died in England; perhaps 10 to 15 percent of urban Flanders and Germany perished; and a large but unknowable segment of rural Europe also succumbed.

Devastating as the Great Famine was, however, it was only a harbinger of things to come.

People who lived through the Black Death took the connection between plague and malnutrition as a given, the way we do the connection between cigarettes and lung cancer. The Florentine Giovanni Morelli attributed the city's 50 percent plague mortality rate to the severe famine that struck central Italy the year before. Not twenty out of a hundred had bread in the countryside, he wrote. "Think how their bodies were affected." The Frenchman Simon Couvin also described malnutrition as a handmaiden of plague. "The one who was poorly nourished by unsubstantial food fell victim to the merest breath of the disease," he observed. However, many modern historians question the link between plague and malnutrition. For every Florence, they point to a counter example where Black Death losses were moderate or light, despite a recent history of famine. Critics also point to another inconsistency. In the years between the Great Famine and the plague, diets actually improved somewhat. If people were eating better, they ask, how could nutrition have been a predisposing factor in the Black Death?

However, it may be that critics have failed to find a connection between plague and malnutrition because they have been looking in the wrong places. The regional outbreaks of the disease that occurred after the Black Death—the epidemics of 1366–67, 1373, 1374, 1390, and 1400—all took place in periods of dearth. More centrally, the profound malnutrition of the Great Famine years may have left millions of Europeans more vulnerable to the Black Death. "A famine of . . . three years is of sufficient length to have devastating long-term effects on the future well being of human infants," says Princeton historian William Chester Jordan, who points out that malnutrition often impedes proper immune system development, leaving the young with lifelong susceptibility to disease.

"By inference," declares Professor Jordan, author of a study on the Great Famine, "the horrendous mortality of the Black Death should reflect the fact that poor people who were in their thirties and forties during the plague had been young children in the period 1315–1322 and were developmentally more susceptible to the disease than those who had been adults during the Famine or were born after the Famine abated."

Dr. Jordan's conclusions are based on animal research, but a recent study by a British researcher, Dr. S. E. Moore, indicates that fetal malnutrition is also a factor in human immune system development. Studying a group of young African adults, Dr. Moore found that subjects born in "the nutritionally debilitating hungry season" (winter and early spring) were four times more likely to die of infectious disease than adults born in the "plentiful harvest season." In the conclusion of her report, Dr. Moore writes, "Other evidence from the literature also favors the hypothesis that intrauterine growth retardation (caused in this case by maternal food shortages) slows cell division during sensitive periods in the development of the immune system. This would provide a mechanism by which early insults could be 'hard wired' such that they [would have] a permanent impact."

The historical evidence also suggests a link between the Great Famine and the Black Death. A connection between the two events should be reflected in plague deaths. Areas that lost large numbers of children in the famine should have suffered less during the Black Death because they had fewer vulnerable adults in the population—adults with congenitally defective immune systems. In medieval Flanders, the mortality pattern fits this paradigm. The region, which lost a great many children to epidemics during the Great Famine, experienced fewer plague deaths than many neighboring regions.

The imbalance between food and population was not the only disease risk factor in the medieval environment. Long before the weather turned and the land gave out and the grain became covered with mold and fungi, the continent was producing more garbage than it could dispose of. Circa 1200, the medieval city was drowning in filth, and in the postboom decades, the situation may have worsened as thousands of dispossessed peasants flooded into urban Europe, animals in tow. By the third decade of the fourteenth century, the amount of refuse on the medieval street was so great, it was literally driving men to murder. One morning in 1326 an irate London merchant confronted a peddler who had just tossed some eel skins into the lane outside his shop.

Pick up the eels, the merchant demanded.

No, replied the peddler.

Fists flew, a knife flashed; a moment later the peddler lay dead on the street.

As the state of public sanitation worsened, public outrage grew. There was a tremendous hue and cry about outdoor slaughterhouses and backed-up street gutters, and an even greater hue and cry about the swarms of black rats that lived off the filth. A fourteenth-century English-French dictionary illustrates just how ubiquitous the rodent was in the Middle Ages. "Sir," goes one passage, ". . . I make bold that you shall be well and comfortably lodged here, save that there is a great pack of rats and mice." Medieval people were also quite aware that the rat was a dangerous animal. Antirodent remedies like "hellebore in the weight of two pence" and "cakes of paste and powdered aconite" were quite popular and widely used. However, what people did not suspect is that *Rattus rattus,* the black rat, was involved in human plague.

This is not quite the ahistorical judgment it sounds like. Premodern peoples had keen powers of observation. During an outbreak of pestilence in Antiquity, the Roman governor-general of Spain offered handsome bounties to local rat hunters. The folklore of medieval and early modern India and China also contains several references to the connection between *Rattus rattus* and *Y. pestis*. An example is the Indian legend of the beautiful Princess Asaf-Khan of Punjab.

Walking through a courtyard one day, Asaf-Khan is said to have seen an infected rat staggering drunkenly. "Throw him to the cat," she ordered. A slave picked up the wobbly rat by the tail and threw it to the princess's pet cat; the cat promptly pounced on the animal, then just as promptly dropped it and fled. A few days later, the cat was found dead outside the princess's bedroom. The following day the slave who picked up the rat died; then, one by one, the rest of Asaf-Khan's slaves died, until only the princess was left alive.

A few centuries later a Chinese poet, Shi Tao-nan, wrote an ode to the relationship between *Rattus rattus, Y. pestis,* and man.

Dead rats in the east,
Dead rats in the west! . . .
Men fall away like . . . walls. . . .

Nobody dares weep over the dead . . .
The coming of the devil plague
Suddenly makes the lamp dim.
Then, it is blown out,
Leaving man, ghost and corpses in [a] dark room.

Europeans first became aware of the biological relationship between *Y. pestis* and *Rattus rattus* during the Third Pandemic of the late nineteenth century, when the rat (along with the flea) was identified as a key agent in human plague. In subsequent years, a great deal has been learned about *Rattus,* including its age and origin. The black rat first evolved in Asia, probably India, sometime before the last Ice Age. At a weight of four to twelve ounces, it is only half the size of its first cousin, the Norwegian brown rat—also an important vector in human plague—but *Rattus* more than makes up for its unprepossessing physical stature with incredible powers of reproduction. It has been estimated that two black rats breeding continuously for three years could produce 329 million offspring, as long as no offspring died and all were paired (fortunately, all very big ifs).

Rattus also has some other remarkable qualities that make it a formidable disease vector. One is great agility. A black rat can leap almost three feet from a standing position, fall from a height of fifty feet without injury, climb almost anything—including a sheer wall, squeeze through openings as narrow as a quarter of an inch, and penetrate almost any surface. The word "rodent" derives from the Latin verb *rodere,* which means "to gnaw," and thanks to a powerful set of jaw muscles and the ability to draw its lips into its mouth (which allows the incisors, or cutting teeth, to work freely), *Rattus* can gnaw through lead pipe, unhardened concrete, and adobe brick.

A wary nature also makes *Rattus* a wily vector; the black rat usually travels by night, builds an escape route in its den, and reconnoiters carefully. This last behavior seems, at least in part, learned. During a foraging expedition, one young rat was observed taking a reconnaissance lesson from its mother. It would scamper ahead a few feet, stop until the mother caught up, then wait as she examined the floor ahead.

Only after receiving a reassuring maternal nudge would the young rat advance. Rats also have another rather unusual, humanlike trait: they laugh. Young rats have been observed laughing—or purring, the rodent equivalent of laughter—when playing and being tickled. *Rattus* is, by nature, a very sedentary animal—usually. A city rat may wonder what lies on the other side of the street, but studies show it won't cross the street to find out. Urban rats live their entire lives in a single city block. The rural rat's range is a not much larger—a mile or so. However, if *Rattus* were phobic about long-distance travel, it would still be an obscure Asian oddity, like the Komodo dragon lizard. Rats do travel, and often for reasons that highlight the role of trade and ecological disaster in plague.

For example, on occasion an entire black rat community will abandon a home range and migrate hundreds of kilometers. Research suggests that what makes the rats override their sedentary impulses is a craving for grain germ—and perhaps more particularly, for the vitamin E in the grain germ. Under normal conditions, rat migrations are infrequent, but under conditions of ecological disaster one imagines that they might become quite common.

For distances beyond the multikilometer range, *Rattus* relies on its long-time companion, man. The stowaway rat is the original undocumented alien. In modern studies, it has been found in planes, in suit-jacket pockets, in the back of long-haul trailers, and in sacks carried by Javanese pack horses. Trade has also been a boon to *Rattus* in another, more subtle but very significant way. In the wild, when rat populations grow unstably large, nature can prune them back with a prolonged period of bad weather and scarce food. The advent of camel caravans, pack horses, ships—and, later, trains and planes and trucks—has weakened this pruning mechanism. Once commercial man appeared, the highly adaptable rat was able to escape to places where food was abundant.

The date of *Rattus*'s arrival in Europe is a source of controversy. Some scholars believe the black rat first appeared in the West during the Crusades, which would mean sometime in the twelfth century. However, this view ignores the Plague of Justinian and the Roman statues of a *Rattus*-like creature, which date back to at least the first century A.D. More credible is the theory of French biologist Dr. F. Audoin-Rouzeau, who dates *Rattus*'s arrival to sometime before the birth of Christ. Given the rat's affinity for trade, its entry point may have been the deserts of the Silk Road or the high mountain passes of Central Asia, where agents of Rome and China met occasionally, or the trading station the Romans maintained on the Indian coast.

Two significant dates in *Rattus*'s European history are the sixth century, when the Plague of Justinian decimated the rodent—and human—population of the Mediterranean Basin, and the year 1000, when a resurgent Christendom began to produce enough food and waste to support a large demographic rebound. Three hundred years later, overcrowding, town walls, and primitive sanitation had turned the medieval city into a haven for *Rattus*.*

Pigs, cattle, chickens, geese, goats, and horses roamed the streets of medieval London and Paris as freely as they did the lanes of rural Broughton. Medieval homeowners were supposed to police their housefronts, including removing animal dung, but most urbanites were as careless as William E. Cosner, a resident of the London suburb of Farringdon Without. A complaint lodged against Cosner charges that "men could not pass [by his house] for the stink [of] . . . horse dung and horse piss."

On the meanest of medieval streets, the ambience of the barnyard gave way to the ambience of the battlefield. Often, animals were abandoned where they fell, left to boil in the summer sun, to be picked over by rats and ransacked by neighborhood children, who yanked bones from decaying oxen and cows and carved them into dice. The municipal dog catcher, who rarely picked up after a dog cull (kill), and the surgeon barber, who rarely poured his patients' blood anywhere except on the street in front of his shop, also contributed to the squalid morning-after-battle atmosphere.

* Woefully inadequate sanitation made medieval urban Europe so disease-ridden, no city of any size could maintain its population without a constant influx of immigrants from the countryside.

Along with the dog catcher and surgeon barber, *Rattus*'s other great urban ally was the medieval butcher. In Paris, London, and other large towns, animals were slaughtered outdoors on the street, and since butchers rarely picked up after themselves either, in most cities the butchers' district was a Goya-esque horror of animal remains. Rivers of blood seeped into nearby gardens and parks, and piles of hearts, livers, and intestines accumulated under the butchers' bloody boots, attracting swarms of rats, flies, and street urchins.

The greatest urban polluter was probably the full chamber pot. No one wanted to walk down one or two flights of stairs, especially on a cold, rainy night. So, in most cities, medieval urbanites opened the window, shouted, "Look out below!" three times, and hoped for the best. In Paris, which had 210,000 residents, the song of the chamber pot echoed through the city from morning to night, intermingling with the lewd guffaws of the prostitutes on the Ile de la Cité and the mournful bleats of the animals going to slaughter at St. Jacques-la-Boucherie on the Right Bank.

No premodern city was clean, but the great urban centers of Antiquity employed a number of ingenious sanitation techniques. The Etruscans, for example, created extensive underground drainage systems to remove garbage and excrement, and the Roman aqueducts carried enough water from the countryside to supply each resident of the city with three hundred gallons per day. The Middle Ages also produced some sanitary wonders, including the privy system in the monastery at Durham, England, which an admiring visitor described thusly: "Every seat and partition in the dormer [dormitory] was of wainscot close, on either side very decent, so that one [monk] could not see the other. . . . [And] there were as many seats of privies as there were little windows in the walls to give light to every one on said seats." The system also had an underground "water course" that drained waste dropped through the privies into a nearby stream.

Though many medieval cities had public sanitation systems, none came close to rivaling Durham's efficiency. The typical urban system began with shallow open gutters in small residential streets; these led into a network of larger central gutters, which, in turn, fed into a central dumping point—usually a large river like the Thames or Seine. Where available, local streams were diverted to provide flushing power, but since urban streams were not widely available, most systems relied on gravity and rainwater. In theory, storms were supposed to flush waste through the downward-sloping gutters to a river dumping point. But dry weather was unkind to theory; large piles of fecal matter, urine, and food would accumulate in the gutters, providing a feast for rats. Storms, when they did come, were not much help. Even a good rain rarely pushed waste much farther than an adjoining neighborhood. However, enough waste matter eventually got to the end points in the system to make the urban river an insult to the senses and an affront to propriety. After a visit to the malodorous Thames, a horrified Edward III expressed outrage at the "dung, lay-stalls and other filth" on the banks.

London supplemented its sewer system with municipal sanitation workers. Every ward in the city had a cadre of inspectors, the Dickensian-named "beadles" and "under-beadles," who probed, peered, sniffed, and questioned their way along the medieval street. Was waste being cleared from housefronts? Were alleys being kept clean? Better-off Londoners often built indoor privies, or garderobes, over alleyways, suspending them "on two beams laid from one house to the other." For the garderobe's owner, the privy meant liberation—no more chamber pots on cold nights—but for his neighbors, it meant piles of dung in the alley, a medley of frightful odors, and swarms of flies (rats do not usually feed on human waste). Beadles and under-beadles also investigated acts of sanitary piracy. The year before the plague arrived in England, two malefactors were arrested for piping their waste into the cellar of an unsuspecting neighbor.

Under the beadles were the rakers, the people who did the actual cleaning up. Rakers swept out gutters, disposed of dead animal carcasses, shoveled refuse from the streets and alleys, and hauled it to the Thames or other dumping points, like the Fleet River.

The beadles and rakers not only had the dirtiest job in medieval London, but the most thankless as well. In 1332 a beadle in Cripplegate Ward was attacked by an assailant who, to add insult to injury, stole the beadle's cart; a few years later, two women in Billingsgate heaped such abuse on a team of rakers, municipal authorities ordered the women arrested. Indeed, judging from contemporary accounts, medieval London seems to have been engaged in a low-level civil war over sanitation. On one side were miscreants, like the foul-mouthed Billingsgate ladies and William E. Cosner, the garbage king of Farringdon Without. On the other side, the king, Edward III, who thundered, "Filth [is] being thrown from houses by day and night"; the nervous mayor, who tried to assuage these royal outbursts with a flurry of widely ignored sanitation ordinances; the much-abused beadles, under-beadles, and rakers; and irate private citizens like the murderous shop owner.

Granaries, fields of oats and barley, and large stocks of domestic animals also made the rat a ubiquitous figure in the medieval countryside, and the architecture of rural Europe may have made the peasant especially vulnerable to its sharp teeth. Most peasant huts were constructed of wattle and daub, a sort of medieval version of wallboard. First, wattle, or twigs, were woven into a lattice design; then the mudlike daub was smeared over the lattice. The combination was so permeable, one unfortunate English peasant was killed when a poorly aimed spear burst through his cottage wall one morning at breakfast.

An early-twentieth-century outbreak of plague in the Egyptian village of Sadar Bazaar highlights another rat-friendly aspect of peasant life. A rat count in the village revealed that families who slept with their domestic animals had more rats per household—the exact number was 9.6—than families who did not: 8.2.

The Greeks, who worshipped the body, considered cleanliness a cardinal virtue, and the Romans considered hygiene so important, their public baths looked like temples. At the baths of Diocletian, "the meanest Roman could purchase, with a small copper coin, the daily enjoyment of [a] scene which might excite the envy of the Kings of Asia," wrote Edward Gibbon. However, early Christians, who thought self-abnegation a cardinal virtue, considered bathing, if not a vice, then a temptation. Who knows what impure thoughts might arise in a tub of warm water? With this danger in mind, St. Benedict declared, "To those who are well, and especially to the young, bathing shall seldom be permitted." St. Agnes took the injunction to heart and died without ever bathing.

Religious suspicions about bathing softened during the late Middle Ages, though not enough to dramatically improve standards of personal hygiene. Catherine of Siena, who was born in 1347, also never bathed, though Catherine's greatest achievement may have been her (reported) ability to go months at a time without a bowel movement. St. Francis of Assisi, who considered God's water too precious to squander, was another infrequent bather. The laity continued to resist the bathtub for less high-minded reasons. Whatever one medieval Miss Manners might say about bathing as a way of being "civil and mannerly toward others," it was easier to wash only your face and hands in the morning, just as it was easier to dump a full chamber pot out the window rather than walk down several flights of stairs. Undressing and changing clothing were also infrequent. Thus, another useful phrase in the fourteenth-century English-French dictionary was, "Hi, the fleas bite me so!"

No doubt, the principal insect vector in the Black Death was *X. cheopis,* the rat flea, but given the state of the medieval body, it is extremely likely that *Pulex irritans,* the human flea, also played an important role in the medieval plague.

From Caffa to Vietnam and Afghanistan, no human activity has been more closely associated with plague than war, and few centuries have been as violent as the fourteenth. In the decades before the plague, the Scots were killing the English; the English, the French; the French, the Flemings; and the Italians and the Spanish, each other. More to the point, in those savage decades, the nature of battle changed in fundamental ways. Armies grew larger, battles bloodier, civilians were attacked more frequently, and property was destroyed more routinely—and each change helped to make the medieval battlefield and the medieval soldier more efficient agents of disease.

Different historians date what is sometimes called the Military Revolution of the Later Middle Ages to different events, but as good a place to start as any is a meadow outside the Flemish village of Coutrai on a steamy July day in 1302. Arriving at the meadow that morning, a large French cavalry force on its way to Coutrai to relieve a group of besieged comrades (Flanders was a French domain in the fourteenth century) found the way blocked by several battalions of resolute Flemish bowmen and pikemen, dressed in wash-bowl helmets and fishnet armor.

Shortly after noon the French commander, Robert of Artois, ordered an attack on the Flemings, and his cavalry—pennants flapping in the summer wind—began advancing across the meadow high grass with all the stately grandeur appropriate to warrior-knights of the "august and sovereign house of France." After fording a small brook midway between the two camps, the French broke into a run. A moment later, an enormous *thwang!* sounded, and the cloudless July sky filled with a thousand steel-tipped Flemish arrows; a few seconds later, there was an even louder *thud!* as several hundred French war horses smashed into the Flemish line at twenty miles an hour. According to conventional medieval military theory, the impact of the charge should have knocked the Flemings to the ground like bowling pins, making them vulnerable to trampling by horse and impalement by rider, but at Coutrai the gods of war rescrambled the rules. Instead of plunging through the Flemish line, the French broke against it like a wave against a sea wall—and dissolved, Humpty Dumpty–style, into a jumble of falling horses and falling men.

The discovery that infantry, well armed and resolute, could defeat cavalry, the queen of the battlefield—a discovery reaffirmed in several subsequent battles—revolutionized medieval military strategy, and like most revolutions, the infantry revolution produced several unanticipated consequences. First, medieval captains upgraded the role of infantry; then, discovering that foot soldiers were much cheaper to field—five or six bowmen and pikemen cost about the same as a single cavalry man—the captains expanded the size of the medieval army; and as armies grew much larger, battles grew much bigger and bloodier. This was partly a matter of numbers, but it also reflected the growing violence of warfare. For one thing, the largely peasant infantry was far less apt to observe the rules of chivalry, particularly in combat with enemy nobles. Since stress, including combat stress, weakens immune system function, arguably one consequence of bigger, more violent wars was a larger pool of disease-vulnerable people. Less arguably, larger armies produced larger concentrations of dirty men and debris, which attracted larger concentrations of rats and fleas.

The *chevauchée,* the second major military development of the fourteenth century, was created to resolve the great military dilemma of the age: how does an army break a siege? "A castle can hardly be taken within a year, and even if it does fall, it means more expenses for the king's purse and for his subjects than the conquest is worth," wrote Pierre Dubois, an influential fourteenth-century military thinker. Dubois's solution to the siege problem, outlined in *Doctrine of Successful Expeditions and Shortened Wars,* was indirection. Attack civilians, Dubois argued, and your opponent will be forced to abandon his fortified position and come out and defend his people. Thus was born the *chevauchée*. The idea of sending large raiding parties on search-and-destroy missions through the enemy countryside was not quite as new as Dubois pretended. The practice had been tried before, including by the Normans against the English in 1066. Civilians had also been targeted before. "If sometimes the humble and innocent suffer harm and lose their goods, it cannot be otherwise," declared Honore Bouvet with a Gallic shrug.

However, the Anglo-French Hundred Years' War—the largest, bloodiest conflict of the Middle Ages—transformed the *chevauchée* into a common and devastating weapon. The war began in 1337, and in the decade before the plague arrived in 1347, the English, who became masters of the *chevauchée,* employed it with lethal effect against Dubois's fellow countrymen. All through the 1340s, flying wedges of English horsemen crisscrossed the French countryside, torching farms and villages, raping and murdering civilians, and looting cattle. In a letter to a friend, the Italian poet Petrarch, a recent visitor to wartime

France, expressed astonishment at the level of destruction. "Everywhere were dismal devastation, grief and desolation, everywhere wild and uncultivated fields, everywhere ruined and deserted homes. . . . [I]n short everywhere remained the sad vestiges of the Angli [the English]."

Even more heartfelt is the account of English terrorism by the French king Jean II. "Many people [have been] slaughtered, churches pillaged, bodies destroyed and souls lost, maids and virgins deflowered, respectable wives and widows dishonored, towns, manors, and buildings burnt, . . . the Christian faith . . . chilled, and commerce . . . perished. . . . So many other evils and horrible deeds have followed these wars that they cannot be said, numbered or written."

While the present is, at best, an imperfect guide to the past, several studies on modern conflict provide some additional insight into how war may have made the medieval world more vulnerable to plague. The subject of the first report, a U.S. Army study, is the old Soviet army, which fought in Afghanistan in the 1980s. Russian combat casualties in the conflict were quite low—under 3 percent—but the Soviet army suffered horrendous rates of illness, especially infectious illness. Three out of four soldiers who fought in Afghanistan—75 to 76 percent of the entire Soviet army in the country—had to be hospitalized for disease. Some soldiers were stricken by bubonic plague, but malaria, cholera, diphtheria, infectious dysentery, amoebic dysentery, hepatitis, and typhus were, if anything, more common.

What caused such a disastrous disease rate? The answer sheds some light on the unchanging nature of soldiering. According to the report, one important factor was military hygiene. The average Russian soldier changed his underwear once every three months, washed his uniform and blankets at about the same rate, drank untreated water, left his garbage unpicked up, defecated near his tent rather than at a field latrine, and, even when involved in kitchen work, only washed his hands after a bowel movement when an officer made him. Significantly, the report says that combat stress may also have played a role in the high disease rate. Stress, as noted earlier, impairs immune system function, lowering resistance to disease. This observation, of course, also applies to civilians menaced by marauding armies.

A second report comes from Vietnam, where approximately twenty-five thousand people, most native Vietnamese, were struck by plague between 1966 and 1974. According to U.S. Army medical authorities, one important factor in the outbreak was the siegelike conditions in large parts of the country. Not atypical is the fire base a team of American doctors visited in the winter of 1967. Twenty-one people were suffering from plague at the base, and an inspection quickly revealed why. Vietcong mortar and artillery attacks had driven life underground. The soldiers—and, in many cases, their families—lived in dirt bunkers whose warm, moist environments perfectly mimicked the rodent burrow; personal hygiene was appalling. No one washed, the bathing facilities being above ground, and no one used the field latrines, for the same reason. The Genoese, who were besieged at Caffa, and the French, who were besieged at Calais on the eve of the Black Death, would have found the fetid Vietnamese bunkers, piled high with human waste, half-eaten rations, and blood-stained battle dressings, quite familiar. Working under appalling conditions, the army doctors managed to save seventeen plague victims, but in four cases the disease was too advanced even for treatment with modern antibiotics.

The uprooting of populations—another feature of war that dates back to the *chevauchee* and beyond—was also of great assistance to *Y. pestis* in Vietnam. In 1969 more than six hundred people—the majority of them children—were stricken by plague in the village of Dong Ha. Again, U.S. Army doctors found rats and fleas everywhere, but this time the infestation was caused by a resource imbalance, not a siege. Little Dong Ha had suddenly become a receiving point for refugees fleeing south from the DMZ (demilitarized zone) and east from Khe Sanh, but the village lacked the sanitary resources to cope with a large influx of dirty people.

Of course, it is impossible to say with any precision how, and to what degree, these three factors—war, famine, and inadequate sanitation—helped to pave the way for the Black Death, but what can be said with some certainty is that by the time *Y. pestis* left Caffa in late 1346 or early 1347, Europe was already up to its chin in water and the tide was still rushing in.

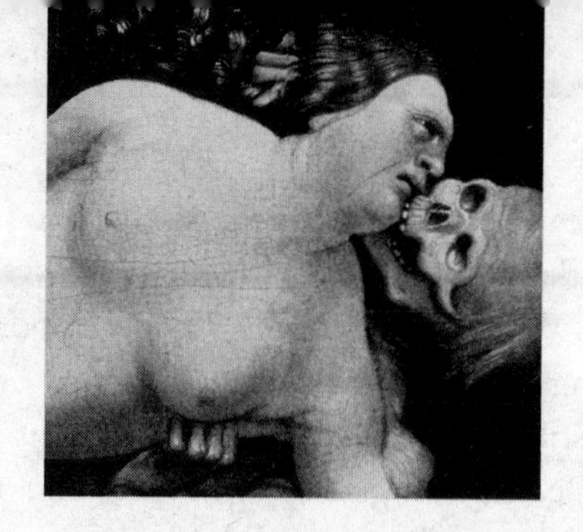

CHAPTER FOUR

Sicilian Autumn

Sicily, October 1347

THE MEDITERRANEAN IS A SEA OF SECRETS. THERE ARE THE SECRETS of its submerged mountains, the lofty ranges that once linked Tunisia to Sicily and Spain to Morocco. There are the secrets of its mysterious ancestor, the Tethys Sea, which, before the birth of Eurasia, flowed across the face of the world into a great eastern ocean. And there are the secrets of the Mediterranean's dead: the Vivaldi brothers, who vanished into the infinity of longitude and latitude, searching for a passage to the

Indies. And the greatest secret of all: the fate of the pestilential Genoese, who fled Caffa with "sickness clinging to their very bones."

"Speak, Genoa. What have you done?" a contemporary demanded on behalf of the plague dead. But Genoa has kept her silence about the ships, and keeps it still. The Caffa plague fleet floats through the literature of the Black Death like a ghost shimmering on a night sea. One account speaks of "three galleys loaded with spices . . . storm driven from the east by the stinking breath of wind"; another of four Genoese ships returning from the Crimea "full of infected sailors"; a third of a Genoese fleet—variously numbered at two to twelve ships—sailing from Asia Minor to the Mediterranean, infecting everything in its wake, including the Black Sea port of Pera, Constantinople, Messina (Sicily), Genoa, and Marseille.

In June, as the apocalypse gathered in the East, few in Europe were aware that something wicked was coming their way. In England the summer of 1347 had some of the sepia-toned glamour of the summer of 1914. Fresh from a glorious campaign season on the plains of northern France, the golden-haired Edward III was in London, enjoying the tournament season, while down the Thames at Westminster Palace, his daughter, Princess Joan, was spending the soft June evenings strolling the palace lawns serenaded by a Spanish minstrel, a gift from the dashing Prince Pedro of Castile. In Siena bootmaker and part-time real estate speculator Agnolo di Tura was at work on a history of the city, eyeing some properties on the Campo, the town square, and doting on his wife, Nicoluccia, and their five children. In Paris the cleric Jean Morellet was updating the building fund at his church, St. Germain l'Auxerrois. The fund received so few donations, Morellet had time for long morning walks along the unembanked Seine, where the river breezes kept the water mills in perpetual motion and the narrow rectangular houses along the bank stood up like "hairs on a multitude of heads." In Thonon, a town on Lake Geneva, a barber surgeon named Balavigny could be found at the town gate most mornings that summer, gossiping with fellow members of the local Jewish community. In Naples, where the warm night air smelled of the "sweet, soft perfume of summer," the most beautiful woman in

Europe and Christendom's reigning Bad Girl—Queen Joanna of Naples and Sicily—was facing accusations of murder.

In July, as the fields around Broughton filled with straw-hatted peasants, and the monks in Avignon glazed their windows to trap in the cool air, in Constantinople hopelessness and despair already reigned. Positioned just below the Black Sea and just above the Dardanelles, gateway to West, the Byzantine capital sat in the middle of a bull's-eye. A Europe-bound ship leaving the Crimea could not reach home without passing the city. Thus, sometime in the spring or early summer of 1347, the plague-bearing Genoese arrived in the harbor, chanting the words of the black-hatted stranger in the American Indian myth: "I am death."

The Venetian scribe who estimated that 90 percent of Constantinople perished in the Black Death surely exaggerated, but no one who lived through the pestilential summer and fall of 1347 in the Byzantine capital ever forgot the experience. "Every day we bring out our friends for burial. [And] every day the city becomes emptier and the number of graves increases," wrote the court scholar Demetrios Kydones. As the mortality intensified, Kydones saw his fellow citizens become twisted by fear and selfishness. "Men inhumanely shun each other's company [for fear of contagion]. Fathers do not dare to bury their own sons; sons do not perform . . . last duties to their fathers."

The plague also left a lasting mark on Ioannes IV, the Byzantine emperor. "Upon arrival . . . [the empress] found [our] youngest . . . dead," wrote Ioannes, in his only known statement about the death of his thirteen-year-old son, Andronikos. After the boy's death the emperor lost his taste for the world. Abdicating the throne, Ioannes retired to the solitude of a monk's cell, to pray and mourn and grieve for the remainder of his life.

From Constantinople, *Y. pestis* followed the trade routes southward into the Dardanelles, the thin vein of blue water that carries the Europe-bound traveler into the Aegean Sea and the Mediterranean beyond. In the summer of 1347, the world divided at the Dardanelles. Immediately to the west lay the green sunlit hills of Europe, still untouched by plague; to the east, the pesilential plains of Asia Minor. Descending through the straits, *Y. pestis* stopped to pay its respects to Xerxes, the Persian king who built a bridge of boats to ferry his army across the waterway. Then, as the Aegean rushed into view, the plague bacillus cloned itself. One strain of the disease swung northward through Greece, Bulgaria, and Romania, in the general direction of Poland; a second strain darted southward across the Mediterranean toward Egypt and the Levant; while a third strain doubled back eastward, striking Cyprus late in the summer. As if trying to repel a malignant body, the island immediately rose up in violent rebellion. First the earth ripped open, and "a tremendous earthquake" uprooted trees, collapsed hills, leveled buildings, and killed thousands. Then the angry sea threw up a tidal wave so enormous, it seemed to scrape the sun as it rushed toward the island. Mothers snatched up their children, farmers ran from the fields; seagulls scattered into the sky, gawking and yakking in alarm; while fishermen, caught between the coast and the wave, whispered a quick last prayer as the luminous Mediterranean light disappeared behind a hard black wall of water. A moment later an enormous crash sent shock waves flying across the water for dozens of miles in every direction; then large parts of coastal Cyprus disappeared beneath a turbulent sea of white foam. "Ships were dashed to pieces on the rocks and . . . this fertile and blooming island was converted into a vast desert," wrote a German historian. Next the air itself seemed to give out. A "pestiferous wind spread so poisonous an odor that many, being overpowered, . . . fell down suddenly and expired in dreadful agonies."* As calamity followed calamity, alarmed Cypriots began to fear that their Arab slaves would rise up in revolt. Closing their hearts to pity, islanders reached for the sword. Day after day that bleak fall, hundreds of Muslim men, women, and children were herded into ruined olive groves littered with puddles and twisted,

* Michael Baillie, a professor in the school of Archealogy and Palaecology at Queens University, Belfast, hypothesizes that the miasma was caused by "outgassing," a rare geological phenomenon in which gas deposits trapped underneath the ocean floor suddenly break free and escape into the atmosphere, corrupting the air. A recent example of the phenomenon may have occurred in 1986, when a poisonous cloud of hydrogen sulfide emerged from Lake Nyos in Cameroon and killed 1,700. Strikingly, survivors said the cloud smelled like "rotten eggs." What released the trapped gas in the lakebed is unclear, but, in the case of Cyprus, the triggering event may have been the earthquake. (M.G.L. Baillie, "Putting Abrupt Environmental Change Back into Human History," in *Environment and Historical change,* ed. Paul Slack [Oxford: Oxford University Press, 1998], p. 68.)

uprooted trees and slaughtered by men who would be dead of the pestilence within a week.

The fourth strain of the plague swept westward across the Mediterranean until it found an island even more tragic and tortured than Cyprus.

Sicily lies only a few miles off the coast of Europe, but it belongs to a different world—a more primitive, insular, and, above all, violent world. There is the violence of the island's sky, which is too blue; of its sun, which is too bright; of its people, who are too passionate; and of its wind, the piercing summer sirocco, which blows northward across the Mediterranean from Tunisia and stings the eyes, burns the throat, and coats the lungs with sand. There is also the violence of Sicilian history, a history so full of duplicity, subjugation, bloodshed, and despair that the sunny Mediterranean island has produced a society of black-hearted fatalists. In Sicily, says native novelist Leonardo Sciascia, "we ignore the future tense of verbs. We never say, 'Tomorrow I will go to the country'; we say, '*Dumani, vaju in compagna*'—'Tomorrow I am going to the country.' How can you fail to be pessimistic in a country where the future tense of the verb does not exist?"

The most tragic moment in Sicily's history begins almost like a child's fairy tale. "In October 1347,* at about the beginning of the month, twelve Genoese galleys put into the port of Messina." The author of these words, a Franciscan friar named Michele da Piazza, does not say from where the galleys originated—Caffa, another Black Sea port, Constantinople, Romania, or somewhere closer—but apparently nothing about the vessels seemed untoward or suspicious. As the ships docked, Messina went about its daily business, enjoying a final

* Friar Michele's account illustrates why historians believe that contemporary descriptions of the Black Death have to be read with caution. Other evidence suggests that the Genoese arrived in Messina in late September, not October, and while their fleet may have contained twelve galleys, Friar Michele also may have selected the number because twelve had a special magical significance for medieval man. Another problem with contemporaneous plague chronicles is plagiarism. Frequently a chronicler would lift an expression, turn of phrase, and sometimes even an entire description whole from an ancient author, two favorite sources being Thucydides, who wrote a classic account of the fifth-century B.C. Plague of Athens and Tacitus, who wrote an equally famous description of the third-century A.D. Plague of Antoine in Rome. (Ole J. Benedictow, *The Black Death 1346–1353: The Complete Story* [Woodbridge, Suffolk: Boydell Press, 2004], p. 70.)

moment of normalcy before the world changed utterly. Fishermen unloaded their catch, old women gossiped from windows, children chased one another across long golden beaches, a soft autumn wind danced up the narrow streets of the town; then anchors dropped, gangplanks came down, and the Genoese crews rolled onto the docks, "carrying such a disease in their bodies that if anyone so much as spoke with one of them he was infected . . . and could not avoid death."

The Black Death had arrived in Europe.*

Almost immediately people began to fall ill, and they fell ill in ways no one in Messina had ever seen before. First, says Friar Michele, "a sort of boil . . . the size of a lentil, erupted on the thigh or arm, [then] the victims violently coughed up blood, and after three days [of] incessant vomiting . . . for which there was no remedy, they died . . . and with them died not only everyone who had talked to them, but also anyone who acquired, touched or laid hands on their belongings."

Friar Michele seems to be describing pneumonic plague *secondary* to bubonic plague—that is, plague that begins in the lymph system (producing the bubo), but metastasizes to the lungs (producing the bloody cough). Messina's rapid infection suggests that if the disease was not pneumonic when it arrived, it rapidly became so—that is, at a certain point, the infection began to spread directly from person to person via airborne droplets.

More puzzling is how the Genoese, ravaged as they were, managed to get to Messina. Some of the crew members may have had a natural immunity to *Y. pestis,* but even assuming that CCR5-Δ32 does heighten resistance to the disease, far from settled scientific fact, could enough crew members have survived to get a fleet from the Crimea or Constantinople to Sicily, both several months' sail away? Given how quickly plague kills, another, perhaps more credible scenario is that the fleet that brought the disease to Messina originated in a port closer to Italy.

Open sea sailing was still very dangerous in the Middle Ages, so mariners rarely sailed anywhere in a straight line. Even during extended

* It is possible that the pestilence slipped into another European port a few days or weeks earlier, but Messina and Sicily are where the plague enters the historical record in Europe.

trips, ships would inch along a coastline like rock climbers on a ledge, stopping every third or fourth day to trade and buy supplies. The practice, called *costeggiare,* would have allowed *Y. pestis* to proceed to Europe in a stepwise fashion, moving from port to port and fleet to fleet, allowing it to kill crews at will.

Messina quickly expelled the Genoese, but the plague had already entered the lifeblood of the city. As the mortality deepened, churches and shops fell silent, beaches emptied, fishing boats lay idle, streets became deserted. Soon Messina, like Constantinople, became two cities, the city of the infected—a municipality of pain and despair—and the city of the uninfected, where fear and hate ruled. "The disease bred such loathing," says Friar Michele, "that if a son fell ill . . . his father flatly refused to stay with him." That autumn many in Messina died, not only absent the consolation of a parent or a child, but without a priest to hear confession or a notary to make out a will. Only Messina's animals maintained the old traditions of loyalty and faithfulness. "Cats and . . . livestock followed their master to death," says Friar Michele.

Soon Messina began to empty out. Friar Michele speaks of crazed dogs running wild on deserted streets, of nighttime fires winking from crowded fields and vineyards around the city, of dusty, sun-drenched roads filled with sweaty, fearful refugees, of sick stragglers wandering off to nearby woods and huts to die. He also describes several incidents of what sound, to a modern sensibility, like magical realism but were probably episodes of panic-induced hysteria. In one, "a black dog with a naked sword in its paw" rushes into a church and smashes the silver vessels, lamps, and candlesticks on the altar.* In another, a statue of the Blessed Virgin comes alive en route to Messina and, horrified by the city's sinfulness, refuses to enter. "The earth gaped wide," says Friar Michele, "and the donkey upon which the statue of the Mother of God was being carried became as fixed and immovable as a rock."

* Rabies seems to have been widespread in medieval Sicily, so this story may have some basis in reality. (Philip Ziegler, *The Black Death* [New York: Harper and Row, 1969], p. 42.)

Not long ago a British historian boasted of English resoluteness in the face of the Black Death. "With his friends and relations dying in droves, . . . with every kind of human intercourse rendered perilous," wrote the historian, "the medieval Englishman obstinately carried on in his wonted way." The assessment is accurate, but English fortitude owes something to good fortune as well as good character. The plague did not suddenly drop from the sky on England one day. Untouched until the summer of 1348, the English had nearly a year to collect intelligence and steady themselves. Without pressing the analogy too far, cities like Messina and Constantinople were in the position of a Hiroshima or Nagasaki. Not only did the plague more or less strike out of the blue, it produced death on a scale no one had ever seen, no one had ever imagined possible—death not in the hundreds or thousands, but in the hundreds of thousands, and in the millions. Moreover, it was a death capable of obliterating whole networks of people in a matter of hours. One day, wrote a contemporary, "a man, wanting to make his will, died along with the notary, the priest who heard his confession, and the people summoned to witness his will, and they were all buried together on the following day."

Faced with catastrophe on such an unprecedented scale, it is hardly surprising that so many in Sicily lost their heads.

The tale of the Black Death in Sicily is also a tale of two cities, Messina and its southern neighbor, Catania. Believing the Messinese to be vain and supercilious, the Catanians had long disliked their swaggering northern neighbors, and when the town became a collection point for refugees from the port, relations between the two cities soured further. "Don't talk to me if you are from Messina," wary townsfolk told the refugees. The Messinese, whose reputation for vanity was not entirely unjustified, did not enhance their standing by promptly asking to borrow Catania's most precious relics, the bones of the blessed virgin St. Agatha. The Catanians were aghast. Even by the standards of Messinese cheek, this was outrageous. Who would protect Catania from the pestilence while St. Agatha was in the north helping the Messinese drive the plague from their native city? Even Friar Michele becomes a little unhinged when he describes the request. "What a stupid idea on the part of you Messinese. . . . Don't you think if she [St. Agatha] wanted to make her home in Messina she would have said so?"

The crisis deepened when Catania's patriarch, Gerard Ortho, experienced a fit of guilt. Under public pressure, the patriarch had agreed to ban Messinese refugees from the city. Now, to appease God and his conscience, not only did he let the refugees talk him into lending them St. Agatha's relics, he promised to carry the relics to Messina himself. Again, Catania was aghast. The patriarch seemed to be imposing a form of unilateral spiritual disarmament on the city. An angry crowd quickly gathered and marched on the cathedral. On every other day, Catanians addressed their patriarch on bended knee and with bowed head, but not on this day, with the city under imminent threat from a horrible illness. On this day the marchers spoke truth to power. Confronting the patriarch inside the cathedral, they told him flatly, "They would rather see him dead before they let the relics go to Messina." A man of some moral courage, Patriarch Ortho insisted on keeping his word to the Messinese. Finally a compromise was struck. Messina would not get St. Agatha's relics, but it would get the next best thing: holy water into which the relics had been dipped—Patriarch Ortho would sprinkle the water over the infected city himself.

Like almost all stories about Sicily in the autumn of 1347, the tale of two cities ends badly. Despite the holy water, the plague continued to rage in Messina; despite St. Agatha's relics, Catania was struck by the pestilence; and despite a close association with two most important symbols of Sicilian spirituality, Patriarch Ortho died a terrible plague death.

The story of Duke Giovanni, Sicily's craven regent, also has an unhappy ending. As the plague spread across the island—while it produced grievous mortalities in Syracuse, in Trapani, in Sciacca, in Agrigento—the duke thought of no one and of nothing but himself. "He roamed here and there like a fugitive," says Friar Michele, "now in the forest . . . of Catania, now at a tower called *lu blancu* . . . now at the church of San Salvatore. . . ." In 1348, certain that the plague was abating, the duke emerged from hiding and settled "in a place called

Sant' Andrea." Hearing of his reemergence shortly before leaving Sicily, *Y. pestis* paid a call on the duke at his new home and killed him.

Near the end of his chronicle, Friar Michele throws up his hands in despair and declares, "What more is there to say?"

Very little—except that by the autumn of 1348, when the pestilence finally burned itself out, the dead had come to inhabit Sicily as insistently as the living. Human remains could be found everywhere on the island: on the desolate volcanic wastelands of the interior, in the soft green valleys near the coastal plains, and along the island's golden beaches. A third of Sicily may have died in the plague; no one knows for sure.

Genoa, November-December 1347

In a disquisition on Genoese "character," a local cleric likened his fellow citizens to "donkeys." "The nature of a donkey is this," he explained. "When many are together . . . and one is thrashed by a stick, all scatter, fleeing hither and thither."

The Genoese expelled from Messina behaved in character. Scattering "hither and thither," they began to infect other ports, but the expelled galleys were almost certainly not the only agents of plague in the Mediterranean in the desperate fall of 1347. From Caffa, the pestilence had spread around the Black Sea, then to Constantinople, Romania, and Greece, producing a panicked flight west. By November there must have been twenty or more plague ships off the southern coast of Europe, some sailing toward the western Mediterranean, others toward the Adriatic, each armed with the equivalent of a large thermonuclear device, and most, if not all, captained by men whose innate greed was greatly enhanced by the fact that they had a personal financial stake in the cargoes their ships were carrying. With holds full of dead and dying, many infected vessels continued to sail from port to port, selling their wares. One contemporary account speaks of three infected ships expelled from French and Italian ports "heading toward the Atlantic along the Spanish coast . . . to conclude their trade." France, Spain, Egypt, Sardinia, Corsica, Malta, and Tunis were all infected via the traditional Mediterranean trade routes, as was mainland Italy, which, in the autumn of 1347, may have been the most vulnerable region of Europe.

For years nothing had gone right on the Italian peninsula. Even the land and sky had seemed to fall into disorder. In 1345 the heavens burst open, and it rained torrentially for six months, flooding fields, washing out bridges, and producing famine on an epic scale. "In 1346 and 1347," says a contemporary, "there was a severe shortage of basic foods . . . to the point where many people died of hunger and people ate grass and weeds as if they had been wheat." In Florence the terrible plague spring of 1348 was preceded by the terrible famine spring of 1347. In April of that year much of Florence was surviving on a municipal bread ration. As if sensing the approaching horror, the Italian earth also began to tremble. Major earthquakes rocked Rome, Venice, Pisa, Bologna, Naples, Padua, and Venice,* perhaps releasing poisonous gas into the atmosphere as in Cyprus. In several places vintners complained that the air in their wine casks had grown turbid. Everywhere on the peninsula, there was also war and rumor of war.

Through hunger and rain, flood and earthquakes, Italians persisted in killing one another. Genoa was at war with Venice, the papacy with the Holy Roman Emperor, the Hungarians with Naples, and in Rome the aristocratic Colonna family and the aristocratic Orsini family were slitting each others' throats with the happy abandon of Mafia clans.

"What was true of late medieval Europe as a whole was *a fortiori* true of Italy," says historian Philip Ziegler. "The people were physically in no state to resist a sudden and severe epidemic, and, psychologically, they were attuned to . . . a supine acceptance of disaster. . . . To speak of a collective death wish is to trespass into the world of

* Research on modern outbreaks of plague illustrates how the ecological unheaveals in Italy may have paved the way for *Y. pestis*. According to Dr. Kenneth Gage, chief of the Plague Division at the U.S. Centers for Disease Control, an earthquake was the triggering event in a 1994 outbreak of plague in India. The quake destroyed rodent burrows in nearby plague foci, forcing the wild rodent community to relocate closer to areas of human habitation, where they exchanged fleas with *R. rattus*. In Africa, which has the highest incidence of plague in the modern world, a frequent triggering event in the disease are cycles of torrential rain and parching drought, says Dr. Gage. The rodent population increases in the rainy years when food is abundant; then, when drought strikes and food becomes scarce, the hungry rodents flee toward cities, towns, and villages in search of food.

metaphysics, but if ever there was a people with a right to despair of life, it was the Italian peasantry of the mid-fourteenth century."

If Italy was the most vulnerable region of western Europe, the most vulnerable region of Italy may have been Genoa, a handsome city with a "fine circuit of walls, . . . beautiful palaces" set against a majestic mountain backdrop. In addition to sharing the afflictions of its neighbors, Genoa also bore the special burden of its hubris and ambition. Having become the center of an eastern trading empire, the city exerted an almost magnetic attraction on anything coming out of Asia, whether it be Sungware from China, spices from Ceylon, ebony from Burma, or death from the Mongolian Plateau.

Perhaps sensing the city's vulnerability, the Genoese exercised great vigilance in the autumn of 1347. Contemporary accounts say that Genoa was infected on December 31, 1347, but a reconstruction of the timeline that fall suggests that *Y. pestis* made a first run at the city eight to ten weeks earlier. In this version of events, on a late October morning three or four galleys, probably members of the expelled Messina fleet, appear in Genoa harbor and are promptly driven away. As the fleet scatters "hither and thither" again, one stray vessel makes its way north along the French Mediterranean coast to Marseille, and infects the unsuspecting city; then upon being expelled for a third time (Messina and Genoa are one and two), the stray meets up with two companions and sails into history as part of the pestilential fleet last seen heading "toward the Atlantic along the Spanish coast."

However, the prompt action by authorities in October only bought Genoa a little extra time. In late December a second infected fleet appeared in the winter sea off the city. It is unclear where the vessels, also Genoese in origin, came from—Messina, Constantinople, the Crimea, or somewhere else—but their visit appears to have been in the nature of a death ride. The crews were "horribly infected," and may have wanted to gaze a final time upon their native city and its "fine circuit of walls" and "beautiful palaces." Again the ships were driven away by "burning arrows and other engines of war," but, in this instance, too slowly. During this second visit, which may be the December 31 episode described in the chronicles, the plague got into the city.

Thereafter, Genoa falls silent. Almost alone among major Italian cities, she failed to produce a plague chronicler. The only record we have of the city's experience in the winter and spring of 1348, the period of the Black Death, is a report of a famous visitor and a few accounts of individual acts of heroism and self-sacrifice.

One such act was performed by a woman named Simonia, who, in late February 1348, nursed her friend Aminigina through the final days of a bitter plague death. Ignoring the danger to herself, Simonia remained at Aminigina's side, changing her soiled nightshirts, holding her hand when she cried, wiping the blood and spittle and vomit from her lips. On February 23, 1348, the dying Aminigina rewarded Simonia with a small monetary bequest. On that same day, in another part of Genoa, in an office above winter streets dancing with plague, a notary named Antonio de Benitio was making out wills. De Benitio and his colleagues, Guidotto de Bracelli and Domenico Tarrighi, both of whom also stayed in the city, are not obviously heroic figures in the mold of a Patriarch Ortho and Simonia. But in a period of mass death, when enormous amounts of money and property were suddenly being orphaned, notaries, who made out wills and other legal documents, played an essential role in the maintenance of civic order. Without notarial records, there could be no orderly transmission of societal resources from the dead to the living; and without such a transmission, chaos and disorder would result.

During the pestilence Genoa also received a brief visit from the most notorious woman in Christendom, the willful, beautiful Queen Joanna of Naples and Sicily. Joanna, a combination of Scarlett O'Hara and Lizzie Borden, was in trouble again. On a September evening three years earlier, her husband, eighteen-year-old Andreas of Hungary, was discovered in the Neapolitan moonlight, dangling from a balcony with a noose around his neck. The queen and her lover, Luigi of Taranto, a man of such extraordinary physical beauty that a contemporary called him as "beautiful as the day," were suspected of plotting the murder. The queen's visit to pestilential Genoa was occasioned by her angry Hungarian in-laws, who had just invaded Italy in pursuit of her and Luigi and anyone who had helped them kill the eighteen-year-old Andreas. In March 1348, while notary de Bracelli was at his desk making out wills, and bodies piled up along Genoa's "fine circuit of walls," the lovely Joanna, in the breathless tradition of the romance novel heroine, was boarding a fast ship in Genoa harbor. Soon Christendom's most beautiful couple would reunite in Avignon, where the Neapolitan queen would participate in a trial so notorious, for a while it upstaged even the plague.

Little else is known about Black Death Genoa, though it is said that if you stand under the statue of Columbus in the harbor on a summer night, you can hear the plague dead speak—but, of course, the voices are just the moans and creaks of small pleasure craft pitching in the night wind. Along with a magnificent harbor, the winds were nature's great gift to Genoa. They blow to the south and west—just the directions the medieval Genoese wanted to go—until the day they arrived at the place where the winds end and discovered what lay in wait.

It is thought that a third of the city's eighty thousand to ninety thousand citizens died of plague but, as with Sicily, no one knows for sure.

Venice, January 1348

The Genoese cleric who likened his fellow citizens to donkeys also had some thoughts about the character of Genoa's great rival, Venice. "The Venetians are like pigs," the cleric declared, "and truly they have a pig's nature, for when a multitude of pigs . . . is hit or beaten, all draw close and run unto him who hits it."

The cleric could have added vanity to his list of Venetian traits. The "ruler of half and a quarter of the Roman empire" liked to boast that it had the sleekest ships, the wealthiest bankers, the most beautiful women, and the most intrepid merchant adventurers. "Wheresoever water runs," declared a local chronicler, Venetians can be found buying and selling. The most famous—and, in the view of most non-Venetians, the most shameless—display of Venetian narcissism was the all-day parade the city gave itself upon the installation of Lorenzo Tiepolo as doge, or city ruler.

The parade began with a glorious morning sail-by. The entire Venetian fleet, fifty magnificent vessels—the decks and masts of each crowded with cheering sailors; sails billowing out like puffed cheeks—glided across the mouth of the harbor, looking as stately as a procession of cardinals. Then, in the sparkling light of an Adriatic afternoon, the city's guilds marched out of St. Mark's Square behind two lines of gaudily attired trumpeters. Behind the musicians came the master smiths, with garlands in their hair, and above their heads, brightly colored banners flapping in the wind. Next came the furriers, in samanite and scarlet silk with mantles of ermine and vair; then the master tailors, in white robes with crimson stars; the gold workers, dressed in a shiny gold fabric; and finally the wicked barbers, ogling and gawking at the scandalously dressed slave girls marching in front of them.

But the Genoese cleric was right: narcissistic the Venetians might be, but in moments of crisis, they did band together—and that trait served them well when the plague slipped out of a silvery January dawn in 1348. Unlike the fatalistic Sicilians, who accepted the pestilence as an act of God, the competent, vigorous Venetians displayed what psychologists call agency. In the context of the times, the city's response to the Black Death was well organized, intelligent, and ruthless in its insistence on the maintenance of public order. On March 20, in an atmosphere of grave crisis, Venice's ruling body, the Great Council, and the doge, Andrea Dandolo, appointed an action committee of leading nobles; the committee's recommendations would form the basis for a reasonably coherent municipal response to the pestilence. Public health would be born in the tortured cities of northern and central Italy in the winter and spring of 1348, and Venice would be in the forefront of the new field.

Under instruction from municipal authorities, all ships entering Venice were boarded and searched; vessels found harboring foreigners and corpses (citizens of Venice being transferred home for burial) were set ablaze. To maintain public order, drinking houses (inns) were shut down, and the gaily colored wine boats that sailed the canals were ordered out of the water. Anyone caught selling unauthorized wine was fined, his goods confiscated and emptied into the canals. On April 3, with the warm weather approaching, the Grand Council issued a new directive. A few days later, a fleet of stiletto-shaped municipal gondolas appeared in the canals, their boatmen shouting, *"Corpi morti, corpi morti,"* as they navigated between the shuttered buildings. "Whoever had such dead in his house had to throw them down into barges under heavy penalty," says a contemporary.

As May settled over the Venetian lagoons, corpse-laden convoys shuttled back and forth through the choppy gray seas to the windswept islands of San Giorgio d'Alega (St. George of the Sea Weed) and San Marco Boccacalame. The convoys carried the poor, collected from streets, canals, hospitals, and charitable institutions. As a reward for being a citizen of a state that ruled "half and a quarter of the Roman Empire," each cadaver got a grave dug exactly five feet deep, a last view of Venice, and a final prayer from a priest. The same rules governed internment at San Erasmo, a mainland burial site under one of the modern city's most famous districts, the Lido.

By summer, with black-draped mourners everywhere, public morale became a grave concern. Venice was becoming the Republic of the Dead. On August 7, so as to avoid a further deepening of feelings of "affliction" in the city, the Grand Council banned *gramaglia,* or mourning clothes. The tradition of laying the dead in front of the family home to solicit contributions was also ended. The practice, popular in poorer neighborhoods, was deemed inappropriate in a time of plague. A new municipal clemency program was also instituted. To fill the empty canals and streets, the prisons were opened and municipal authorities softened their stance on the readmission of debt exiles; a right of return was granted to those who agreed to pay a fifth of what they owed.

Mass flight did occur in Venice, but the city clamped down vigorously on refugees. On June 10, with the death rate approaching six hundred a day, the authorities issued an ultimatum to absent municipal workers: return to your post within eight days or lose your position.

Caffa is often mentioned as the source of Venice's infection. But, leaving aside the problem of how a crew could have survived such a long journey, if Caffa—or even Constantinople—were the source of infection, Venice, on the east coast of Italy, should have been infected around the same time or even a little sooner than Messina, instead of months later. Ragusa (present-day Dubrovnik), a Venetian colony on the Balkan side of the Adriatic, which was visited by a Crimean fleet late in 1347, is a more likely source of infection. Contemporaries put the municipal death toll at a hundred thousand, but, with a population of roughly 120,000, that would give Venice a preposterously high mortality rate of almost a hundred percent. A historian of the city, Frederic C. Lane, thinks the plague killed about 60 percent of Venice, roughly 72,000 people, extraordinary enough.

One thing even the Black Death could not damage was Venetian self-esteem. On being awarded an annuity for his valorous services to the city during the pestilence, the municipal physician Francesco of Rome declared, "I would rather die here than live anywhere else."

The forceful Venetian response to the Black Death proves the point of *Disaster and Recovery,* the U.S. Atomic Energy Commission study on thermonuclear war. In the worst years of the mortality, Europeans witnessed horrors comparable to Hiroshima and Nagasaki, but even when death was everywhere and only a fool would dare to hope, the thin fabric of civilization held—sometimes by the skin of its teeth, but it held. Enough notaries, municipal and church authorities, physicians, and merchants stepped forward to keep governments and courts and churches and financial houses running—albeit at a much reduced level. The report is right about human resiliency: even in the most extreme and horrific of circumstances, people carry on.

Central Italy, Late Winter–Early Spring 1348

"At the beginning of January [1348], two Genoese galleys arrived from Romania, and when [the crews] reached the fish market someone began to talk with them and immediately . . . fell ill and died."

The incident in the fish market in Pisa marks a new stage in the spread of the Black Death. Behind the coastal city lay a dense network of rivers, roads, and trade routes, and at the other end of the network lay the urban centers of Tuscany. As if scenting fresh blood, the previously seaborne plague suddenly changed direction and thrust inland with the ferocity of a feral animal. In Florence, eighty-one kilometers to the east, alarmed municipal authorities frantically made preparations for the coming onslaught. Citizens were exhorted to keep their homes and streets clean and butchers to observe municipal restrictions on the slaughter of animals. Prostitutes and sodomites—medieval Florence had a reputation as a hotbed of sodomy—were expelled, and a 500-lire fine was imposed on visitors from already afflicted Pisa and Genoa. In early April, with the city not appreciably cleaner, local officials established a special municipal health commission with quasi-military powers. Commission agents were authorized to forcibly remove "all putrid matter and infected persons from which might arise . . . a corruption or infection of the air."

To the north, Florence's neighbor Pistoia issued a series of extraordinary public health directives. As a warm late spring breeze wafted across the town square, a municipal official announced that, henceforth, "bodies . . . shall not be removed from the place of death until they have been enclosed in a wooden box and the lid of the planks nailed down"; that "each grave shall be dug two and a half arms length deep"; that "any person attending a funeral shall not accompany the corpse or kinsmen further than the door of the church"; and that "no one shall dare or presume to wear new clothes during the mourning period." And "so that the sound of bells does not trouble or frighten the sick, the keepers of the campanile shall not allow any bells to be rung during funerals."

However, at least one new measure would have had a familiar ring to listeners. The official decreed that "it shall be understood that none of this applies to the burial of knights, doctors of law, judges, and doctors of physic, whose bodies can be honored by their heirs . . . in any way they please."

In Perugia, to the south of Florence, anxious local authorities turned to Gentile da Foligno for help. A leading physician of the day, Gentile was already famous for his paper on human gestation. After studying the vexing question of why human gestation tends to be more variable than that of the elephant (two years), the horse (twelve months), and the camel (ten months), Gentile concluded that one factor in the variability was the tendency of humans to become excited while having sex. In another renowned paper, Gentile examined a second vexing contemporary question: Is it better to suck poison from a wound on an empty stomach, as the great authority Serpion held, or on a full stomach, as the equally famous Maimonides and George the German had concluded? Gentile sided with Maimonides and George the German.

Questioned about the pestilence, Gentile, a professor at the medical school in Perugia, was initially reassuring. His plague tract, prepared at the request of municipal authorities, is so measured in tone—it describes the pestilence as less dangerous than some previous epidemics—that a modern German scholar would accuse him of writing most of it before 1348. However, the calm tone of the tract more than likely reflects geographical distance. Since the plague was still far away when Gentile began writing, he had set pen to paper lacking a firsthand knowledge of the disease. The passages that were added to the tract later, when the pestilence was approaching Perugia and more information was available, show that the "prince of physicians" was quick to appreciate *Y. pestis*'s unique destructive power. These new additions describe the plague as "unheard of" and "unprecedented."

While Florence exhorted its citizens to clean the streets, and Venice burned suspect ships, Siena, as ever, remained preoccupied with *la glorie de Sienne*. Municipal records show that in February 1348, as the pestilence thrust eastward across the wintry Tuscan countryside, Siena's governing body, the Council of Nine, was preoccupied with getting the municipal university upgraded to a more prestigious *studium generale*. The Nine adopted a typically Sienese solution to the problem: bribery. The council instructed its representatives at the papal court—the arbiter of such things—to spend whatever sums necessary to obtain the prestigious *studium generale* designation. If Siena took any precautions to protect itself from the plague, they have been lost.

In Orvieto, eighty miles to the south of Florence, town officials had an even more novel reaction to the approaching danger. They simply ignored it. Examining municipal records for the late winter and spring of 1348, French historian Elizabeth Carpentier found not a single reference to the pestilence. Perhaps Orvieto's Council of Seven, the town's governing body, was concerned about further depressing public morale, already badly shaken by the famines of 1346 and 1347 and a series of bloody and incessant local wars. In such a fraught atmosphere, plague talk could easily produce panic. However, the Seven also seem to have been engaging in a bit of magical thinking. It almost seems as if the council had convinced itself that if the pestilence did not hear its name spoken in Orvieto, it would pass over the town as the Angel of Death passed over the children of Israel who marked their doors with lambs' blood.

When the last of the winter snow had melted and the morning sky was flooded with golden light again, the plague came. The dying started slowly in March and early April, then quickly gathered momentum. On April 11, with local mortality levels approaching Sicilian levels, Florence suspended municipal deliberations; Siena followed suit in early June; and on July fifth, Orvieto. By August 21, six of Orvieto's seven town councilors were dead, and the survivor was recovering from plague. All through the desperate spring and summer of 1348, only once had the word "plague" been spoken in a council meeting, and then not until June, when the pestilence seemed about to swallow the town whole. Contemporaries put Orvieto's death rate at 90 percent, though Professor Carpentier thinks 50 percent a more reasonable estimate. In June, with the summer heat settling over the hills of Umbria, renowned physician Gentile da Foligno died a simple country doctor's death, tending patients in Perugia. A devoted student would later claim the great man died of overwork, but the brief course of Gentile's illness suggests a pestial death. In Pistoia, which gave its name to the pistol, draconian public health measures proved as ineffective as had St. Agatha's relics in Messina. "Hardly a person was left alive," wrote a local chronicler, and while surely that was an exaggeration, a half century later Pistoia's population would be only 29 percent of its mid-thirteenth-century level. In neighboring Bologna, where will making reached record levels on June 8, 1348, the Black Death claimed 35 to 40 percent of the city.

In Florence and Siena, the death rates would be even worse.

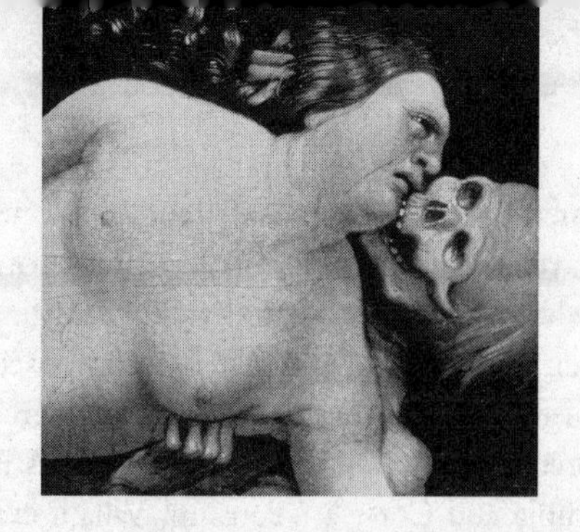

CHAPTER FIVE

Villani's Last Sentence

March 1348

ON A GRAY MARCH AFTERNOON IN 1348, PAST AND PRESENT intersected in the writing room of Florentine Giovanni Villani. As Villani sat at his desk composing a history of the plague, the disease was already in the villages to the west of the city and expected in Florence within days. Walking home from church that morning, the old man had seen dozens of carriages and carts rushing eastward toward the hills behind the city. Many of the shops and homes he passed had already been shuttered. Everyone who could seemed to be fleeing the city, and everyone who could not was deep in prayer. Seventy-two-year-old Villani, former banker and lifelong chronicler of Florence, would seek solace in his writing. He picked up his pen; on the walk home, he had composed a first sentence for his brief history of the pestilence. "Having grown to vigor in Turkey and Greece . . . ," he wrote, "the said pestilence leaped to Sicily, and Sardinia and Corsica." Pausing, Villani examined the sentence. What should come next? Ever since the previous November, Florence had been full of rumors about the plague, but which to believe? Ah yes, the old chronicler recalled, one story had struck him as both true and haunting—full of great courage and great foolishness. It was a report about eight Genoese galleys that had dared the pestilential Crimea. Four ships had returned "full of infected sailors . . . smitten one after the other on the return journey"; the other four vessels were said to be still wandering the Mediterranean, crewed by dead men. As the old man began to recount the fleet's odyssey, the room fell silent. There was just the crackle of embers in the dying fire and the sound of carriages rushing by in the empty street outside. As Villani wrote, the gray afternoon light in the window behind him faded into the lifeless black of a March night.

In his prime, Giovanni Villani had been a revered figure in Florence. A dazzling polymath, the young Villani had seemed capable of anything: computing the city's population from its grain consumption, counting the number of workers in the municipal cloth industry, writing a multivolume history of Florence in the manner of Virgil and Cicero. A wealthy banker at thirty, a civic leader at forty, the elegant Villani had climbed to the very pinnacle of Florentine society with silky ease, serving once as chief of the municipal mint and twice as prior, the city's most important civic office. Florence, more celebrated than any "republic or city state, save the Roman Republic . . ."; Florence, the city that invented eyeglasses and modern banking; Florence, the city Pope Boniface VIII called the earth's fifth element along with earth, wind, fire, and water—Florence had reached its glorious apotheosis in the feline, multitalented Signor Villani. For a time, the chronicler's only fault seemed to be marrying unwisely. When his second wife, the haughty Monna dei Pazzi, ran afoul of the sumptuary laws (dress codes), the rueful Villani grumbled that the "disordinate appetite of women . . . overcomes the reason and good sense of men."

However, by March 1348 the former Florentine wunderkind was an impoverished and disgraced old man—his fortune lost, his good name blackened beyond repair. Ten years earlier, at the age of sixty-two, Villani had endured the double humiliation of bankruptcy and debtors' prison. After his release, the former banker returned to chronicling, a passion that survived all the storms and seasons of life, but postimprisonment Villani showed a new appetite for disastrous and apocalyptic events, as if drawn to situations that mirrored his own embittered old age. And during the 1340s, Florence was happy to provide him with many such episodes.

Even without the plague, the 1340s would have been a desperate decade for the city. In 1340 there was a terrible epidemic; in 1341, a war with Pisa; and in 1343, political upheaval and civil strife, the latter culminating in a monstrous act of public barbarism that had left the old chronicler deeply shaken. "In the presence of the father and for his greater sorrow," Villani wrote of the execution of the city's chief of police and his son, the crowd "first dismembered the son, cutting him into small bits. This done, they did the same with the father. And some were so cruel . . . they ate of the raw flesh." In the mid-1340s ecological upheaval and financial ruin added to the sum of Florentine misery. There were the torrential rains of 1345 and the terrible famine of 1347, and in between there was the financial catastrophe of 1346, when Edward III of England, who was using Florentine money to fight the Hundred Years' War, defaulted on his loans to local banks to the tune of 1,365,000 florins—a sum the horrified Villani described as the "value of an entire kingdom."

However, no catastrophe captured the old chronicler's imagination quite like the plague. In the autumn of 1347, as Catania and Messina fought over St. Agatha's bones, Villani, in an I-told-you-so mood, was writing, "This plague was . . . foretold by the masters in astrology last March. . . . The sign of Virgo and its master . . . Mercury . . . signif[y] death." A series of sinister ecological portents in late 1347 and early 1348 reaffirmed Villani's belief that death would soon be astride the Arno plain. The winter before the plague arrived, the earth ripped open again, and large parts of northern Italy and Germany were rocked by earthquakes; shortly after Christmas 1347 a mysterious "column of fire" appeared over Avignon. Eyewitnesses claimed that the glowing shaft of golden light was a natural phenomenon produced by "the sun's rays like a rainbow," but Villani was having none of that. Even if the column was a natural phenomenon, he insisted that its appearance "nevertheless [is] a sign of future and great events." Whenever Villani wrote "great," he meant "terrible."

The pestilence, when it arrived, was everything that a Lear-like old man could have hoped for. Slipping under Florence's three sets of walls on a grim March day, *Y. pestis* toured the city like a conquering King Death. Stopping here and there to admire "views that resemble paintings," it infected Florence's "beautiful streets, beautiful hospitals, beautiful palaces and beautiful churches." Demonstrating a new ferocity, it burst into homes and churches and leaped upon the inhabitants "with the speed of fire racing through a dry or oily substance." The plague killed eighty Dominicans at the monastery of Santa Maria Novella and sixty Franciscans at Santa Croce del Corvo; it killed vast numbers of Florence's eight thousand to ten thousand schoolchildren, thirty thousand wool workers, six hundred notaries and lawyers, and sixty physicians and surgeons. And, as if emboldened by its success, with each passing day the pestilence killed with ever-mounting ferocity. It killed through a gray, wet April and a sunny May; and when the summer came and the July sun baked a thousand tangerine rooftops, it killed with an even greater ferocity, as if killing was the only happiness it knew.

In early spring, as the pestilence was taking hold in Florence, Villani completed his history. After following *Y. pestis* from its origins to the present moment, the chronicler wrote, "And the plague lasted until . . ."—then put down his pen, apparently expecting to pick it up again after the disease had burned itself out. It was an uncharacteristic act of optimism on the old pessimist's part, and, as it turned out, an unwarranted one.

Seven hundred years later, Villani's last sentence still awaits completion.

In the opening scene of *The Decameron,* Giovanni Boccaccio's allegory set in hills above Black Death Florence, several young women, "fair to look upon" and highborn, are attending a funeral in the city. Afterward, sitting in the oppressive darkness of the church nave, the little group falls into a fit of communal gloom. Outside, on the hot, pestilential streets, a world of pain and death awaits them. Suddenly a member of the group—a pretty young woman named Pampinea—brightens. Turning to her companions, she says, "Dear ladies, Here we linger for no purpose . . . [other] than to count the number of corpses being taken to burial . . . If this be so (and we plainly perceive that it is), what are we doing here? . . . We could go and stay together in one of our various country estates. . . . There we shall hear birds singing . . . see fresh green hills and plains, fields of corn undulating like the sea."

While the sparkling twenty-somethings in *The Decameron* are fictional, the account of the plague that precedes their conversation in the church is not. Giovanni Boccaccio lived though the Black Death in Florence,* and his account of the epidemic captures the texture and feel of life in a pestilential city as no other document of the period does.

"It is a remarkable story I have to relate," Boccaccio begins, then offers the reader a small sample of how "remarkable" life in Florence was. "One day," he says, "the rags of a pauper who had died from the disease were thrown into the street where they attracted the attention of two pigs. In their wonted fashion, the pigs first of all gave the rags a thorough mauling with their snouts, after which they took them between their teeth and shook them against their cheeks. And within a short time, they began to writhe as though they had been poisoned, then they both dropped dead to the ground, spread-eagled upon the rags that had brought about their undoing."

* Several scholars have argued that Boccaccio was not in the city during the plague. While that seems unlikely, the author's whereabouts remain a source of controversy.

A visiting Venetian once described Florence as a "clean, beautiful, happy place," but the city Boccaccio describes had become a vast, open-air death pit. "Many dropped dead in the open streets by day and night, . . . whilst a great many others, though dying in their own houses, drew their neighbors' attention to the fact more by the smell of their rotting corpses than by any other means. And what with these, and the others who were dying all over the city, bodies were here, there and everywhere."

Of the plague's divisive effects, Boccaccio writes, "It was not merely a question of one citizen avoiding another; . . . this scourge had implanted such a great terror in the hearts of men and women that brothers abandon brothers, uncles their nephews, sisters their brothers, and in many cases, wives deserted husbands. But even worse, and almost incredible was the fact that fathers and mothers refused to nurse and assist their own children, as though they did not belong to them."

The dying could find no succor outside the family, either. "Countless numbers of people who fell ill, both male and female, were entirely dependent upon . . . the charity of friends (who were few and far between) or the greed of servants, who remained in short supply despite the attraction of high wages out of all proportion to the services they performed." As shocking to Boccaccio as the desertion of the sick was "a practice almost never previously heard of, whereby when a woman fell ill, no matter how gracious or well born or beautiful she might be, she raised no objection to being attended by a male servant, whether young or not. Nor did she have any objections to showing him every part of her body as freely as she would have displayed it to a woman. . . . This explains why those women who recovered were possibly less chaste in the period that followed."

According to Boccaccio, "A great many people died who would perhaps have survived had they received some assistance. . . . And, hence, what with the lack of appropriate means for attending the sick and the virulence of the plague, the number of deaths reported in the city whether by day or night was so enormous that it astonished all who heard tell of it, to say nothing of the people who actually witnessed the carnage. And it was perhaps inevitable that among the citizens who survived, there arose certain customs that were quite contrary to the established tradition."

Boccaccio is referring here to the way the plague changed the grand opera that was the Florentine Way of Death. "It had once been the custom . . . ," he writes, "for the women relatives and neighbors of the dead man to assemble in his house in order to mourn in the company of the women who had been closest to him; moreover, his kinsfolk would foregather in the front of his house along with his neighbors and various other citizens, and there would be a contingent of priests, whose numbers varied according to the quality [that is, status] of the deceased; his body would be taken thence to the church in which he had wanted to be buried being borne on the shoulders of his peers amidst the funeral pomp of candles and dirges. But as the ferocity of the plague began to mount, this practice all but disappeared. . . . Not only did people die without having many women about them, but a great number departed this life without anyone at all to witness their going. Few indeed were those to whom the lamentations and bitter tears of their relatives were accorded; on the contrary, more often than not, bereavement was the signal for laughter, and witticism and general jollification—the art of which the women, having for the most part suppressed their feminine concerns for the salvations of the souls . . . , had learned to perfection."

Even sadder to see, according to Boccaccio, were the pathetic little trains of mourners who followed the dead through the summer streets. It became "rare for bodies . . . to be accompanied by more than ten or twelve neighbors to church, nor were they borne on the shoulders of worthy or honest citizens but by a kind of grave digging fraternity newly come into being and drawn from the lowest orders of society. These people assumed the title of sexton and demanded a fat fee for their services, which consisted of taking up the coffin and hauling it away swiftly not to the church specified by the dead man in his will, but usually to the nearest at hand."

Boccaccio's grave diggers are the sinister *becchini,* who circled Florence like vultures during the plague. Adopting the death's-head motto, "Those who live in fear die," these rough country men from the hills above the city earned an unsavory reputation not only for their cavalier attitude toward death, the way they seemed almost to condescend to it, but also for their swashbuckling behavior. In a city swollen with grief and loss, the *becchini* drank and wenched and caroused and stole like happy buccaneers. As spring became summer, the terrors of life in Florence grew to include a front door bursting open in the dead of night and a group of drunken, shovel-wielding grave diggers rushing into the house, threatening rape and murder unless the inhabitants paid a ransom.

The greatest blow to the Florentine Way of Death, however, was not the *becchini* or the cynicism of the female mourners, but the plague pits. "Such was the multitude of corpses . . . ," wrote Boccaccio, that "huge trenches were excavated in the churchyards, into which new arrivals were placed in their hundreds, stored tier upon tier like ships' cargo, each layer of corpses being covered with a thin layer of soil till the trench was filled to the top."

Professor Giulia Calvi was describing the psychological effects of the pits during a recurrence of the plague in the city, but her words apply equally well to Black Death Florence. "Nothing," she writes, "was more senseless, uncommon and cruel [to Florentines] than to be buried . . . far from the family vault, one's own church . . . from the texture of family life and neighborhood life . . . naked, mutilated by animals, a victim of the elements." For many, the pits held another, even greater terror. The modern idea of a personal death, of "my death," is a product of the European Middle Ages. In Antiquity and the early medieval period "death, at least as described in epic and chronicle, was a public and heroic event," says historian Caroline Walker Bynum. "But in the later Middle Ages death became increasingly personal. In painting and in story, [it] was seen as the moment at which the individual, alone before his personal past, took stock of the meaning of his life." The plague pit was the antithesis of this idea; it made death anonymous, casual, animal-like, and left the individual unrecognizable "even for future resurrection."

However, one thing even the Black Death could not change was human nature. Florentines responded to the pestilence in ways that still sound familiar. "Some people," says Boccaccio, "were of the opinion that a sober and abstemious mode of living considerably reduced the risk of infection. They, therefore, formed themselves into groups and lived in isolation from everyone. Having withdrawn to a comfortable abode, . . . they locked themselves in and settled down to a peaceable existence, consuming modest quantities of delicate food and precious wines and avoiding all excesses.

"Others took the opposite view and maintained that an infallible way to ward off this appalling evil was to drink heavily, enjoy life to the full . . . gratifying all one's cravings . . . and shrug the whole thing off as one enormous joke." Members of this group visited "one tavern after another drinking all day and night to immoderate excess or . . . they would do their drinking in various private houses, but only in ones where the conversation was restricted to subjects that were entertaining or pleasant. . . . People behaved as though their days were numbered and treated their belongings and own persons with abandon."

A third group of citizens "steered a middle course . . . living with a degree of freedom sufficient to satisfy their appetites, and not as recluses. They therefore walked about, carrying in their hands flowers or fragrant herbs or divers sorts of spices, which they frequently raised to their noses, deeming it an excellent thing thus to comfort the brain with such perfumes because the air seemed to be everywhere laden and reeking with the stench emitted by the dead and dying, and the odors of the drugs."

A fourth group reacted like Pampinea and her friends. "Some again," says Boccaccio, "the most sound perhaps . . . affirmed that there was no medicine for the disease superior or equal in efficiency to flight; following which prescription, a multitude of men and women, negligent of all but themselves, deserted their city, their houses, their estates, . . . and went into voluntary exile, or migrated to the country, as if God . . . would not pursue them."

The chronicle of another Florentine, Marchione di Coppo Stefani, helps to flesh out the picture of life in the pestilential city. Though he wrote several decades after the plague, Stefani employed a perhaps more resonant metaphor to convey the horror of pits. He says the dead were laid out, "layer upon layer just like one puts layers of cheese on lasagna." The chronicler also offers a fuller explanation of how friends and relations would desert the dying. As night fell, plague victims "would plead with . . . relatives not to abandon them," and to avoid an ugly scene, often the relatives would agree to stay. "'So you don't have to wake me during the night,' they would tell the victim at bedtime, 'take some sweetmeats, wine or water, they are on the bedstead by your head.'" According to Stefani, this supposed act of kindness was usually a ruse. "When the sick person fell asleep [the relative] left and did not return." Frequently these little dramas of abandonment and betrayal had an even darker second act. The next morning, awakening to find herself deceived and abandoned, the plague victim would crawl to a window and cry out for help, but since "no one . . . wished to enter a house where someone was sick," the call would go unheeded, leaving the victim to die alone in the warm morning light in a pool of her own blood and vomit.

Stefani's account also contains a description of the mordant dinner parties that became popular in Florence during the mortality. These often had the aspect of the game Ten Little Indians. "The [pestilence] was a matter of such great discouragement and fear," says the chronicler, "that men gathered . . . to take some comfort in dining together. And each evening one of them provided dinner to ten companions and the next evening they planned to eat with one of the others." But often when the next night arrived, the guests would find that the host "had no meal planned because he was sick. Or if [he] made dinner for ten, two or three were missing."

For many Florentines, one of the strangest aspects of the Black Death was the eerie stillness that fell over the streets and squares. Normally church bells echoed through the city morning, noon, and night, but during the plague the heavy thud of the gloomy bells became too much for people to bear and municipal authorities ordered them silenced. "They could not sound bells . . . nor cry out announcements, because the sick hated to hear this, and it discouraged the healthy as well." Human nature being what it is, greed flourished. "Servants, or those who took care of the ill, charged from one to three florins a day, and the cost of things grew as well," says Stefani. "The things that the sick ate, sweetmeats and sugar, seemed priceless. Sugar cost from three to eight florins a pound, . . . capons and other poultry were very expensive, and eggs cost between twelve and twenty-four pence each. . . . Finding wax was miraculous. A pound of wax would have gone up more than a florin if there had not been a stop put [by the municipal government] to the vain ostentation that the Florentines always make over funerals. . . . The mortality enriched apothecaries, doctors, poultry vendors, *beccamorti* [literally, vultures, and another name for the *becchini*], and greengrocers, who sold poultices of mallow, nettles, mercury, and other herbs necessary to draw off the mortality."

The Black Death's visit to Florence is unusually well documented. We know that the mortality claimed roughly fifty thousand lives, a death rate of 50 percent in a city of about a hundred thousand. We also know that while public order held, anarchy and disorder were common. Major riots were avoided, but flight was general and greed ubiquitous. During 1348, municipal officials stole 375,000 gold florins from the inheritances and estates of the dead. We know, too, that in Florence victims often developed two buboes instead of the one characteristic of modern plague. We know as well that many animals died; along with Boccaccio's pigs, there are reports of dogs and cats and apparently even chickens being stricken by the *gavocciolo,* or plague boil.

What remains a source of contention, however, is why the plague was so severe in the city.

This question goes to the heart of an even larger and far more contentious and puzzling question: Why was the plague in late medieval Europe so much more catastrophic than the plague of the Third Pandemic? Victorian scientists arrived in late-nineteenth-century India and China expecting to encounter a ravaging, galloping, all-consuming monster. Journalist William Seveni warned readers of the British *Fortnightly Review* to brace themselves. "We must not deceive ourselves [for once] the dreadful scourge . . . obtains a foothold [it] will be a far greater danger than when the terror-stricken Romans cried, 'Hannibal ante portas!'"

However, while millions died, the plague of the Third Pandemic proved to be a far more manageable disease than the plague of the Black Death. In early-twentieth-century India, *Y. pestis* traveled at an average of eight miles *per year;* in South Africa, a little faster: eight to twenty miles annually. By contrast, in Black Death Europe, *Y. pestis* covered the eighty-one kilometers between Pisa and Florence in two months—January to March 1348. In France and England the disease also moved with alacrity. Between Marseille and Paris, it traveled at a rate of two and a half miles *per day;* between Bristol and London, at two miles per day.

Contagion rates were also strikingly different. When the plague season arrived in Third Pandemic India, people would simply move two hundred yards from the family home, encamp, and wait for the disease to burn itself out. Indeed, *Y. pestis* proved so lethargic, British doctors in the colonial medical service joked that the safest place to be during the pandemic was the plague ward of a hospital. The pestilence of the Black Death, in contrast, was the pathogenic equivalent of a piranha. Boccaccio's description of the two pigs who fell dead after shaking an infected blanket was not literary hyperbole. The medieval plague spread so quickly, several medieval medical authorities were convinced the disease was spread via glance. "Instantaneous death occurs," wrote a Montpelier physician, "when the aerial spirit escaping from the eyes of the sick man strikes the healthy person standing near and looking at the sick." The medieval plague also produced symptoms uncommon in the modern disease, including a gangrenous inflammation of the throat and lungs; violent chest pains; a hacking, bloody cough; uncontrollable vomiting; a foul body odor; and a rapid course. Like Friar Michele da Piazza, chronicler Villani noted that most victims were dead within three days.

Most striking of all is the difference in mortality rates. In the Black Death, mortalities of 30 and 40 percent were common, and in the urban centers of eastern England and central Italy, death rates reached an almost unimaginable 50 to 60 percent. Historian Samuel K. Cohn claims that in the worst years of the Third Pandemic, death tolls never exceeded *3 percent*. While that estimate is open to question, no one challenges Cohn's contention that, overall, the mortality rates of the Third Pandemic were dramatically lower than those of the Black Death.

In the 1980s these discrepancies gave rise to a new theory of the medieval plague. A group of scholars began to argue that the Black Death was not a plague at all, but an outbreak of another disease—possibly anthrax, possibly an Ebola-like illness called hemorrhagic fever. The arguments of the Plague Deniers, as the group might be called, will be examined in the afterword. Suffice it to say here that recently DNA from *Y. pestis* has been found in several medieval plague sites. Moreover, it is not necessary to reinvent the Black Death to explain the discrepancies between the Second and Third Pandemics.

Microbiologist Robert Brubaker thinks that many of the differences between the two outbreaks dissolve if the vast differences between medieval and Victorian medicine are factored into the equation. Another possible explanation for the differences may lie in the unique impact of a disease like plague on a premodern society with no access to a relatively sophisticated colonial medical service. Unlike viral infections, which often left behind a large core of immune survivors to care for the ill and harvest the food the next time an epidemic struck, plague spared no one. Despite the findings about CCR5-Δ32, the best available current evidence is that *Y. pestis* does not produce permanent immunity in victims. During the Black Death, this biological quirk may have produced an enormous *secondary* mortality. As both Boccaccio and Stefani suggest, many people seem to have died not because they had particularly virulent cases of plague, but because the individuals who normally cared for them were either dead or ill themselves. In addition, the medieval streets may have become even dirtier and more rat infested because the street sweepers were all dying, and the malnourished even more malnourished because the farmers who grew the food and the stevedores who carted it to the city were also being decimated by plague.

A third possible explanation for the supermortality of the Black Death is that the medieval pandemic was caused by an unusually virulent strain of *Y. pestis,* marmot plague, which spread to rats as it moved toward Europe, while the Third Pandemic was an outbreak of less lethal rat plague that, for the most part, stayed in rats.

In the view of many Russian scientists, what makes marmot plague more virulent than, say, rat or gerbil plague, is a long-shared evolutionary history. Having cohabitated together, perhaps since the plague bacillus first evolved on the steppe, marmots have had more time than other rodents to develop resistance to the bacillus. Thus, to survive in marmots, biologically speaking, *Y. pestis* has been forced to go nuclear by adopting a strategy of hypervirulence, which it has done by, for example, evolving a tropism for the lungs.

Many Western microbiologists question whether the virulence of the plague bacillus varies from one rodent species to another. However, unlike most existing explanations of the Great Mortality, the Russian marmot theory has the virtue of simplicity. It provides a single, coherent explanation for several aspects of the medieval plague that continue to puzzle historians and scientists, including the high mortality rates and the apparently very high incidence of pneumonic plague even during the warm Italian springs and summers.

One or several of these factors may well have been responsible for the fearful death tolls in Florence, and a few months later, in Siena, a few dozen kilometers to the south. Except in Siena the mortality was even greater and the song of death that arose from its stilled summer streets, even more haunting than Boccaccio's mournful dirge.

Siena, April–May 1348

"La mortalità cominciò in Siena di maggio"—the mortality commenced in Siena in May, wrote Agnolo di Tura. Other sources place the plague's arrival in mid-April. All that is known for certain is that in Siena, as in Florence, men and women grew fearful as spring gathered in the countryside. The rustics, the rough countrymen who brought oil into Siena, stayed home; shops and stores were shuttered; the mu-

nicipal courts fell silent; and the wool industry, a sinkhole for so much Sienese wealth and pride, shut down. However, one aspect of Sienese life went on as usual as death arrived with spring in the fateful year of 1348.

As they had every day for a decade, the Sienese continued to awake each morning to the shouts of laborers mounting the scaffolding around the city cathedral, and to walk home each evening in the aura of the loveliest vista in Tuscany, the cathedral's white marble pinnacles and statues flushing rose red under a twilight blue Tuscan sky. The decision to pour thousands of lire into transforming the main town church into a majestic Tuscan St. Peter's was typically Sienese. "That vain people," Dante mocked in the *Inferno,* but a more accurate description of the medieval Sienese might be "deluded." Feeling overshadowed by larger, wealthier Florence, little Siena spent the better part of the thirteenth century huffing and puffing to make itself look bigger than it was, usually with disastrous results. Deciding the road to municipal glory lay in naval power, in the manner of Genoa and Venice, landlocked Siena frittered away a fortune trying to turn a malarial coastal village called Talamone into a major seaport. A few decades earlier, deciding cloth-making was the key to municipal glory and wealth, waterless Siena (water was essential to the manufacture of cloth) had frittered away another fortune digging up the rocky hill under the town in search of a mythic underground river called Diana.

In Agnolo di Tura, the city of dreams found its perfect spokesman. Reality has its place in Agnolo's journals, but not so prominent a place that it interfered with Siena's march toward the "broad sunlit uplands" of civic glory. Thus, in 1324, when the town wall is expanded, Agnolo boasts that Siena "has . . . grown in population such that the . . . walls [have to be] extended at Val di Montone." And in 1338, when the City Council decides to expand the town church, Agnolo is so enthusiastic you can already see the new cathedral floating above the Arno plain like a great gothic ship asail on a pumpkin-colored sea. "Siena," he writes, "is in a great and happy condition. Accordingly a great and noble enlargement of the city cathedral has begun." Even in 1346, a year of torrential rain and widespread famine, Agnolo remains, as ever, upbeat. After visiting the Campo, the town's main square, he declares that it is "more beautiful than any other piazza in Italy."

Like Giovanni Villani, Agnolo di Tura was a town chronicler, but there the resemblance between the two men ends. The older Villani was a scion of a wealthy mercantile family: well educated, urbane, and, before his financial embarrassment, an important civic figure. Agnolo was an everyman, albeit an ambitious everyman. He seems to have begun life as a humble shoemaker. There is still a bill of sale for some molds and other tools he bought in January 1324. The rest of Agnolo's early life remains unknown, except for his mother's name, Donna Geppo, and where he grew up, the Orvile section of Siena. However, Agnolo's literary career suggests some early education, and his habit of signing himself Agnolo di Tura del Grasso, or Agnolo the Fat, suggests that he liked to eat.

Agnolo seems to have thought big as well. Everything else we know about him points to a young man eager to rise in the world. The dowry of his wife, Nicoluccia, for example, indicates that the humble shoemaker managed to marry a bit above himself. Three hundred and fifty lire was a tidy sum for a craftsman's wife to bring into a marriage. Other surviving evidence also points to dreams too big for a cobbler's bench. One is a set of receipts for some gifts Agnolo bought the wife of a high official in the Biccherna, Siena's treasury department, where he worked part-time as a tax collector. It is unclear whether the gifts were purchased at the man's behest or on Agnolo's initiative; either way, the purchases were clearly meant to curry favor.

The fact that Agnolo convinced the tax office to reimburse him for the presents a few months later suggests that he was also shrewd about money, an impression reinforced by Nicoluccia's dowry and Agnolo's real estate dealings, which include the sale of a piazza near a place called Fontebradda for the handsome price of twelve gold florins in 1342. Real estate, tax collecting, shoemaking—would even an ambitious young man wear so many professional hats? It may be that Agnolo was not the only di Tura eager to rise in the world. In the medieval version of the "Take my wife" joke, the wife never tires of reminding her husband that her family is higher born than his. Some of this dynamic may have been at work in the di Tura marriage. During the 1330s and 1340s so many houses are listed under the name Agnolo di Tura, historians have speculated that Siena had several men of that name. But there may be another explanation. The residences all belonged to the same Agnolo di Tura, who kept moving his family into ever bigger houses in hopes that one day his higher-born wife would stop reminding him of how much she had sacrificed to marry a lowborn shoemaker.

The other salient fact about Agnolo is that he and Nicoluccia had five sons.

Agnolo mentions the children only once in his chronicle, but he compresses so much emotion into the single reference, the door flies open to the di Tura family's life—to the Christmas visits of Donna Geppo's, the Sunday outings at the Campo, the evening walks through the little squares that blink out from Siena's converging streets like an eye to a keyhole: the five di Tura boys running across the square, scattering a flock of birds into a vermillion-colored sky, an out-of-breath Agnolo chasing after them, and Nicoluccia shouting for everyone to stop, especially Agnolo, who is too *grasso* to run.

In the early summer of 1348, this happy life ended in a field near the cathedral.

The first official Sienese reaction to the pestilence came in early June. On the second, the City Council shut down the civil courts until September. A week and a half later, with plague pits already beginning to appear in the city, council officials resorted to a favorite municipal strategy—bribery. To appease a wrathful God, on June 13 a thousand gold florins were appropriated for the poor and gambling was banned "forever" in the city. On June 30 money was appropriated for the purchase of *"torchi e candele"* for a great religious procession.

The Palazzo pubblico, the building where the City Council met, was familiar to Agnolo. In his chronicle, he mentions a 1337 renovation there: "Rooms were constructed for Signori [Lords] and their staff [above the council chamber] and scenes from Roman history . . . were painted outside them." In the terrible May and June of 1348, one sees Agnolo, a big, heavyset, sad-faced man, walking the halls of the palace offering condolences to bereaved colleagues, listening to debates on how to dispose of the corpses accumulating in the sweltering, malodorous streets, comforting the dying on the new second floor. However, if Agnolo did any of these things, he never wrote about them. The impression one gets from the chronicle is that summer—the summer of the plague—Agnolo became a walker in the city.

"In many parts of Siena, very wide trenches were made and in these, they placed the bodies, throwing them in and covering them with but a little dirt."

One sees Agnolo walking to the cathedral, the workmen gone from its roof and walls now; its nave lit by the glow of a thousand candles—

"After that they put in the same trench many other bodies and covered them also with earth and so they laid them layer upon layer, until the trench was full."

—and through the Plaza de Campo, where he and Nicoluccia would take the children on Sundays—

"Members of a household brought their dead to a ditch as best they could without a priest, without divine offices."

—and the little squares, where the di Tura children would scatter birds into the evening sky with their shouts and charges.

"Some of the dead were . . . so ill covered that the dogs dragged them forth and devoured many bodies throughout the city."

Here a final image suggests itself: Agnolo standing at the a burial site near the unfinished cathedral on a soft April Sunday in 1356 or 1357. The field is featureless—in the pestilential summer of 1348 there were too many dead for individual graves and headstones. There is just a marker noting that during the Great Mortality many people of Siena were buried here. Agnolo places a bouquet of flowers under the marker and says a prayer. Later, walking home, he begins to relive the day he first visited the plague pit: the smells and the sights of that day—the hungry dogs snarling at one another while they pawed at the loose earth, the burly rustic gravediggers stripped to the waist in the summer heat, the wails of the grieving mothers and fathers, the piles of greasy white corpses in the shallow plague pits, and himself—full of fury, grabbing a shovel from a rustic and digging a separate, deeper grave.

One imagines that this was the day Agnolo added the concluding sentence to his chronicle for 1348:

"And I, Agnolo di Tura, called the fat, buried my wife and five children with my own hands."

The mortality in Siena was grave. According to Agnolo, "52,000 persone" died in the city, including 36,000 "vecchi"—old people; for the countryside, he gives a figure of 28,000. In a region with a preplague population as potentially high as 97,000, that works out to a death rate of 84 percent, a figure most modern historians consider improbable. Current estimates put the mortality in Siena at between 50 and 60 percent. The plague produced one additional casualty. While the eternal ban on gambling was revoked within six months (Siena was broke again), seven hundred years on, the cathedral renovation awaits completion.

Rome, Summer 1348

In August the pestilence traveled southward from Orvieto to Rome, where one of the most extraordinary of all of medieval Italy's little municipal dramas was playing itself out.

Imagine Mussolini three times as handsome and four times as preposterous, and you have the drama's hero, Cola di Rienzo, self-proclaimed tribune of Rome, fantasist extraordinaire, and local hero. For smashing the Mafia-style rule of Rome's old noble families, the *populus romanus* were willing to forgive their handsome Cola almost anything, including the fantasy that he was the bastard son of a German emperor instead of the peasant son of a barkeep. But when Cola knighted his own son in the blood of another man, even the Roman crowds recoiled in horror.

The second major character in the drama was Cola's nemesis, eighty-year-old Stefano Colonna, Rome's most powerful aristocrat and an authentic natural wonder. "Great God, what majesty is in this old man," wrote a contemporary. "What a voice, what a brow and countenance, . . . what energy of mind and strength of body at such an age!" When Cola's supporters killed Stefano's son, grandson, and nephew, the old man refused to mourn, saying, "It is better to die than [to] live in servitude to a clown."

The third major character in the drama was the urbane and learned Pope Clement VI, a shrewd sybarite with a Friar Tuck–sized waistline and an X-rated libido. One story has it that, when rebuked for wantonness, Clement would plead *ex consilio medicorum*—he was following the advice of his physicians; another, that he would confront the chastiser with a list of other libidinous popes he kept in a "little black book," then wonder out loud why it was that the church's greatest leaders had also been among its greatest philanderers.

The last major player in the drama was Francesco Petrarch, literary celebrity and early practitioner of radical chic. "I feel I have met a God, not a man," Petrarch wrote, after encountering the handsome Cola, thereby proving that he was a better poet than a judge of character.

The *deus ex machina* that brought these figures into play was the intolerable state of medieval Rome. By 1347 the great capital of Antiquity had shrunk to a squalid little ruin; it was a city of buildings without walls, arches without roofs, pedestals without statues, fountains without water, columns without arches, and steps that led from nowhere to nothing. From a half million—and perhaps many more—residents during Antiquity, the population had fallen to a pitiful thirty-five thousand, and with no other visible means of support, medieval Romans survived by cannibalizing the decaying city. The rich pilfered marble and brick from the imperial ruins to build their gloomy castles and fierce towers; the poor, to erect their stinking hovels. Even the great palaces on Palatine hill, the baths of Diocletian, and the Julian basilica were torn down and thrown into kilns to make lime. Materials the Romans could not use themselves, they happily sold to others.

Many of Italy's great cathedrals and even London's Westminster Abbey were, in part, built from imperial rubble. On summer mornings medieval visitors could still see women hurrying across the Ponte Sant' Angelo, balancing bundles on their heads, and fishermen bent over their pots and nets along the banks of the Tiber. But beyond the river, now malodorous and polluted, and the center city, where the poor lived in streets so narrow the afternoon sun never fell below the tiled rooftops, there was nothing but lumpy grassland, broken buildings, and cow pastures all the way to the Aurelian wall, the boundary of the old imperial city.

Life on the Roman street mirrored the city's physical condition. The Hobbesian state of all against all, characteristic of medieval Italy generally, reached its apotheosis in the gangsterism of medieval Rome. The city's ruling class—the great aristocratic families like the Colonna and the Orsini—engaged in a perpetual war against one another, and beneath the violence of the highborn there was the violence of the gutter: of robbers and muggers and streetcorner toughs. In 1309, when the papacy, the last bastion of municipal authority, fled to the safety of Avignon, civic order collapsed entirely. With "no one to govern," wrote a contemporary, "fighting [was a] daily occurrence, robbery was rife. Nuns—and even children—were outraged; wives were torn from their husbands' beds. Laborers on their way to work were robbed at the very gates of the city . . . ; priests became evil doers, every sin was unbridled. There was only one law—the law of the sword."

Cola di Rienzo's emergence as the self-appointed savior of a bleeding Rome owed something to civic patriotism, something to personal grievance—a Colonna henchman killed Cola's brother—and something to a romantic imagination inflamed by constant rereadings of Seneca, Livy, and Cicero. Sometimes, after studying the Great Ancients, the dreamy young Cola would stand in a cow pasture at twilight and, surveying a broken column or arch, wonder out loud, "Where are those good old Romans? Where is their lofty rectitude? Would that I could transport myself back to the time when these men flourished." Lacking a time machine, Cola did the next best thing: he invented a fantasy version of himself. Long before his rise to prominence, he took to signing himself Cola di Rienzo, "Roman Consul and sole legate of the people and of the orphans, widows and the poor." He also began telling people that he was the illegitimate son of the German emperor Henry II, who had seduced his barkeep mother on a visit to Rome.

Cola first rose to public attention in 1343, on a visit to Avignon. He was by then a notary—one of the few careers open to poor, bright boys—and prominent enough to be appointed to an important commercial delegation. To revive tourism in popeless, lawless Rome, municipal authorities wanted Clement to declare 1350 a Jubilee Year and offer a special indulgence—a forgiveness of sin—to pilgrims who visited the city. A similar celebration in 1300 had attracted more than a million tourists. However, when the delegation met with the pope, Cola used the occasion to launch into a furious denunciation of the Roman nobility and their gangsterism. Horrified by the outburst, Cola's fellow delegates tried to shush him, but Clement, already on record as declaring the nobles "horse thieves, murderers, bandits and adulterers," was impressed by the ardent young notary. Before leaving Avignon, he put Cola under papal protection and gave him a new title, rector of Rome.

During the same visit, Cola also met the poet Petrarch, another lively fantasist and sly self-promoter, who already had a good part of literate Europe trying to guess the name of the mystery woman who appeared in his love poetry, the luminous Laura.

Love and I, stood agape; we marveled how
no wonders ever amazed the human sight
like the speaking lips and laughing eyes alight
of our lady.

"You say that I have invented . . . Laura to have something to talk about and to have everyone talking about me," Petrarch wrote to a friend, who, like many others—in the poet's own time and since—have thought Laura the artful fabrication of an artful fabricator. But the mystery woman in the poems was real enough. Her full name was Laura de Sade, she was related by marriage to the infamous eighteenth-century Marquis de Sade, and Petrarch loved her as deeply and truly as he claimed, though perhaps not as chastely. He had children with at least two other women.

An international celebrity as well as a poet, Petrarch dined with the aristocratic Colonna, walked the beaches of Naples with the beautiful Queen Joanna, attended audiences with Clement VI—if there had been a fourteenth-century *People,* the fish-eyed poet would have been on the cover under the headline, "The Fabulous Francesco!" Normally, lowborn notaries were of no interest to the celebrity poet, but besides sharing a reverence for Rome's past, the bookish Petrarch was infatuated by Cola's bold man-of-action stance. After their first meeting, the poet gushed to his new hero in a letter, "When I think of our earnest, sanctified conversation . . . I feel afire, as if an oracle had issued from the recesses."

In May 1347, as the plague was sailing west, Cola, who had been building a power base in Rome since his return from Avignon, launched a coup.

On the nineteenth, a sleepy Saturday, forces loyal to the notary seized the buildings in the municipal district. The next morning, as church bells echoed through the streets, the gates of Rome flew open and Cola strode into the city the way he must have imagined a thousand times in his dreams: dressed in full knight's armor, with the red banner of freedom and the white banner of justice flapping overhead and the papal vicar at his side. Ahead of Cola marched a phalanx of blaring trumpeters.

"Cola! Cola! Cola!" the crowd shouted. The notary, who looked especially handsome in shining armor, raised a hand in acknowledgment; a child emerged from the crowd; Cola took the bouquet she offered and gave her a kiss. Then the trumpets blared again and the notary's little ship of state sailed off through flower-strewn May streets to the palace of the capital, where the coup ended with a remarkable oration. As thousands of voices shouted approval, Cola pledged himself ready to die for Rome, swore to restore the city to its former glory, and promised to devote himself to the cause of equal justice for all.

"I think of you day and night," an excited Petrarch wrote his hero shortly after the coup. "And . . . O Tribune, if you must die in battle do so courageously for you will assuredly be rapt in heaven. . . . Unfortunately," the poet added in a postscript, "my . . . circumstances prevent me from joining in your holy war." Happily, no one had to die in Cola's cause, not even Cola.

Stefano Colonna flew into a rage when he heard about the notary's coup. "If this fool provokes me further, I shall throw him out the window," the old man shouted. The rest of Rome's aristocrats, however, chose to sit on their swords and await events. When Cola summoned the nobility to the capital to swear allegiance to him, everyone came and solemnly placed a hand over his heart. Many in the illustrious gathering could remember when the notary had been the pet bad boy of Rome's elite, the impudent provocateur whom fashionable hostesses would invite to dinner parties to shock and titillate highborn guests. Perhaps Cola was a still a fool.

While the great bided their time, Cola collected titles and tried on clothes. On the last day of July 1347, the notary, dressed in a white silk robe embroidered with gold and accompanied by a matching pennant emblazoned with the sun, led a procession to the Baptistry of St. John. There Cola, who had just had himself proclaimed a knight, took a ritual bath in the tub where Emperor Constantine was said to have been cured of leprosy. The next day, now dressed in scarlet, Cola appeared on the balcony of the Lateran Palace. Declaring Rome the capital of the world and all Italians Roman citizens, the new knight unsheathed his sword and made three cuts in the air—one to the east, one to the west, and one to the north. The enthusiasm of the crowds was enhanced by a nearby statue of Constantine; the emperor's horse spewed free wine from the nostrils.

On August 15, three months after seizing power, Cola gave himself a new title, tribune of Rome. After receiving a silver wreath, a scepter, and an orb, a symbol of sovereignty, the newly crowned tribune stepped to the podium and reminded the audience that he was thirty-three, the same age as Christ when He died on the cross for the sins of humanity.

On the feast of Saints Peter and Paul, Cola, now dressed in green and yellow velvet and carrying a scepter of glittering steel, rode to St. Peter's Basilica on a white charger. In addition to fifty spearmen, the new tribune was accompanied by a horseman holding a banner with his coat of arms, a knight throwing gold coins into the crowd, a chorus of trumpeters blaring through long silver tubes, and a chorus of cymbalists clanging silver cymbals together. On his arrival at the basilica, an assemblage of bowing dignitaries greeted the tribune to the strains of *"Veni Creator Spiritus."*

After a summer full of comic opera events, support for Cola began to evaporate. In September, with the plague now only a few weeks' sail from Sicily, the pope denounced the tribune as a usurper and heretic. Petrarch, fearful of alienating his powerful friends in the Roman elite, expressed disquiet and doubt, and the Roman crowds, believing their presumptive savior was acting the fool, faded away.

Sensing the change in mood, in mid-September Cola organized a second coup. He invited all Rome's great barons to a banquet. At dinner, Cola had to endure old Stefano Colonna's sarcasm about his magnificent attire, but after dessert the notary had his revenge. As the guests prepared to leave, Cola ordered the arrest of seven leading nobles, including five members of the powerful Orsini family, and impudent old Stefano. You will be executed in the morning, the tribune informed the prisoners.

When a priest came to Stefano's cell the next morning, the old man snarled and waved him away, saying he had no need to confess; Stefano Colonna would never die at the hands of a nothing like Cola di Rienzo. The remark proved prophetic. A few hours later, as bells tolled for the condemned men, the tribune lost his nerve. Executing the Colonna and Orsini prisoners, members of the most powerful aristocratic families in the city, might provoke the other nobles. A few minutes later, a chastened Cola stepped out onto a balcony and, reminding the crowd of the biblical adage "Forgive us our trespasses," announced that he had decided to pardon the prisoners.

In November, as the plague was arriving in Marseille, Cola alienated his few remaining supporters with an act of unimaginable barbarism. During an attack on the city by a group of nobles, twenty-year-old Giovanni Colonna, old Stefano's grandson, was cut to pieces by Cola's cavalry. The morning after the assault, the tribune brought his son Lorenzo to the spot were Giovanni fell. As a crowd watched in horror, Cola unsheathed his sword, dipped it in Colonna's blood, then placed the red-tipped sword upon his son's head, proclaiming him knight Lorenzo of Victory.

A few weeks later, as the last of Cola's support was ebbing away, Petrarch wrote to the tribune, "I cannot alter matters, but I can flee them. . . . A long farewell to thee also, Rome, if these stories are true."

When the plague arrived in the city in August 1348, Cola was safely out of harm's way. Unseated the previous December in a countercoup, he was living in disgraced exile with the Celestine monks in Abruzzi. That was too far away to hear the tremendous earthquake that rocked Rome at the start of the pestilence, or to catch the scent of burning flesh rising above the Coliseum, where many of the plague dead were cremated, or to see the piles of corpses lined up along the Tiber like a levee wall.

"I am overwhelmed," wrote a contemporary, in the sad, desperate voice that became the voice of Europe in the summer of 1348. "I can't go on. Everywhere one turns there is death and bitterness. . . . The hand of the Almighty strikes repeatedly, to greater and greater effect. The terrible judgment gains in power as time goes by."

CHAPTER SIX

The Curse of the Grand Master

ON A WINTRY PARISIAN AFTERNOON IN 1314, SEVERAL DOZEN people stood huddled together on a windswept island in the Seine, awaiting an execution. Against the low gray light, the spectators' faces looked like petals on a rain-soaked bark. But when the jobbers of Les Halles, the butchers of St. Jacques-la-Boucherie, and the prostitutes of Marché Palu arrived to swell the crowd, the petals would merge into a frenzied pastel mass, two thousand faces whipped to high color by a river wind and the expectation of death in the sharp March air.

The crowd had gathered on the island to witness the execution of Jacques de Molay, former grand master of the Templars, until recently one of the most powerful religious orders in Christendom. Earlier in the day, de Molay had caused a great furor outside Notre Dame. In return for a life sentence, the old man had agreed to publicly confess to "crimes that defile the land with their filth," including sodomy, idol worship, and spitting on the cross. But this morning in front of the cathedral, de Molay had surprised everyone. With half of Paris looking on, the old man boldly asserted his innocence and denounced the charges against him and the Templars as a lie and a crime against heaven. Emboldened, one of his lieutenants, Geoffroi de Charney, had done the same.

The king was infuriated, the Church embarrassed, the Parisian mob titillated. The Templars affair promised to end as dramatically as it had begun seven years earlier, when agents of the French Crown swooped out of an October dawn and arrested two thousand unsuspecting, mostly elderly Templars in a series of nationwide raids. Dazed members of the order were pulled from their beds and their prayers, pushed into carts, and hauled off to royal prisons. By the end of the day—a Friday the thirteenth, the superstitious noted—if there was a Templar alive in France who had not been charged with having intercourse with demons; spitting on Christ's image; urinating on the cross; administering the "kiss of shame" to the penis, buttocks, and the lips of the order's prior; or engaging in other homosexual acts, it was because he was hiding in a haystack or under a bed. The Templars' crimes were "a bitter thing, a lamentable thing, a thing which is horrible to contemplate, terrible to hear . . . a thing almost inhuman, indeed set apart from all humanity," declared the author of the charges, the grand master's former friend, Philip the Fair, King of France and "more handsome than any man in the world."

Philip's construction of the Templars affair as a case of odious blasphemers (the Templars) versus a "watch-tower of regal eminence" (himself) was an early example of the Big Lie. Money, not sin, drove the king. The ambitious Philip was an architect of the modern nation-state; his great dream was to transform feudal France, a patchwork of regions with different traditions and practices, into a unified nation, bound by a single set of institutions and laws and answerable to a single authority, the French Crown—which is to say, to himself. To a significant degree, he succeeded. Increasingly in the France of Philip the Fair, the king's peace became "the peace of the whole kingdom; and the peace of the kingdom . . . the peace of the church, the defense of all knowledge, virtue and justice."

With tax revenues uncertain, however, the new, assertive French state lacked a firm financial foundation. The Templars, who possessed the largest treasury in northern Europe, could provide the king with a lucrative new revenue stream. As a potential target, the order also had the additional advantage of being both loathed and feared. Part secret society—members were rumored to practice black magic—and part international bank, the Templars were viewed as a sinister organization peopled by powerful, shadowy figures. Every *éminence grise* in Christendom was thought to wear a Templar cross. The only thing the order lacked was culpability. The society's crimes might be legion, but none had been directed against the French Crown. However, Philip and his ministers also anticipated another aspect of the modern nation-state, the false confession. The "august and sovereign house of France" was quite adept at constructing fanciful crimes and torturing the innocent until they agreed to confess.

Templar knight Gerard de Pasagio testified that after his arrest he was tortured by the "hanging of weights on his genitals and other members." Other Templars were strapped to the rack, their ankles and wrists dislocated by a winching device that—slowly—pulled joints from sockets. Another popular torture was called *strappado*. The prisoner was pulled to the ceiling by a rope that suddenly went slack, his fall broken at the last moment by a violent jerk. Sometimes weights were attached to the testicles and feet to make the jerk more violent and painful. One Templar, Bernard de Vaho, had his feet smeared with fat and placed in an open flame. A few days later, when de Vaho tried to walk, "the bones in his feet dropped out." Other popular tortures included yanking teeth and tearing out fingernails one by one.

By early 1314 the Templars were in tatters. The order had been disbanded by papal bull, most of its treasury was in the hands of the French Crown, and its leadership either dead, in prison, or gone mad. All that remained was to end the affair on an appropriate note of dignity.

But the only part of the finale that went according to schedule was the date, March 18, 1314. On learning that the grand master and his lieutenant, de Charney, had defied the Crown, Philip overrode the objections of Church officials who wished for a day to deliberate the men's fate and ordered both "burned to death" immediately.

Shouts of "heretic" and "blasphemer" greeted the condemned men as they arrived at the execution site, the desolate Ile des Javiaux in the Seine, late on the afternoon of the eighteenth. Someone in the mob picked up a stone and threw it. The raw river wind had put the crowd in an ugly mood, but there was also more than a hint of expectation in the air. People were hoping for a last great surprise from the grand master, something like the repudiation in front of Notre Dame that morning. And, according to legend, de Molay did not disappoint. As he disappeared into a plume of flame and smoke, the old man is supposed to have thrown back his head and called down a curse on the King of France and on all the king's descendants unto the thirteenth generation.

Stories about the grand master's curse spread as far as Italy—Giovanni Villani mentions it in one his chronicles—but no one seems to have taken de Molay's words very seriously. And, indeed, why should they have?

Surveying Philip's France in the year 1314, the chronicler Jean Froissart described it as "gorged, contented and strong." Foreigners might complain of "prating Frenchmen always sneering at nations other than their own," but when left alone those foreigners would exclaim to one another, "Oh, to be God in France!" In the early fourteenth century, few would have challenged the assertion of Jean de Jardun that "the government of the earth rightfully belongs to the august and sovereign house of France."

Stretching from Flanders and Picardy in the north to the Pyrenees in the south, Philip's France was the largest state in Europe, with the largest population—variously estimated at between 16 million and 24 million. The country had the richest farmland, and its fief, Flanders, contributed the most important industry in medieval Europe: cloth making. Paris, the French capital, had grown into a bustling metropolis of 210,000* souls with a half dozen or more paved streets, including a cobblestoned marvel, the Grande Rue, the city's main thoroughfare and the axis of Paris's northward march from the Ile de la Cité and the Seine in the twelfth and thirteenth centuries. In the military arena, France also stood supreme. No other monarchy in Europe could routinely field armies of 20,000 to 25,000.

Almost everywhere, medieval culture was also by and large French culture. Medieval men and women wore French fashions, emulated French manners, imitated French chivalric traditions, mimicked French troubadours, read French sagas, and prayed in French-inspired Gothic cathedrals. Of another Gallic architectural wonder, the nation's abbeys, the chronicler Joinville likened them to illuminations of a manuscript in azure and gold. Of the University of Paris, a young Irish scholar exclaimed, it is "the home and nurse of theological and philosophical science, mother of liberal arts, mistress of justice and the standard of the morals, the mirror and lamp of all theological virtues."

Against all of this, what was the grand master's curse? Nothing.

And yet . . .

A month after de Molay's execution, Pope Clement V, Philip's reluctant ally in the Templars affair, died suddenly. Then, in November, the forty-six-year-old king was killed by a stroke. The following year, 1315, the Great Famine arrived.

In 1316 Philip's successor and oldest son, Louis X, died after a brief eighteen-month reign. A few years later, the Pastoureaux, a peasant rebel movement, swept through northern France, storming castles and abbeys, burning down town halls, ravaging farms, and killing Jewish moneylenders.

And still worse was to come.

* Estimates of medieval Paris's population, like many medieval demographic estimates, range all over the lot. Some historians think the city may have had only a hundred thousand residents.

In 1323 Philip's second son and Louis's successor, Philip V, died prematurely. A mysterious series of epidemics followed, then more famine and another royal mortality; in 1328 Charles IV, the last of Philip the Fair's three sons, died; none of his heirs had survived to age thirty-five or ruled for more than six years. And since Charles left no male heir, the Capetian dynasty, rulers of France since 987, died with him. In the succession crisis that followed, Edward III of England, Philip the Fair's grandson (through Edward's mother), invaded France to lay title to the French throne, igniting the bloodiest conflict of the Middle Ages, the Hundred Years' War.

And still worse was yet to come for "the august and sovereign house of France."

Marseille, November 1347

In Marseille, where a street fight could leave thirty dead and where even the clerical crime rate was a scandal, the plague arrived before the winter rains and may have killed half the population. Yet it failed to shake the city's resolve or to destroy a centuries-long tradition of tolerance.

Any history of fourteenth-century Marseille would have to include its energetic, sharp-elbowed population of 20,000 to 25,000, its reputation as a medieval Big Easy, and its commercial importance. Marseille was a principal entry point for goods coming from Spain and the Levant, and a principal disembarkation point for Crusaders. This latter role led to the city's involvement in one of the most notorious incidents of the Middle Ages. In the early thirteenth century, thousands of members of the Children's Crusade—youngsters who believed that pure prepubescent Christian hearts, not swords, held the key to liberating the Holy Land—descended on Marseille, seeking passage to the Levant. A number of local ships were booked, and the young Crusaders duly taken aboard, but, character being destiny in Marseille, en route to the East, the captains had a change of heart and instead sold the young Crusaders in the slave markets of the Muslim Levant.

Any biography of medieval Marseille would also have to include the sloping hill the city sat on, and its odd three-tiered wedding-cake structure. At the top of the hill was the *ville-episcopal*—bishop's town; beneath it, the administrative buildings of the *ville de la prevote*—provost's town; and at the bottom of the hill, the *ville-bas,* or lower town. A rats' maze of thoroughfares, the *ville-bas* was where medieval Marseille lived and worked and played. Inside the quarter's shops, drapers, fishmongers, and box and barrel makers bent over workbenches, cutting, tearing, and banging, while outside on sinewy streets illuminated by a sliver of blue sky, money changers shouted out the latest exchange rates, drunken mariners ogled broad-hipped women in dresses cut so low the necklines were called "windows of hell," and tanners poured vats of steaming hot chemicals into piles of mud and human waste. With ventilation limited to a breeze from the harbor, on most days the *ville-bas* had the pungent odor of a mermaid with loose bowels.

In spring and summer, when the torrid Mediterranean heat settled over the city, making Marseille's stone buildings sweat on the inside and hot to the touch on the outside, local magistrates, notaries, and lawyers would abandon the gloomy precincts of Hopital du St. Esprit, the center of the municipal legal system, for the outdoor courts of the Place des Accoules. Passing through the plaza on a spring morning in 1338, a visitor could have heard a young woman named Guilelma de Crusols giving testimony in the case of Bonafos v. Gandulfa.

At issue in the case was Madame Gandulfa's wandering drainpipe. According to M. Bonafos, the madam had moved the pipe, which was supposed to be equidistant between their two homes, closer to his house, so waste would drain onto his property. Young Mlle. de Crusols, testifying on M. Bonafos's behalf, told the court that she had tried to reason with the difficult Madame Gandulfa. "I went . . . and asked [her] why she had moved the drain. [But] all she would say is that if the drain was moved back to its original place she would move it again." Mlle. de Crusols also told the court that during the conversation, Madame Gandulfa kept waving a piece of drainpipe back and forth in her hand.

Ten years later, the Place des Accoules makes another brief appearance in Marseille's history, although in a context that would have been incomprehensible in 1338. A reference to the square appears in an entry a notary named Jacme Aycart made in his casebook on April 30, 1348. The father of Silona and Augeyron Andree had just died of plague, and notary Aycart had been called to the Change, a new outdoor court near the harbor, to draw up a transfer of guardianship for the children. In a dotal act it was customary to record the location as well as the date. Accordingly, M. Aycart wrote down "the Change," then, realizing that would sound odd since the outdoor court usually met on the Place des Accoules, Aycart added a clarifying note. He wrote that the court had changed venues *"ob fecarem mortuorum terribilem de symmeterio ecclesie Beate Marie de Acuis"*—"on account of the terrible stench of the dead from the cemetery of Notre Dame des Accoules" (next door to the Plaza).

The plague came to medieval Marseille the way most things did—by sea. The infecting agent was apparently one of the pestilential galleys Genoa had expelled in late October 1347. Marseille's last day of normalcy was November 1—All Saints' Day, as it happens. Sometime during the day, the stray galley appeared outside the harbor like a shark fin circling in the water; the iron chain guarding the entrance was lowered, and the ship sailed past La Tourette, a fort manned by the Knights of St. John, and docked. The prompt action of the Genoese authorities suggests that many coastal towns in southern Europe were already on high alert for pestilential ships, but, strangely, the galley, already "driven from port to port," aroused no suspicion in Marseille. Perhaps the previous expulsions had made the captain crafty. Aware that the physical condition of the crew would cause alarm, he may have waited for evening light or for a thick fog or heavy rain before making a run at the harbor. Whatever ploy he used, it worked; the arrival of the plague seems to have taken Marseille by surprise.

"Men," says a contemporary, "were infected without realizing it and died suddenly and the inhabitants thereupon drove the galley away." But, as in Messina and in Genoa, all *Y. pestis* needed to establish itself was that narrow hour between not knowing and knowing. As the expelled galley vanished back into the Mediterranean, a silvery stream of disease was already slithering thorough the cavernous streets of the *ville-bas,* stopping here and there to admire the piles of waste and refuse in front of the four- and five-story buildings before heading north to La Juiverie, the Jewish quarter, and the Palais de Marseille, seat of government. Meanwhile, outside the harbor, the expelled ship joined two companions, and the three galleys faded into the autumn light somewhere "along the Spanish coast" heading toward the Gibraltar Gap. Of the little fleet, a contemporary wrote, "The infection that these galleys left behind their whole route . . . particularly in coastal cities . . . was so great that its duration and horror can scarcely be believed, let alone described."*

Marseille's historic commercial ties to the Levant and to Asia Minor made it a natural target for any disease coming out of the East. In A.D. 543 the Plague of Justinian took a terrible toll on the city, and in 1720 Marseille would become one of the last cities in Europe to experience a major outbreak of pestilence. However, the 1347 outbreak was particularly severe. In April 1348 Louis Heyligen, the musician who lived in the papal city of Avignon to the north, wrote to friends in Flanders that "in Marseille . . . four of five people died." The true figure was probably closer to one out of two, but that was enough for abbot Gilles li Muisis to describe the city's suffering as "unbelievable."†

Another entry in M. Aycart's casebook captures the casual pervasiveness of death in a city where half the population of twenty thousand to twenty-five thousand disappeared within a year. Dated April

* Economic desperation probably kept the little fleet going. The collapse of the Crusades at the end of the thirteenth century and the weakening of the Mongol empire a few decades later produced a great maritime depression. In 1248 the casebook of a single Marseille notary contained more than a thousand commercial acts. In the ten years prior to 1348, all Marseille's notaries *combined* recorded only 147 such acts. (Daniel Lord Smail, "Mapping Networks and Knowledge in Medieval Marseille, 1337–1342," unpublished Ph.D. thesis [Ann Arbor: University of Michigan, 1994], p. 22.)

† Contemporary accounts put the mortality in Marseille at 56,000, but since that is more than twice the city's medieval population, the figure is meaningless. The estimate of something like a 50 percent mortality rate is based on the fact that the plague probably arrived in Marseille, as it did in Messina and Genoa, in—or about to burst into—its highly contagious and lethal pneumonic form.

10, 1348, the entry concerns the court appearance of a crafty old peasant named Jacme de Podio. Jacme was trying to get his hands on the dowry of his daughter-in-law Ugueta, a recent plague fatality. The purpose of the hearing was to establish the legitimacy of the old man's claim. Normally, when a woman died intestate, as Jacme's daughter-in-law had, her dowry automatically went to her daughter. But it was Jacme's contention that his granddaughter was dead, too, and despite the cascade of death in the streets, he had managed to talk several of Ugueta's neighbors into appearing in court to corroborate his claim. One told the magistrate that, yes, she had seen the girl on the streets a few times after her mother's death; then she vanished, too.

How do you know the girl is dead? the magistrate asked.

The neighbor said that one day, by chance, she happened to see the girl's corpse in one of the wagons that carry the dead to the cemetery of Notre Dame des Accoules. The next person in the line of inheritance was Jacme's son, Peire, Ugueta's husband. The magistrate asked about Peire's fate. Dead, too, Jacme said. Again, the old man produced a neighbor to corroborate his claim. The neighbor said Peire died right after his wife and daughter, maybe two days later, the neighbor couldn't be sure—so many people were dying these days, it was hard to keep track. A week after the hearing, M. Aycart made a final note about the case in his ledger: Old man Jacme was dead now, too.

Venality like Jacme's was not uncommon in Marseille. At a hearing after the plague ended, a young woman named Uga de Bessa gave testimony in the case of a man who had spent hours on the pestilential streets of the city, searching for a notary for his dying wife. At first the story sounded like an act of selfless spousal devotion: a caring husband risks death to give his dying wife the small final comfort of putting her affairs in order. But then the magistrate asked if the victim had indicated how she wished to dispose of her estate. Yes, replied Mme. de Bessa. Since no notary could be found, the dying woman had called several witnesses to her bedside and in their presence testified that "she was leaving a hundred florins to her husband Arnaut."

Yet this story, like so many other stories about the plague, has a twist ending.

And where is Arnaut now? the magistrate asked.

Dead, replied Mme. de Bessa. He died of the plague, too.

If venality was common in Marseille, so was a kind of dogged, undemonstrative resolve. Though it was struck soon after Sicily, Marseille did not collapse into panic or social breakdown. No doubt, there were cases of desertion—parental, clerical, and civic—but not enough to find their way into local chronicles. There also do not seem to have been many instances of mass flight. City "residents accommodated the effects of the plague," says historian Daniel Lord Smail, author of an illuminating study of Black Death Marseille. "Municipal institutions . . . did not fold up. . . . [People] stayed by their kinfolk, friends and neighbors." Sometimes avarice can have a steadying effect. During March, the worst month of the pestilence, scheming old Jacme de Podio, his greed undimmed, was knocking on doors in his daughter-in-law's neighborhood, looking for witnesses to testify in his court case. April, another terrible month, not only found the wealthy merchant Peire Austria still in Marseille, but he and two colleagues, Franses de Vitrola and Antoni Casse, were planning a new business venture. April was also the month that the solicitous husband, Arnaut, was searching the pestilential streets for a notary to document his dying wife's wish to leave him a hundred florins. People even continued to marry. In May Antoni Lort attended the wedding of his friend Pons Columbier.

Sharp-elbowed Marseille was also not without its acts of human kindness. Taking pity on a client who had lost everything in the Black Death, Jewish moneylender Bondavin de Draguignan told the man that he could continue to work the garden he had offered in payment of his loan. The moneylender also told the man that when the debt was paid off, he could have his garden back.

"If the plague had a profound impact on the residents of medieval Marseille, it was not a blow that led to despair," says Professor Smail.

* * *

Marseille's experience in the Black Death is also noteworthy in another respect.

On Palm Sunday night, April 13, 1348, the Christian residents of Toulon, a quiet seaside village to the east of Marseille, attacked the local Jewish quarter. Doors were smashed, windows broken, furniture overturned; men, women, and children were hauled from beds and hurled into the nighttime streets to be jeered, taunted, kicked, and spat upon. Homes were torched, property looted, money stolen, forty Jews killed. Parents were cut down in front of sons and daughters, husbands in front of wives, brothers in front of sisters. The next morning, the bodies of dozens of dead Jews were hanging from poles in the town square.

Within days the pogrom spread to neighboring villages: to Digne, Mezel, Apt, Forcalquier, Riez, Moustiers, and La Baume. In some places, the Jews were offered the option of conversion; most chose death. "The insane constancy shown by [the Jews] . . . was amazing," wrote a chronicler. "[M]others would throw their children into the flames rather than risk them being baptized and then would hurl themselves into the fire . . . to burn with their husbands and children." On May 14 Dayas Quinoni, a La Baume Jew who had been in the papal city of Avignon when the pogroms broke out, returned home to find his family dead and the local Jewish quarter burned and deserted. "There is no one left but me," M. Quinoni wrote that night. ". . . I sat down and wept in the bitterness of my soul. Would that the Lord in his mercy allow me to see the consolations of Judah and of Israel . . . and permit me and my descendants to rest there forever."

Easter week violence against the Jews was a tradition in the Middle Ages. The season, with its echoes of Jewish "complicity" in the Crucifixion, perhaps inevitably stirred hatred in Christian hearts. But the outbursts in Toulon and La Baume were quickly superseded by a new and far more venomous form of anti-Semitism. As the plague swept eastward across France, Germany, and Switzerland in the summer of 1348, rumors began to spread that the mortality was a Jewish plot. In the earliest iterations, the rumors were just that: vague accusations. Christians, it was said, were dying because their wells were being contaminated with a Jewish plague poison.* But during the fall, as the pestilence worsened, the rumors grew increasingly elaborate, detailed, and bizarre—until they constituted a medieval version of the *Protocols of the Elders of Zion*. By November 1348 every well-informed citizen in eastern France understood that the plague was not the act of a vengeful God or of infected air, but of an international Jewish conspiracy aimed at achieving world domination. "It is our turn now," one well poisoner is alleged to have told his Christian interrogators.

Authorities in the Swiss town of Chillon played an important role in promoting the rumors. The confessions they obtained from local Jews in September 1348 gave the plot a persuasive patina of fact. The confessions spoke of a mastermind, a sinister Rabbi Jacob, formerly of Toledo, Spain, now living in eastern France—and a network of agents who purportedly delivered packets of plague poison to Jews throughout Europe. The Chillon conspiracy theorists even created names and personalities for the agents. There was the bullying Provenzal, the kindhearted merchant Agimetus, the maternal Belieta, the compliant barber-surgeon Balavigny, and a clever youngster known simply as "the Jewish boy." The theorists also created a list of supposedly "contaminated" sites. One was said to be a certain fountain in the German quarter of Venice; another, a public spring in Toulouse; a third, a well near Lake Geneva.

Even the poison used to contaminate the Christian water supply was described in meticulous detail. It was "about the size of an egg," except when it was the "size of a nut" or a "large nut," "a fist" or "two fists"—and it came packaged in "a leather pouch," except when it was packaged in a "linen cloth," "a rag," or a "paper coronet"; and the poison was variously made from lizards, frogs, and spiders—when it was not made from the hearts of Christians and from Holy Communion wafers.

* During a wave of epidemics in the 1320s, a variant of this rumor appeared. At that time it was lepers who were alleged to be poisoning the wells, but they were said to be in the pay of the Jews and their ally, the Muslim Caliph of Granada (see chapter 10). The association between well poisonings and the pestilence is ancient. During the Plague of Athens in the fifth century B.C., it was said that Athenians were dying because the Spartans were poisonings the wells. (Barbara W. Tuchman, *A Distant Mirror: The Calamitous 14th Century* [New York: Ballantine Books, 1978], p. 109.)

Jews who were questioned in connection with the well-poisoning plot were required to swear to a special "Jewish" version of the interrogation oath. "If what you say is not true and right," the interrogator would say to the prisoner, "then, may the earth envelop you and swallow you up . . . and may you become as leprous as Naaman and Gehazi, and may calamity strike you that the Israelite people escaped as they journeyed from Egypt's land. And may a bleeding and flowing come forth from you and never cease as your people wished upon themselves when they condemned God, Jesus Christ."

Special "Jewish" tortures were also available to interrogators. One technique was to place a crown of thorns on a prisoner's head, then smash it into the skull with a mailed fist or a blunt instrument. Another was to place a rope of thorns between a Jewish prisoner's legs and then yank it up into the crotch and scrotum.

Between the summer of 1348 and 1349, an unknown but large number of European Jews were exterminated. Some were marched into public bonfires, others burned at the stake, still others barbecued on grills or bludgeoned to death, stuffed into empty wine casks and rolled into the Rhine. In some localities, the killings were preceded by show trials; in other cases, there were no legal proceedings—sometimes not even an accusation. Jews were killed simply as a prophylactic measure.

The pogroms around Marseille not only pointed to this new form of plague-related anti-Semitism, they also heralded an important change in the nature of French anti-Semitism. Traditionally, Langue d'Oc—roughly, Mediterranean France—was the land of the troubadour; it was cosmopolitan, romantic, poetic, sensual, and tolerant. Jews had a long, mostly happy relationship with the south. Langue d'Oui—roughly, Atlantic France—was the land of the knight; it was ambitious, aggressive, resolute, and intolerant. The pogroms in the southern villages of La Baume, Apt, and Mezel were a signal that this historic division was coming to an end—that the north, which had long had political designs on the tolerant, more cosmopolitan south, was beginning to absorb the southern region culturally as well militarily. The torchbearing citizens of Toulon and La Baume were acting in the grand tradition of Atlantic French anti-Semitism, a tradition that included the 1240 trial of the Talmud—Parisians celebrated its conviction for blasphemy and heresy by burning fourteen cartloads of Talmudic works—a mass expulsion of the Jews in 1306, and the violent pogroms of the post–Great Famine era, which ended with nearly every Jew between Bordeaux and Albi dead.

The singular achievement of Black Death Marseille was to resist the wave of anti-Semitism and remain true to its Mediterranean heritage of tolerance. During the plague, the local Jewish community of 2,500 experienced no harassment or attack. Moreover, as the pogroms mounted in ferocity, Marseille gained a reputation as a haven for Jews fleeing persecution elsewhere.

Avignon, January 1348

In Avignon—where there were seven churches, seven monasteries, seven nunneries, and eleven houses of ill repute—the plague arrived in bitter January of 1348, filled the local cemeteries to capacity, and further damaged the already tarnished reputation of the papacy.

In February 1300, when Boniface VIII, the last of the great medieval popes, stepped onto a Vatican balcony and proclaimed 1300 a Jubilee Year, "in order that . . . [Rome] be more devoutly frequented by the faithful," the papacy had seemed invincible. But the aura of omnipotence was an illusion. Even as Boniface stood basking in the adulation of the faithful that wintry February morning, history was working against the Church, and if the pope was still oblivious to the fact, his longtime nemesis Philip the Fair was not. Within a decade "the august and sovereign house of France," the new power in Europe, would humiliate the papacy not once but twice. In 1303 Philip's agents arrested Boniface at his summer palace, an experience the aged pope found so shocking, he dropped dead a few weeks later. Then, in 1308, Boniface's successor, the pliant, jolly Gascon Clement V, was bullied into acting as Philip's surrogate in the Templars affair.

After fifty-four members of the order had been executed in the squares of Paris for drinking the powdered remains of their illegitimate children and for other crimes "most wicked" and of a "burning shame to heaven," Clement, with Philip at his side, announced that the bulk of the Templars' treasury would be awarded to the French king, in recognition of his efforts in bringing the order to justice.

At his execution, Grand Master de Molay did not forget to thank Clement for his participation in the order's demise. Legend has it that as he went up in flames, the grand master invited the pope to join him and Philip in hell.

Nothing the French Crown did to the papacy, however, was as damaging as what the papacy did to itself in Avignon. The concept of the pope-out-of-Rome was not new when Clement V fled to the Provençal countryside in 1308. Between 1100 and 1304 popes had spent more time out of the Holy City than in it. But the Avignon exile was different. First, there was the suspicion that Clement would not come to Rome because he was unwilling to leave his beautiful French mistress, the Countess of Perigord. Secondly, there was Avignon itself: full of burned-out houses—a legacy of the thirteenth-century Albigensian Crusades—and crooked little streets, swept by violent winds, and surrounded by crumbling walls, the town had all Rome's decrepitude, discomfort, and filth, but none of its historic glamour and authority or its infrastructure. The close proximity of the French Crown—Provence was still nominally independent—also enhanced the impression that the pope was becoming a French puppet.

The most damaging aspect of the Avignon papacy, however, was its utter lack of moral seriousness. Clement V and his successors transformed the Church into a spiritual Pez dispenser. The fertile minds at the curia had managed to create an indulgence for every imaginable situation and every imaginable sin. For a price, an illegitimate child could be made legitimate, as could the right to trade with the infidel, or marry a first cousin, or buy stolen goods. Dispensations were also created for special niche markets such as nuns who wished to keep maids, converted Jews who wished to visit unconverted parents, and people who wanted to be buried in two places (a wish that required cutting the deceased in half). The opulent lifestyle of the Avignon popes added further to the air of moral squalor that hung over the town. "The simple fishermen of Galilee" are now "clad in purple and gold," complained Petrarch.

A dinner party Clement V gave in 1308 is characteristic of the imperial style of the Avignon papacy. Under exquisite Flemish tapestries and silk hangings, a staff of four knights and sixty-two squires served thirty-six papal guests a dinner of nine courses on plates of silver and gold. Each course consisted of three elaborate *pieces montées,* or centerpieces, such as a pastry castle made of roast stag, roebucks, and hares. Between the fourth and fifth courses, the guests presented the pope with a magnificent white charger valued at 400 florins (one florin could buy a man a good sheep) and two rings, one with an enormous sapphire, the other with an enormous topaz. To show his appreciation, Clement gave each guest a special papal ring. During a second interval between the fifth and sixth courses, a fountain spouting five different kinds of wine was rolled out. The margins of the fountain were garnished with peacocks, pheasants, partridges, and cranes. At an interval between the seventh and eighth courses, guests were treated to an indoor jousting match. Following the ninth course and a concert, dessert was served. It consisted of two edible trees; one, silver-colored, bore gilded apples, peaches, pears, figs, and grapes; the other, garden green, was laden with candied fruits. The evening concluded with another round of entertainment. A pair of hands clapped, and the chef and his staff of thirty came racing out of the kitchen to perform a dance for the papal guests.

John XXII, Clement V's successor, was more frugal, but only because the spindly, pinch-faced John preferred counting his money to spending it. In an idle moment one scholar calculated that John's personal fortune of twenty-five million florins weighed ninety-six tons.

Benedict XII, John's successor, returned the Avignon papacy to the tradition of opulent magnificence. On a country walkabout in 1340, Benedict's papal party was led by a white charger surrounded by several grooms; next came a chaplain, squires holding aloft three red hats on poles, two pontifical barbers carrying red cases containing papal vestments and tiaras, a subdeacon with a cross, and a mule with the Corpus Christi. In the middle of the procession rode Benedict, mounted on a white horse, shielded from the noonday sun by a canopy held aloft by six nobles, and followed by a squire with a papal mounting stool, should the pope wish to dismount. The tail of the procession was made up of assorted chamberlains, stewards, prelates, abbots, and, at the very rear, like an ambulatory exclamation mark, a papal almoner, tossing coins to the crowd.

However, compared to his successor, even Benedict looked parsimonious.

"No sovereign exceeded him in expenditure, nor bestowed his favors with greater generosity," wrote an observer of Cola di Rienzo's former patron, Clement VI. "The sumptuousness of his furniture, the delicacy of his table, the splendor of his court, filled with knights and squires of the ancient nobility, was unequalled." And that was barely the half of it. Clement VI had a personal wardrobe of 1,080 ermine skins, delighted in games of "chance and in horses," owned "the finest stud to be procured," and, despite clucking tongues, kept "his palace . . . open to the fair sex at all sorts of hours."

On misty mornings, the magnificent papal palace at Avignon rose above a surrounding belt of oak-filled and dew-splashed meadows like a spectral presence: a stately jumble of rocket-shaped turrets, wandering rooftops, and pyramid-shaped chimneys floating atop a pigeon gray cloud. *"Valde misterioseum et pulcrum"*: very mysterious and beautiful, declared one visitor. "Of solemn and wondrous beauty in its dwellings and of immense fortitude in its towers and walls," declared another. The solemn magnificence of the palace was enhanced by its setting on a rock above the Rhône and by vast, cathedrallike corridors with vaulted windows, where red-hatted cardinals glided across checkerboard patterns of shadow and light like living chess pieces.

Daily, live saltwater fish from Marseille, freshwater fish from the Rhône, sheep and cattle from the Alpine pastures, and fowl and vegetables from the Provençal countryside flowed to the papal dining table. The palace also had a staff of more than four hundred, who worked in several kitchens, dining halls, money chambers; a papal steam room with a boiler; a zoo for the papal lion and the papal bear; and a large contingent of papal relatives, most of whom dressed in expensive brocade and fur and were usually accompanied by a knight or two.

When asked why he was more profligate than his predecessor, Clement VI replied haughtily, "My predecessors did not know how to be popes."

If everyone above the rank of bishop lived in opulence in Avignon, nearly everyone below that rank lived in squalor. As scholar Morris Bishop has noted, moving the enormous papal bureaucracy, the curia, to semirural Avignon was akin to moving the United Nations to a small New England town. The almost overnight influx of thousands of new residents strained the local infrastructure, then broke it. Petrarch, a sometime resident, complained that Avignon was "the most dismal, crowded and turbulent [town] in existence, a sink overflowing with all the gathered filth of the world. What words can express how one is nauseated by the rank-smelling alleys, the obscene pigs and snarling dogs, . . . the rumble of wheels shaking the walls, and the carts blocking the twisting streets. So many races of men, such horrible beggars, such arrogance of the rich!" The mistral, the Provençal version of the sirocco, was another bane; it scattered papers, flared up skirts, stung eyes, and left everything covered with a fine coat of dust, but few complained because, as a local saying went: *"Avenio, cum vento fastidiosa, sine vento venenosa"*—"Avignon, unpleasant with a wind, poisonous without it."

Papal bureaucrats suffered most from the lack of adequate infrastructure, says Professor Bishop. They shivered through winters in unheated, drafty buildings; sweltered through summers in shuttered rooms—to protect piles of paper from the disruptive mistral; and worked in semidarkness all year round. Beeswax candles were too expensive for routine use, tallow candles smelled awful and required constant trimming of the wick, while oil lamps lacked sufficient illuminating power for office work. Each evening, joints aching and eyes strained, the bureaucrats of the curia would arise from their stools and descend into Avignon's streets, and, with no sights to see or friends to visit, head for the local taverns to drink and wench. Residents boasted that while the Holy City had only two whorehouses, Avignon had eleven.

"A field full of pride, avarice, self-indulgence and corruption," declared St. Birgitta of Sweden. "The Babylon of the West," Petrarch agreed.

If it was fashionable to criticize Avignon, it was also fashionable to come and gawk.

Crossing the bridge at Avignon on a Sunday morning in the spring of 1345, a visitor might encounter any number of celebrities, including Petrarch himself. As his graceful figure emerges from an avenue of oaks near the Rue des Lices, where papal employees and the native Avignonnais live, the poet looks as his friend Boccaccio described him: light and agile of step, cheerful of gaze, and round and handsome of face. Vain about his supposed lack of vanity, Petrarch, in *Letter to Posterity,* writes, "I can't boast of remarkable good looks"; then, a few sentences later, he boasts about his "brown, sparkling eyes" and "high complexion neither light nor dark" to the reader.

On such a pretty morning, Petrarch could be thinking about almost anything, but probably he is thinking about Laura, the companion of his soul. The poet may be on his way to the studio of Simone Martini, who is painting a pocket portrait of Laura for him or planning a nocturnal visit to her home to swoon under the balcony—or a stroll through the gardens, where the pair sometimes walked and where once they quarreled.

There is no peace; I am too weak for war,
fear and hope; a burning brand, I freeze.
. . . This is my state my lady; It's your doing.

Linger on the bridge for a while longer, and a visitor might see Laura herself. Like other fashionable Avignon women, she is wearing a *baudea*—a silk veil—but her dress, high-necked and loose fitting, is chaste by the standards of the day, which are quite daring. "Watching a woman undress," complained one contemporary fashion critic, "is like watching a skinning."

The morning light suits Laura. The sun adds luster to the golden hair on her forehead and a tint of pink to her snowy complexion. Accompanying her this morning is her very proper-looking husband, chevalier Hugues de Sade. In a neat twist of history, M. de Sade is an ancestor of Petrarch's most eminent eighteenth-century biographer, the Abbé J. F. X. de Sade, who, in turn, is an uncle of the diabolical marquis of the same name. The de Sades are a prominent Avignon family. Wealthy gentry, they own several spinning mills in the region. The Pont d'Avignon has borne the de Sade family coat of arms since 1177.

No doubt, de Sade would be shocked to learn that a few hours earlier the most famous poet in Christendom had crossed the same bridge, fantasizing about his wife. But would the chevalier feel threatened? M. de Sade "cannot have taken [Petrarch] very seriously," says Professor Bishop; otherwise he would never have tolerated the poet's relationship with Laura. Besides, adds the professor, "de Sade knew very well the Provençal tradition of the infatuate poet suppliant. Whenever Petrarch went too far, he would lock up poor Laura, but otherwise, if the poet wanted to sigh at dawn beneath his wife's window, there was no great harm done."

Linger longer on the bridge, and the visitor might encounter another friend of Petrarch's, the musician Louis Heyligen. The glamorous Italians regard northerners as crude country bumpkins, but a decade in Avignon has given the Flemish-born Heyligen more than a little southern panache. Emerging from a crooked street in a vapor of cologne, Heyligen looks like an advertisement for Avignon's most voguish tailors and barbers. His hair is cut fashionably short; his mustaches, which twirl upward like the toes of an elf's shoe, are fashionably long; and his clothes are fashionably tight. This morning Heyligen has squeezed his upper body into a short, particolored, form-fitting jacket, and his buttocks, crotch, and legs into the male equivalent of the "windows of hell," a pair of hose so tight they leave hardly anything to the imagination. Head thrown back, chin thrust forward, and shoulders squared, the former scholarship boy from rural Beerigen in Flanders glides across the Pont d'Avignon like the king of France himself. His is the walk of a man who has known success in life. And, indeed, in Avignon, a city full of brilliant musicians, Heyligen is regarded as the most brilliant.

Heyligen, however, would be horrified to hear himself described as "creative." Personal expressiveness and intuition had no place in medieval music, which was regarded as a branch of mathematics. Like every other aspect of the universe, music was thought to possess inherent structures. Musical structures were the fixed ratios between various notes and chords. The more accurately a musician could calculate the ratios with mathematical formulations, the more likely his music was to duplicate the "aural sound of God."

Currently Heyligen is employed by the erudite and handsome young Cardinal Giovanni Colonna, one of old Stefano's sons and a patron of Petrarch. On Sunday mornings the musician can be found conducting the choir in the cardinal's private chapel. Heyligen is probably coming from there now, going home to prepare next Sunday's musical program.

As Heyligen disappears behind a shuttling cart, the surgeon Guy de Chauliac lumbers into view. "Guigo," as his friends call him, bears a passing resemblance to the French actor Gerard Depardieu (if contemporary portraits are to be believed). He is a big swarthy bear of man, with a very French kind of earthy masculinity. The surgeon looks as if he should have dirt under his fingernails, gold under his bed, and garlic on his breath. A casual observer would declare the surgeon a peasant, and the observer would be half right. Guigo is another bright scholarship boy. Born to a poor farming family in the Languedoc, he would still be pushing a plow but for a pair of "magic hands." Legend has it that when he was a boy, Guigo's skills at suturing wounds and setting bones earned him a reputation as a medical prodigy; he is said to have once saved the leg of a young noblewoman badly hurt in a fall.

There may be some truth in the story. We know the surgeon's education was paid for by a local baron; the subsidy may have been a gesture of gratitude for a life-saving act. From Bologna, where he studied anatomy and surgery, Guigo went to France to study and teach at the University of Paris, then south to Avignon to become personal physician to Benedict XII and John XXII, and now to Clement VI, who, true to his belief that his "predecessors did not know how to be popes," employs a medical staff of eight physicians, four surgeons, and three barber surgeons. As chief papal physician, Guigo has the task of monitoring the papal bowels—stool along with urine analysis was a key diagnostic tool of medieval medicine. Guigo records the number of papal bowel movements made each day and examines the odor and form of each stool for signs of pollutants.

Surgeon de Chauliac could also be thinking about almost anything on this lovely Sunday morning: *Chirurgia magna,* his masterwork, which will influence medical thinking for the next two hundred years; an irregular papal bowel movement; the pretty blond woman who passed by a moment ago. But what the surgeon could not be thinking about—what he could not even imagine on this fine spring day in 1345—is what Avignon will look like three years hence.

No one could.

Plague! The word conjured up . . . fantastic possibilities . . . Athens a charnel-house reeking to heaven and deserted even by the birds; Chinese towns cluttered with victims silent in their agony, the convicts at Marseille piling rotting corpses into pits; men and women copulating in the cemeteries of Milan; London's ghoul haunted darkness. . . . A picture rose before him of the red glow of pyres mirrored on a wine dark slumberous sea, battling torches . . . thick, fetid smoke rising toward the watchful sky. Yes, it was not beyond the bounds of possibility.

Unlike Dr. Rieux, the hero of *The Plague,* Albert Camus's novel of modern pestilence, the people of Avignon knew nothing about the history of *Y. pestis* in 1347. But in the fading weeks of the year—as rumors floated up the Rhône from Marseille, Sicily, and Genoa about sulfurous rains and poisonous winds and great walls of fire, and about a contagion so supple it spread by glance—in some dark recess of the mind the Avignonnais, too, must have thought: "Yes, it was not beyond the bounds of possibility."

"From the outlying districts . . . a gentle breeze wafted a murmur of voices, of smells . . . a gay perfumed tide of freedom sounding on its way": in the modern Oran of *The Plague,* the pestilence arrives gently, inconspicuously, like an odorless, tasteless poison. But the disease that Camus described was already old and enfeebled, drained of its most virulent poisons by centuries of battles in the streets of Sicily, the towns of China, and soot-stained cities of Renaissance Europe. The pathogen that struck Avignon in 1348 was still in the full vigor of youth.

Emerging from the half light of a January dawn, the pestilence fell upon fleshy, wicked Avignon, killing relentlessly, unceasingly, with skills honed on the windswept plains of Mongolia, the shores of Lake Issyk Kul, the twisted olive groves of Cyprus, and the tormented roads between Messina and Catania.

"They say that in the three months from 25 January to the present, 62,000 people have died . . . ," musician Heyligen wrote to friends in Flanders on April 27. "Within the walls of the city [there are] more than 7,000 houses where no one lives because everyone in them has died, and in the suburbs one might imagine there is not one survivor." In Avignon people fell in the streets, in churches, in homes, and in palaces; they fell from workbenches and under carts, and they fell in such astonishing numbers that throughout the cold wet winter and spring of 1348, the thud of the gravedigger's shovel never ceased. The dying was so furious that the rustics who buried the dead—"half-naked men with no fine feeling," Heyligen called them—could hardly keep up with the work. By March 14, eleven thousand people were already interred in the new cemetery that Clement VI had bought the city, and, said Heyligen, this was "in addition to those buried in the churchyards of the Hopital de Saint-Antoine and the religious orders and in many other church yards." When Avignon ran out of ground, Clement consecrated the Rhône; each morning that plague spring, hundreds of rotting corpses would flow down the stream like a mysterious new species of sea creature. Passing Aramon, Tarascon, and Arles, Avignon's dead would flood out into the open Mediterranean, where, under the low gray light of a sea dawn, they would gather in communion with the dead of Pisa and Messina, Catania and Marseille, Cyprus and Damascus.

Bonfires were set to ward off the pestilence and guards were posted to keep strangers out of the city. "If powders or unguents were found . . . the owners . . . were forced to swallow them," wrote surgeon de Chauliac. However, little was done to protect the hastily buried dead from the twitching snouts of Avignon's pigs. Each night, as the city slept a fretful pestilential sleep, the pigs would gather in the darkness and descend on the local graveyards, rooting through the loose, damp, corpse-laden ground until dawn; then, satiated, sleepy, and caked with cemetery mud, they would return home in the morning light.

In local churches, preachers did what preachers do when confronted with inexplicable, senseless human tragedy. They told the faithful that the pestilence was a blessing from God, part of the "small still flame in the dark core of human suffering [which] reveals the will of God in action, unfailingly transforming evil into good." But in the winter streets outside, another kind of transformation was taking place. As the plague wore on, along with the snorts of "the obscene pigs and the snarling dogs," increasingly Avignon echoed with the shouts of pursued Jews, the crackle of streetcorner bonfires, and the harsh, hacking sound of hemmorhagic lungs. Residents were quick to recognize the uncontrollable cough of pneumonic plague, the violent, spasmodic tattoo that threw people against walls or doubled them over in the streets, left chins and shirtfronts stained with bloody mucus, and produced a rattling noise in the lungs that sounded like a heavy iron chain being dragged across cobblestone. In April, only a few months after *Y. pestis* arrived, Heyligen wrote, "It [has been] found that all those who died suddenly had infected lungs, and had been coughing up blood. And this form is the most dangerous . . . which is to say that it is the most contagious."

In Avignon, as elsewhere, the plague also illuminated the complexity of the human condition. There was the familiar story of abandonment, of people dying "without any mark of affection, piety or charity. . . . Priests do not hear confession . . . or administer sacraments. . . . Everyone who is still healthy looks only after himself," complained Heyligen. There was also an alarming new series of attacks on the Jews. "Some wretched men . . . were accused of poisoning the wells," the musician wrote to Flemish friends. "Many were burnt for this and are being burnt daily." Avignon, however, was not bereft of heroism. The monks and brothers of La Pignotte, the municipal almshouse, displayed selfless devotion, feeding the hungry and tending the sick, swabbing oozing pustules, cauterizing painful buboes, bandaging cracked, gangrenous feet, washing bloodstained floors. But, alas, in a time of pestilence almost no good deed goes unpunished. The sick and dying who flocked to La Pignotte with contagious pneumonic plague made the almshouse a death trap. "Whereas at La Pignotte, they normally [go] through 64 measures of grain a day with one measure making 500 loaves of bread," Heyligen noted, "now no more than one measure and sometimes only half is needed."

On May 19 Petrarch, who was in Parma, received a letter from Heyligen. After reading it, the poet made a notation on the flyleaf of his most beloved book, a copy of Virgil. "Laura," he wrote, "illustrious for her virtues and long celebrated in my poems, first appeared to my eyes, in my early manhood, in the Church of St. Clare in Avignon, in the 1327th year of Our Lord, on the sixth of April, at the early morning service. And in the same city in the same month of April, on the same sixth day, at the same hour in 1348 her light was subtracted from the world. . . . That very chaste and loving body was laid to rest in the church of the Franciscan brothers on the very day of her death at evening. But her soul has, I am persuaded, returned to heaven whence it came. . . . I have decided to make this record in this place . . . which often falls under my eyes . . . that I may reflect that there can be no more pleasure for me in this life . . . now that the chief bond has been broken."*

The same day—May 19, 1348—Petrarch may have written these lines:

She closed her eyes; and in the sweet slumber lying
her spirit tiptoed from its lodging place.
It's folly to shrink in fear, if this is dying;
for death looked lovely in her face.

In the early stages of the pestilence, Avignon attempted every imaginable protective measure. People stopped eating fish, maintaining that they have been "infected by infected air," and spices, "for fear they had been carried on Genoese galleys." They tried making bonfires, and then they burned Jews—until Clement issued a bull denouncing the murders. Next, Avignon took to the streets in bloody, semihysterical candlelit marches. Some were "attended by 2,000 people from all the region . . . ," says Heyligen, "men and women alike, many barefoot, others wearing hairshirts or smeared with ashes . . . [some] beat themselves with cruel whips until the blood ran."

After all alternatives had been exhausted, Avignon did what other pestilential towns did: it fell into the state of stuporous resignation Camus described in *The Plague*. "None of us was capable of exalted emotion [any longer] . . . ," says the narrator of the novel. "People would say, 'It's high time it stopped.' . . . But when making such remarks we felt none of the passionate yearning or fierce resentment of the early phase. . . . The furious revolt of the first few weeks had given way to a vast despondency. . . . The whole town looked like a railway waiting room."

In the midst of its suffering, Avignon did have one miraculous moment.

On March 15, 1348, as dawn broke over the city, cooks from the papal palace; scribes from the Holy See; chamberlains from Cardinal

* Petrarch probably invented the matching deaths for literary effect. Life is rarely so heat.

Colonna's palace; and stewards, clerics, and servants from everywhere jostled one another in Avignon's crooked, malodorous little streets. Above the excited crowds, walls and windows were decorated with flowers and silk drapings, and on terraces above the walls stood the "most fair and noble ladies . . . dressed in those costly garments of ceremony which are passed from mother to daughter for many generations."

Around nine a.m. a chorus of silver trumpets sounded, and the wintry morning burst into glorious Technicolor. As a thousand excited spectators oohed and aahed, a parade of brilliantly dressed notables marched through the city. Leading the procession were the smiling bishop of Florence and the cap-waving chancellor of Provence. Behind them marched eighteen cardinals dressed in scarlet robes of the finest cloth, and behind the cardinals marched the most magical couple in Christendom and the reason the crowd got up early this morning. There was Luigi of Taratino, dressed in the latest Spanish fashion—short hair and tight jacket—and looking "as beautiful as the day," and, walking a few paces ahead of him, twenty-three-year-old Queen Joanna of Naples and Sicily, draped in a gold and crimson robe and bearing a scepter and orb. The queen's lovely blond head was protected from the pale winter light by a brilliantly colored canopy held aloft by nobles of the highest rank. The crowd was rapturous. "Her figure," says one admirer, was ". . . tall and nobly formed, her air composed and majestic, her carriage altogether royal, [and] her features of exquisite beauty."

Joanna's fair skinned, blond beauty—a combination troubadours called "snow on ice"—was one of the great wonders of the medieval world. "Fair and goodly to look upon" is how Giovanni Boccaccio described the young queen. "Exquisite and enchanting," declared Petrarch. "More angelic than human," added the chevalier de Brantome. For the gallant young cavalier Galeazzo Gonzaga of Mantua, words alone failed to describe Joanna's loveliness. After a single dance with the Neapolitan queen, the cavalier fell to his knees and vowed to "go through the world until I have overcome in battle two knights whom I swear I will present to you in recompense." Presently, two Burgundian knights arrived in Joanna's court, accompanied by a note from ardent young cavalier Galeazzo.*

In character, the young queen was typically Neapolitan, which is to say that both her subjects—who believed Joanna to be kind and good-hearted—and her brother-in-law, King Louis of Hungary—who called Joanna that "great harlot that . . . ruleth over Naples"—had a point.

The queen's visit to pestilential Avignon was occasioned by her involvement in one of the most sensational celebrity murders of the Middle Ages. On a late summer evening three years earlier, Joanna's husband—and Louis's younger brother—eighteen-year-old Prince Andreas of Hungary, was found hanging from the balcony of a Neapolitan abbey. According to contemporary accounts, the young prince was still alive when a maid discovered him, but when she let out a scream a mysterious figure suddenly emerged from the darkness, grabbed the dangling prince by the ankles, and yanked down hard, breaking his neck and killing him.

On hearing of the murder, Joanna was reportedly inconsolable. The next morning, whenever Andreas's name was mentioned, she would sob, "My murdered man!" Declaring, "I have suffered such intense anguish at the murder of my husband . . . I well nigh died of the same wounds," a few days later the queen offered a reward for any information pertaining to the crime. The Neapolitans were touched—the queen was so young and lovely and her grief so great. The Hungarians were suspicious—the circumstances surrounding Andreas's death were unusual. There was first of all the fact that on the night of the murder, Andreas was summoned from the royal bedchambers by one of Joanna's maids, a young woman named Mabrice di Pace. Mabrice knocked on the bedroom door late in the evening and told the prince an adviser wished to speak to him. There was the added circumstance that when Andreas left the bedroom, where Joanna was supposedly asleep, someone locked the door from the inside. And there was also the fact that one of the men who assaulted the prince in the darkened abbey

* The laws of chivalry dictated that when a knight defeated an opponent, the opponent became his property. Joanna gave her two "gifts" their freedom.

hallway outside the royal bedchamber was Raimondo Cabani, husband of the queen's childhood tutor. After wrestling Andreas to the ground, Cabani and two accomplices shoved a glove into the prince's mouth, slipped a noose around his neck, then dragged him to a balcony and threw him over the rail.

The Hungarians were also quick to point out that Joanna had good reason to desire Andreas dead. There was the queen's semipublic affair with the handsome Luigi, rumored to be the father of the child she was carrying at the time of the murder. Added to that was the queen's well-known dislike for Andreas, who was said to have been a plump, dull boy, and for Andreas's older brother King Louis, who Joanna believed to have designs on her Kingdom of Naples and Sicily. "I am a queen in name only," she told Petrarch one day.

For months after the murder, informed opinion in Europe was divided about the queen's complicity. Luigi predictably insisted that Joanna was innocent, and Petrarch, a relatively unbiased observer, came to the same conclusion. Boccaccio could not make up his mind. In the first of several thinly disguised accounts of the murder, he described a Joanna-like character as a "pregnant she-wolf." But in a later version of the story, the author changed his mind and transformed the queen's character into a beautiful maiden in distress. Louis of Hungary suffered no such agonies of doubt. Shortly before the plague arrived, he wrote to Joanna, "Your former ill faith, your impudent assumption . . . the vengeances you have neglected to take [on Andreas's alleged assassins], the excuses made for it, all prove you to have been accessory to the death of your husband. . . . Be sure, however, that none ever escape vengeance for such a crime."

When Joanna visited Genoa in March 1348, she was fleeing the armies of the avenging Louis, who had just celebrated his conquest of Naples by decapitating a cousin of Luigi's on the balcony where Andreas was hanged.

To Joanna's supporters, her decision to risk the pestilential papal city to clear her name in a Church trial was *prima facie* evidence of her innocence. "To quietly triumph before the world, would for her, outweigh the risk of a hundred pestilences," an admiring biographer wrote later. To detractors, however, the visit only proved that the queen was more afraid of ending up like Luigi's cousin than she was of the plague. Joanna, they charged, was driven to Avignon by the need to cut a deal with Clement VI, one of the few figures powerful enough to protect her from the vengeful Hungarians. Both views would find support in the queen's trial, which was held the same day as her arrival in Avignon, March 15, 1348, in the great Hall of the Consistory.

As Joanna and Luigi took wine and refreshment in an antechamber, the court assembled inside the hall. Seated on a papal dais that put him two steps above everyone else was the presiding judge, Clement VI, who was wearing a bejeweled triple tiara, white robes of exquisite hand-woven silk, and linen slippers with delicate little gold crosses embroidered on the toes. Seated below the pontiff were a semicircle of cardinal-judges, and standing in front of the cardinals were Joanna's accusers, the glowering prosecutors of the Hungarian Crown. Leaning against the walls of the consistory was the most fabulous audience in Christendom. Ignoring the plague, "prelates, princes, nobles, and ambassadors of every European power" had gathered in Avignon to witness the trial.

The evidence against Joanna was incriminating in the extreme, and the Hungarians made the most of it. They reminded the papal court of the deep and well-known animosity that had existed between the queen and her prince consort, of the many plots Joanna's advisers had launched against the unsuspecting Andreas, and, most incriminating of all, of the queen's proximity to the young prince the night of the murder. Only a few inches of bedroom door had separated the royal couple; surely the queen must have heard her husband's cries for help? Though the Hungarian case was powerful—perhaps even overwhelming, as one historian later noted—"The Queen of Naples could seduce the Areopagus itself."

Entering the all-male court, Joanna "came pale and slowly in her beauty, the open crown of Naples set softly upon her bright wavy hair; her long fur-fringed azure mantle . . . strewn with fleur-de-lys." Advancing through an avenue of glittering nobles and cardinals, the

Neapolitan queen fell to her knees in front of the papal dais and kissed the papal feet. Clement VI ordered Joanna to rise, kissed her on the mouth, then invited the young queen to sit by his side.

When called upon to answer the charges against her, Joanna rose; one admirer wrote: "a woman, a mother, and a queen, three voices in one." The queen conceded that, yes, her marriage to Andreas had been lacking in sentiment, but she insisted that shortly before the prince's murder, a reconciliation had been effected. Joanna also reminded the court that when the prince was slain, she had lost not only a dear husband but a dear childhood companion; the royal couple had played together as boy and girl. The young queen then spoke movingly of the horrors of widowhood and exile, and passionately about the cruelties of her horrid Hungarian in-laws, who had snatched away her infant son. "Proclaim to the world at large, the innocence of a persecuted orphan and injured queen," Joanna begged the court.

The papal court did that and more. The judges pronounced Joanna not only innocent but "above suspicion of guilt." Embracing the exonerated queen, Clement declared her his "blameless and beloved daughter." As Joanna and Luigi exited the great Hall of the Consistory, church bells echoed through the pestilential streets of Avignon.

A few months later, it was announced that Clement had purchased Avignon from the queen, who, as countess of Provence, held title to the city. The selling price, eighty thousand gold florins, was deemed very reasonable by most observers—indeed, perhaps even a bit low for what was, after all, the capital of Christendom. Nonetheless, rumors persisted for years that no money ever changed hands in the sale.*

As March turned to April and April to May, Avignon continued to die. Shops and businesses shut down; people fled to the countryside;

* Alas, Hungarians, like elephants, have long memories. On May 22, 1382, thirty-seven years after Andreas's murder, agents of the Hungarian Crown slipped into the chapel where Joanna, then on her fourth husband, was kneeling in prayer and strangled her to death. As with all stories about the queen—who continued to seduce historians, biographers, novelists, and playwrights long after death—this one is clouded in ambiguity. Another account has Joanna being poisoned; a third, smothered by a pillow; a fourth, starving herself to death. (St. Clair Baddeley, *Queen Joanna I, of Naples, Sicily and Jerusalem, Countess of Provence, Forcalquier and Piedmont: An Essay on Her Times* [London: W. Heinemann, 1893], p. 295.)

astrologers warned that the pestilence would last a decade. In April Heyligen told friends that the departure of the pope was expected daily, and if Clement left Avignon, he would leave, too. "They say that my lord [Cardinal Colonna] follows the Pope and that I am to go with him. Since that place looks towards Mount Ventoux, where the plague has not yet come, it is the best place to be, or—anyways—so they say."

Clement's departure in May did not arouse much public comment. Almost anyone who could was fleeing the city, and the pope had done what he could for Avignon. He had bought the city a new cemetery, granted a blanket absolution to the dying, waived the ban on autopsies so that physicians could explore the cause of the disease, condemned the attacks on the Jews in a strongly worded bull, even appointed a commission to calculate the number of plague fatalities worldwide—the commission came up with a figure of almost twenty-four million dead. But the pestilence wore Clement down, as it did nearly everyone else. Between Joanna's trial in March and late spring, when he fled to the papal retreat at Etoile-sur-Rhône, the pope spent a great deal of time seated between two roaring fires in the papal chambers. The fires were the idea of surgeon de Chauliac, who believed heat would cleanse the papal chambers of infected air, thought to be the cause of the pestilence. And the treatment worked, though for reasons that would have surprised the surgeon: the fires kept the papal chambers free of infected fleas.

It would be unfair to heap obloquy on Clement; if he can be accused of anything, it is the sin of being ordinary in an extraordinary time. The pope did what he thought he ought to do, and some of what he did was meritorious. He marched with the fearful and purchased cemeteries for the dead. It can even be argued that Clement was a bolder defender of the Jews than Pius XII, the pope who presided over the Church during World War II. However, in a situation that called for a leader with a Gandhi-like spiritual authority—someone who could both give comfort and inspire—Clement acted like a head of state. He was responsible but, in the end, unimaginative and self-preserving. Surgeon de Chauliac is a more inspiring figure. When the pope fled to Etoile-sur-Rhône, taking Cardinal Colonna and Heyligen with him, the surgeon chose to stay behind in Avignon. One sees Guigo walking through the wintry streets, a big lumbering man, calm of manner, with observant eyes and oversized peasant hands so gentle they could make a feverish child stop trembling. "To avoid infamy I dared not absent myself," de Chauliac wrote of his decision to remain in Avignon, but he never explained whose infamy he feared. Indeed, the surgeon was generally reticent about his personal experiences during the plague. "Toward the end of the mortality," he says, he was stricken himself. "I fell into a continuous fever, with a tumor in the groin. I was ill for nearly six weeks and in such great danger that all my associates thought I would die, but the tumor being ripened and treated . . . I survived."

Surgeon de Chauliac's only other reference to his personal experience comes in the context of a scientific observation. He noted that between winter and spring, the plague changed character in Avignon, altering from a pneumonic to a bubonic form. "The mortality," he wrote, "began with us in January and lasted for seven months. It had two phases. The first was for two months and with continuous fever and the spitting of blood, from which victims died within three days. The second phase lasted for the remainder of the period [that is, spring and early summer] and patients also had continuous fever. In addition, abscesses and carbuncles—i.e., buboes—formed in the extremities, namely in the armpits and groin."

It is only a conjecture, but perhaps the infamy the surgeon feared was the infamy of scientific conscience, of failing to stand in the whirlwind and try to understand and tame it with human reason.

How many people died in Avignon?

A contemporary puts the mortality in the city at 120,000, but this figure is as suspect as Marseille's death toll of 56,000. Whenever a medieval observer used large numbers, what he meant to say was not, "This is the how many bodies were counted," but "A great many people died." Given the lethality of pneumonic plague and the number of contemporary accounts that describe Avignon's losses as severe,

Philip Ziegler's estimate of a mortality in the 50 percent range sounds about right.

In *On Thermonuclear War,* another landmark of Cold War literature, theorist Herman Kahn describes a casualty rate of 50 percent in a nuclear exchange as unacceptably high. In little Avignon in the year of 1348, one out of every two people died.

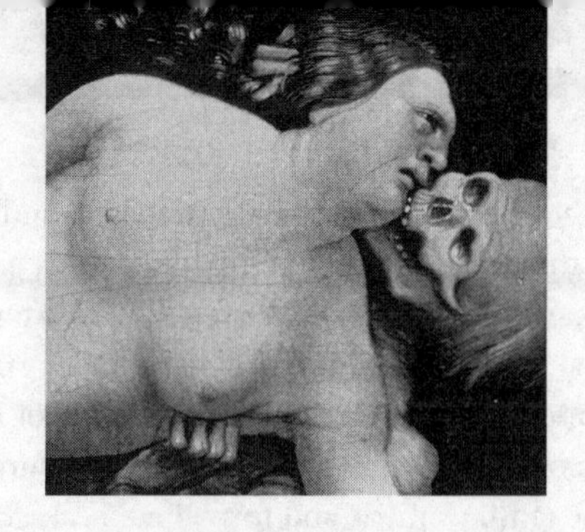

CHAPTER SEVEN

The New Galenism

Paris, Summer 1348

IN PARIS, WHERE THERE WERE EIGHTY-FOUR PHYSICIANS, TWENTY-six general surgeons, and ninety-seven barber surgeons, the plague arrived on a summer day, lingered for over a year, and prompted the forty-six masters of the Paris Medical Faculty to write one of the most renowned scientific works on the Black Death.

Medieval medicine is often regarded as a branch of medieval torture, but the plague caught the field at an important turning point. Outside the monasteries, where a few surviving classical texts on phlebotomy (bloodletting), midwifery, and the pulse were still taught, in the early Middle Ages medicine was a mixture of folk wisdom, magic, superstition, and craft. To the extent that a ninth-century medical practitioner engaged in professional reflection, he thought of himself as a craftsman like a carpenter or butcher—which he was. Specialized terms like "physician" and "surgeon" were unknown until the tenth century, and formal medical schools did not exist until almost the thirteenth century. In the early medieval period, the only remotely scientific tool available was what we might call urinalysis. A healer would sniff and eyeball a patient's urine, then make a diagnosis. One German healer became so adept at the procedure that when the Duke of Bavaria tried to pass off the urine of a pregnant servant girl as his own, the healer announced that "within a week the Lord will perform an unheard of miracle, the Duke will give birth to a son!"

In comparison to his unlettered predecessors, the medical practitioner of the fourteenth century was a model of enlightened scientific professionalism. As Chaucer noted in *The Summoner's Tale*

. . . Nowhere a better expositioner
On points of medicine and pathology.
For he was grounded in astrology;
Treating his patients with the most modern physic
Dependent on his skill in natural magic;
He knew which times would be most propitious
For all his cures to be most expeditious . . .

Not only that, but:

Old eminent authorities he knew,
Some Greek, some Roman, some Arabian, too.
He'd read both Aesculapius the Greek
And Dioscorides, whose drug critique
Was current still. Ephesian Rufus, too
Hippocrates and Haly, all he knew.

Galen, Serapion and Rhazes, all
Their text books he could instantly recall . . .

Chaucer's medical professional was, like the merchant and notary, a product of the new towns, where prosperity and a growing population created a brisk demand for medical care. What made the "better expositioner" something new under the medical sun was the "scientific" training Chaucer poked fun at. The New Galenism, as it was often called, was based on a reinterpretation and expansion of classical medicine by Arab physicians like Avicenna (*Canon*), Haly Abbas (*Pantegni*), and Rhazes. For European scholars accustomed to the ad hoc, craftlike structure of Western medicine, the works of the Arab master physicians, which began to be translated in the late eleventh century, were an astonishment. Drawing on Aristotle, Hippocrates, and especially Galen, the Arabs transformed medicine into a sophisticated intellectual discipline. Like the ancient Western fields of law and theology, Arab-Greco medicine had a unifying set of philosophical principles, a logical, coherent structure, and an intellectual consistency. In the hands of the masterful Arabs, the theory of the four humors created by Hippocrates and expanded and elaborated upon by Galen could explain just about anything, from ulcers to pestilences to the dangers of hot, moist air. The clever Arabs also introduced the West to a number of exciting new diagnostic techniques, including the signature tool of the medieval physician, astrology. In an early-thirteenth-century tract called *De urina non visa*—On Unseen Urines—William the Englishman told his colleagues that they could now dispense with urinalysis; astrology had made the technique obsolete. With a knowledge of the stars, declared William, you could tell what was in a patient's urine without examining it.

For all its sophisticated intellectualism, however, the New Galenism contained some serious flaws. The most pronounced was a typically medieval reverence for authority, especially ancient authority, over observable fact. In practice, this meant that, while Chaucer's new medical professional knew a great deal about the "old eminent authorities" and how more recent masters like "Avicenna, Averroes, Damascan John and

Constantine (just dead and gone)" had intepreted them, the professional's knowledge of medicine was, in effect, based on one- and two-thousand-year-old ideas. Medieval medical students learned very little that was new, practical, or the result of direct scientific observation. Courses in anatomy were offered in many medical schools, but since autopsies were frowned on by the Church, students had to learn about human anatomy by watching a pig being dissected.

The medical schools themselves were a by-product of the New Galenism. The first formal academic medical training was offered in the southern Italian town of Salerno, where the Arab masters were first translated; a hundred fifty years later, Montpelier, Bologna, Oxford, Cambridge, Padua, Perugia, and Paris also had medical schools. And while each institution had its own individual style—Paris had a renowned astrology department; Montpelier, a large number of Jewish students—all the schools required five to seven years of study and taught a curriculum based on the New Galenism.

Along with extensive training, during the thirteenth century Chaucer's new medical professional also acquired another attribute of the modern physician, a license—obtained, as today, by examination. Promoters of medical professionalization described licensure as an essential public health measure. It would—decreed the University of Paris Medical Faculty, an aggressive promoter of licensure—prevent the "shameful and brazen usurpation" of the profession by the untrained and the ill-lettered. But licensure was as much about professional and economic hegemony—it would allow physicians to dominate the old-style healers, who still provided the bulk of medical care—as it was about public health.

A small landmark in the physician's climb to professional dominance was the 1322 trial of a Parisian healer named Jacqueline Felicie. Despite a lack of formal training, Madame Felicie, one of medieval Paris's many female practitioners, apparently had a natural gift for medicine. During her trial, several former patients came forward to testify on her behalf, including a John of St. Omar, who told the bishop's court that during a recent illness, Madame Felicie had visited him several times and refused payment until he had been cured. John described this as unprecedented in his experience of physicians. Other defense witnesses included a second John, John Faber, who testified that Madame Felicie had healed him with "certain potions, one green in color"; and a servant girl named Yvo Tueleu, whose fever had resisted the ministrations of several university-trained physicians. Mademoiselle Tueleu told the bishop's court that after a careful physical examination, Madame Felicie prescribed "a glass of very clear liquid which acted as a purge." Shortly thereafter, the young woman's mysterious fever vanished.

The prosecution's chief witness was John of Padua, a crusty old former medical adviser to Philip the Fair. Judging from his testimony, John seemed to think that Madame Felicie ought to be convicted on the grounds of gender alone. Women were already barred from law, John thundered; how much more urgent, then, to keep them out of a serious profession like medicine!

On November 2, 1322, Madame Felicie was convicted of violating an ordinance that prohibited unlicensed healers from visiting, prescribing medications, or performing other duties for a patient, except under the guidance of a university-trained and licensed physician. The conviction was a major victory for the Paris Medical Faculty, a principal architect of the new medical pecking order, which had a pyramid-like shape. At the pinnacle was a relatively small coterie of the university-trained physicians; they practiced what we would call internal medicine. Beneath them were the general surgeons, who usually lacked academic training, although that was changing. By the early fourteenth century surgery was beginning to find a place in the medical schools. A surgeon could treat wounds, sores, abscesses, fractures, and other disorders of the limbs and skin. Beneath the general surgeon was the barber surgeon, a kind of paramedic, who could perform minor operations, including bleeding, cupping, and applying leeches, as well as cutting hair and pulling teeth; next came the apothecary and the empiric, who usually specialized in a single condition, like hernias or cataracts. At the base of the pyramid were thousands of unlicensed healers like Madame Felicie.

To reflect their new eminence, in the decades prior to the plague, physicians began to adopt a more professional—that is, authoritative—

demeanor and code of behavior. A cardinal "don't" in the new medical etiquette was: don't jeopardize your professional dignity by visiting patients to solicit business. "Your visit means you are putting yourself in the patient's hands," warned William of Saliceto, "and that is just the opposite of what you want to do, which is getting him to express a commitment to you." A cardinal "do" in the new etiquette was to conduct a comprehensive physical exam on a first visit; the exam should include not just urinalysis, but a detailed medical history and an analysis of the patient's breath odor, skin color, muscle tone, saliva, sweat, phlegm, and stool. Some physicians also cast a patient's horoscope on the first visit. Another cardinal "don't" in the new etiquette was to admit to diagnostic uncertainty. Even when in doubt, said Arnauld of Villanova, a physician should look and act authoritative and confident. For the uncertain physician, Arnauld recommended prescribing a medicine, any medicine, "that may do some good but you know can do no harm." Another strategy was to "tell the patient and his family that [you are] prescribing this or that drug to cause this or that condition in the patient so that [they] will always be looking for something new to happen." A third "don't" in the new etiquette was volubility. Reticence conveyed authority, especially when combined with a grave manner; besides, said one savant, the physician who discusses his medical reasoning with the patient and his family risks letting them think that they know as much as he does, and that may tempt them to dispense with his services.

What made the university-trained physician such an impressive figure to laymen, however, was not only his authoritative bedside manner but his mastery of the arcanae of the New Galenism. Its signature principle was the theory of the four humors. For the ancient Greeks, whose thinking shaped so much medieval medicine, the number four was, like the atom, a universal building block. Everything, the Greeks believed, was made out of four of something. In thc casc of thc physical world, the four elements were earth, wind, water, and fire; in the case of the human body, the four humors were blood, black bile, yellow bile, and phlegm. An important element in the humoral theory were the four qualities of all matter: hot and cold, wet and dry. Thus, blood was said to be hot and moist; black bile, cold and dry; yellow bile, hot and dry; and phlegm, cold and wet.

In *On the Nature of Man,* Hippocrates wrote that "health is primarily that state in which [the four] constituent elements [that is, the four humors] are in correct proportion to each other, both in strength and quality, and are well mixed. Pain occurs when one of the substances presents either a deficiency or excess or is separated from the body and not mixed with others."

The theory of corrupt or infected air also played an important part in the New Galenism. Bad air was dangerous because it could disrupt the balance of bodily humors, and particularly dangerous was hot and humid air because both heat and humidity corrupted the life force around the heart. Contagion was a by-product of this corruption. People became ill by inhaling not airborne germs but the corrupt vapors emanating from diseased bodies.

In the new medical schools like the University of Paris, students also learned that earthquakes, unburied corpses, decaying crops, stagnant water, poor ventilation, and even poisons could infect the air; but in the case of epidemics, which affected hundreds of thousands of people in widely separated places, infection was thought to result from a global disturbance, like an unfavorable planetary alignment. The movement of the moon clearly controlled tides; ergo, reasoned medieval (and ancient) man, air quality must also be affected by planetary movements and cycles.

The *Compendium de epidemia per Collegium Facultatis Medicorum Parisius,* the plague treatise of the Paris medical masters, offers an example of how the new medicine used the theories of astrology and infected air to explain the origins of the pestilence.

According to the *Compendium,* "the first cause of this pestilence was and is [the] configuration of the heavens [which occurred] in 1345, at one hour after noon on 20 March, [when] there was a major conjunction of three planets in Aquarius." In the masters' view, the conjunction caused "a deadly corruption in the air," and Mars and Jupiter, two of the three planets in the conjunction, played a particularly important role in the corruption. "For Jupiter, being wet and hot draws up evil vapors from the earth and Mars, because it is immoderately hot and dry, then ignites the vapors and as a result there were lightning sparks, noxious vapors and fires throughout the air."

The second chapter of the *Compendium* explains how these astrological changes led to the plague. "What happened," explained the masters, "was that many of the vapors . . . corrupted at the time of the conjunction . . . then mixed with the air and [were] spread abroad by frequent gusts of wind in the wild southerly gales. . . . This corrupted air, when breathed in, necessarily penetrates to the heart and corrupts the substance of the spirit there and the heat thus destroys the life force."

Like many contemporaries, the Paris masters believed that the extraordinary ecological upheavals of the 1330s and 1340s—the succession of earthquakes, floods, tidal waves, heavy rains and winds, and unseasonable weather—played an important role in the plague. "Experience," declared the masters, "tells us that for some time the seasons have not succeeded each other in the proper way. Last winter was not as cold as it should have been with a great deal of rain. . . . Summer was late, not as hot as it should have been and extremely wet. . . . Autumn, too, was very rainy and misty. It is because the whole year here—or most of it—was warm and wet that the air is pestilential. For it is a sign of pestilence for the air to be warm and wet at unseasonable times."

How could people protect themselves against the plague?

Facing its first great public health crisis, the New Galenism was not lacking in ideas. Between 1348 and 1350, twenty-four plague tracts were written, most by university-trained medical professionals. Like the *Compendium* of the Paris masters, some of the tracts took a big-picture view; others dispensed practical advice on how to stay healthy; two examples of the latter are *Description and Remedy for Avoiding the Disease in the Future,* by Ibn Khatimah, a Muslim physician who lived in Granada, and *Consilia contra pestilentium,* the tract of the Italian master physician Gentile da Foligno. One plague work, Simon of Covino's *Concerning the Judgment of the Sun at the Banquet of Saturn,* was written in verse, while others, such as *A Very Useful Inquiry into the*

Horrible Sickness and *Is It from Divine Wrath That the Mortality of These Years Proceeds?* convey some of the terror of the time.

To varying degrees, most plague authors agreed with the Muslim physician Ibn al-Khatib, who described the pestilence as "an acute disease, accompanied by fever in its origin, poisonous in its material, which primarily reaches the vital principal [the heart] by means of the air, spreads in the veins and corrupts the blood, and changes certain humors into a poisonous character, whence follow fever and blood spitting."

There was also agreement that the best defense against plague was to remain healthy, and above all, this meant avoiding infected air. But how? One way, said the Paris masters, was to avoid marshes, swamps, and other bodies of stagnant water where the air is dense and turgid; another way was by keeping windows with a northern exposure open to let in good air—that is, cool and dry air—and keeping windows with a southern exposure shut to keep out bad air—that is, warm, humid air. To be extra safe, the Paris masters further recommended glazing, or putting a wax cloth over windows with a southern exposure. According to Ibn Khatimah, in plague as in real estate, location was everything. Bad to live in, said the Muslim physician, were cities with a southern coast. The reason? The rays of the sun and other stars bounced off the sea, blanketing such cities in warm, damp air. Also to be avoided were cities facing the south pole, particularly if unprotected on the south side. Cities with an eastern or western exposure occupied a middle ground in the risk spectrum, though a western exposure, which favors dampness, was more dangerous than an eastern exposure.

A physician named John Colle emerged as a particularly innovative thinker on the question of corrupt air. Noting that the "attendants who take care of latrines and those who serve in hospitals and other malodorous places are nearly all to be considered immune," John argued that the best antidote to bad air was more bad air. One of the most surreal images to emerge from the Black Death is of knots of people crouched at the edge of municipal latrines inhaling the noxious fumes.

For Muslims like Ibn Khatimah and his fellow Spanish Arab Ibn al-Khatib, the issue of contagion posed a special problem. According to Islam, God's will determined who lived and who died in an epidemic. Choosing to play it safe, Ibn Khatimah described contagion as a phenomenon "in which the Arabs in their ignorance" [in other words, before Islam] used to believe but no longer do." Ibn Khatimah probably did not believe what he wrote, but he knew it would keep him out of trouble. Braver, Ibn al-Khatib said what he thought: contagion's role in the spread of plague was "firmly established by experience, research, mental perception, autopsy and authentic knowledge of fact." In 1374, when Ibn al-Khatib was dragged from a prison cell and murdered by a Muslim mob, there were those who said that one factor in the physician's undoing was his disregard of Islamic teaching during the plague.

Untroubled by theological dilemmas, Christian writers were free to concentrate their energies on preventive stratagems. To protect against infection indoors, a number of authors recommended burning dry and odiferous woods, such as juniper and ash, vine and rosemary, oak and pine. In winter, this regimen could be supplemented with aromatic substances, such as wood of aloes, amber, musk, laurel, and cypress, and in summer with fragrant flowers and plants sprinkled with vinegar and rosewater. Out of doors, people were advised to carry a smelling apple, a kind of personal scent that, like a gas mask, would protect against noxious fumes. A physician named John Mesue said a very good smelling apple could be made from black pepper, red and white sandal, roses, camphor, and four parts of bol armeniac (Armenian bole). Gentile da Foligno, who liked to keep things simple, said a pleasant-smelling herb—of the type many Florentines used during the plague's visit—would do. Gentile also recommended streetcorner bonfires, a public health measure already in use in Avignon and other cities.

Changes in lifestyle could also protect against infected air. The Swedish bishop Bengt Knutsson, for example, recommended avoiding both sex and bathing because "where bodies have open pores as is the case of men who abuse themselves with women or often have baths . . . they are the more disposed to this great sickness." What if abstinence proved impossible? Ibn Khatimah recommended regular bleedings to purge excess heat and impurities from the body. An energetic self-bleeder himself, the Muslim lost eight pounds via phlebotomy. Also to be avoided—or taken in moderation—was exercise, another pore opener.

Antidotes were also a popular preventive. For those who liked tasty antidotes, a prebreakfast snack of fig, filbert, and rue was suggested. Pills of aloe, myrrh, and saffron were also recommended by many physicians; but theriac, mithridate, bol armeniac, and terra sigillata—all traditional poison remedies—were the most popular antidotes. In Gentile da Foligno's view, however, no antidote could rival the power of an emerald ground into a powder; Gentile's assertion that the remedy could "crack a toad's eyes" suggests that the "prince of physicians," a university-trained professional, harbored a secret weakness for black magic.

Since a good diet kept the four humors in balance, many tracts also stressed the importance of proper eating. Specific dietary recommendations varied according to age, sex, season, and circumstance, but in general, wise to avoid were foods that spoiled easily, like milk, fish, and meat. If meat was one's choice, then fowl, lamb, and kid were best, and should be tender and digestible. As for preparation, the Paris masters took a firm stand for roasting and against boiling. A little cheese helped digestion, said Ibn Khatimah. The Moor's stand on eggs, like his stand on contagion, was subtle. Good were eggs dipped in vinegar, bad were eggs dipped in garlic. The anonymous author of *First About the Epidemic* could barely contain himself on the subject of hard-boiled eggs; they were a garden of dangers, he fulminated.

Everyone liked bread, though good flour and proper baking were important. Wine was another universal recommendation. Best against the pestilence, said Gentile da Foligno, was white wine, preferably, old, light, and aromatic, mixed with water. However, fruits and vegetables produced finger-pointing and argument. The Paris masters took a dim view of lettuce, but Gentile da Foligno was unshakable in his defense of the vegetable. "Cabbage is good for you," said the anonymous author of *First About the Epidemic*—but not if eaten with eggplant and garlic, warned Ibn Khatimah. Some agreement emerged on figs, dates, raisins, and pomegranates, but the only universally recommended fruit was the filbert.

The six non-naturals, factors such as personal habits, behaviors, and emotional states were also stressed by the authors of many plague tracts. Thus, the Paris masters urged people to avoid "accidents of the soul," a far more felicitous description of emotional upset than our dreary modern clinical terms. To be especially eschewed were fear, worry, weeping, speaking ill of others, excessive cogitation, and wrath, which, according to Gentile da Foligno, "overheated the members." Sadness, which cools the body, dulls the mind, and deadens the spirit, also predisposed a person toward plague. Good, in Ibn Khatimah's view, was stupidity, which lowered the risk of pestilence; bad was intelligence, which raised it.

Galen, who lived through the third-century Plague of Antonine* (a devastating outbreak of measles or smallpox) believed that little could be done to cure the pestilence once a person was infected, and his pessimism was reflected in the plague tracts of his medieval disciples. Only a handful of authors even suggested cures, and aside from symptomatic relief, most suggestions revolved around bleeding, which was thought to draw poisons and corrupt humors away from vital organs such as the heart, liver, and brain. With phlebotomy, said John of Penna, a professor at the University of Naples, speed was essential. At the first sign of plague symptoms, the patient should be bled on the same side of the body as the pain; even better than bleeding, said John, were purges. Gentile da Foligno, another advocate of bleeding, recommended a venesection of the median vein when the course of the disease was unclear. If the bubo was on the neck, Gentile advised a venesection of the cephalic vein; if under the arms, of the pulmonary vein. Gentile said the bleeding should continue until faintness developed.

Ibn Khatimah, who, unlike many tract authors, seems to have actually treated plague patients, believed that in the bubonic form of the

* Unlike Gui de Chauliac, Galen, who in addition to being a medical theorist and sports doctor (he treated gladiators) was a Roman celebrity physician, showed no interest in acquiring a first-hand knowledge of the plague. When the Plague of Antonine struck Rome, he fled.

disease, the crisis point arrived on the fourth day. After that, the evil vapors begin to detach from the heart, he said. In the recovery period, the Muslim recommended applying an ointment to the buboes to hasten ripening and then surgical excision on the seventh day.

How did the New Galenism perform against the plague?

Some of the advice in the plague tracts was plainly sensible, but, alas, not a great deal—and usually what worked, worked for reasons that would have surprised the tracts' authors. Diet, for example, was useful because it enhanced immune system function, not because it balanced the four humors, fire because it drove away fleas. For all his careful training and his command of the Arab and Greek masters, the best advice Chaucer's new medical professional could offer his patients was the commonsensical admonition to "run far and run fast." And by the time the plague neared Paris in the summer of 1348, even that advice was becoming ineffective since the pestilence was now everywhere, from the Mongolian Plateau to the coast of Greenland.

In May 1348, as the plague slithered north through a misty French countryside, Paris was enveloped in a sense of déjà vu. Scarcely two years earlier the English king, Edward III, desiring "nothing so much as a deed of arms," had landed a force of ten thousand archers and four thousand foot solders on the windswept Cotentin Peninsula, next door to the D-day beaches of 1944; within a month, the English were standing astride the approaches to Paris. The enemy "could be seen by anyone . . . who could mount a turret," wrote the chronicler Jean de Venette, who described Parisians as "stupefied [with] amazement" by the proximity of the danger.

But the English threat at least had been comprehensible, and the spectacle and glamour of war had provided a tonic for turbulent souls and unquiet minds. Paris in the summer of 1346, like Paris in the summer of 1914, crackled with electricity: shouts and cheers echoed through streets and squares and marketplaces as the *arrière-ban*—the general summons to arms—was read. On August 15, there had been the thrilling spectacle of the greatest knights in the realm, led by the dashing Count of Alençon, the king's brother, rushing out to meet the enemy, accompanied by a brigade of Genoese bowmen and by blind King John of Bohemia. All day long, the thunderous sound of horse hooves had echoed across the cobblestoned Grand Rue. There was also the thrilling news that the king, Philip VI, had challenged Edward to personal combat (Edward refused), and the heartening sight of Philip, sitting on his horse like a simple knight, addressing the humble folk of Paris before marching out to battle. "My good people," the portly Philip declared, "doubt ye not, the Englishmen will approach you no nearer than they be."*

Paris in the summer of 1346 had flags to put out, trumpets to blare, war drums to beat. Paris in the summer of 1348 had nothing to do but visit churches, light candles, listen to rumors, think, and wait. "Everyone in our neighborhood, all of us, everyone in Paris is frightened," wrote the physician Peter Damouzy, who lived to the north of the city. Damouzy, a former member of the Paris medical faculty, tried to occupy his mind by writing a treatise on the pestilence, but the approach of the plague kept breaking into his thoughts. "I write without benefit of time," he scribbled at one point, and, later, with even more urgency, "I have no time beyond the present to say or write more."

Physician Damouzy's account is one of the few reports we have of what, by the summer of 1348, was becoming a common experience—waiting. Though the plague was moving with great swiftness, often advancing several miles in a single day, the sense of shock had evaporated. Most localities had several days' to several weeks' advance notice of its arrival. Enough time to think and wonder and worry.

Eight months later, a waiting Strasbourg would vent its anxiety by killing Jews, nine hundred of them. "They were led to their own cemetery into a house prepared for their burning and on their way were stripped naked by the crowd which ripped off their clothes and found much money that had been concealed," wrote a local chronicler. Paris had no Jews to burn, having banished them all; but in the long rainy weeks of May, June, and July 1348, there were prayers to be said and

* The king was half right: the English won at Crécy, one of the most important early battles of the Hundred Years' War, but Paris was not besieged.

rumors to listen to, many of them filled with "stupefied amazements." From Normandy in the west, from Avignon in the south, and from points between came stories of church bells echoing through deserted streets, of black plague flags flying above villages, of abandoned countryside, where the only sound to be heard was the banging of a farmhouse door in the wind. The pestilence's magisterial pace also gave Parisians ample time to contemplate the meaning of love and duty and honor in a time of plague. What would they do if a loved one was afflicted? What would the loved one do if *they* were afflicted? The fear of contagion makes the psychology of plague different from the psychology of war. In plague, fear acts as a solvent on human relationships; it makes everyone an enemy and everyone an isolate. In plague every man becomes an island—a small, haunted island of suspicion, fear, and despair.

"In August, a very large and bright star was seen in the west over Paris," wrote chronicler de Venette, who believed the brilliance of the star "presaged the incredible pestilence which soon followed . . ." However, since no one is sure when the plague arrived in Paris—estimates range from May to August, with June being the most probable date—its start must have been less spectacular than the chronicler's brilliant star.

On a summer morning when the sky was again heavy with black-bottomed rain clouds and the streets were full of watery light, perhaps a young housewife awoke with a terrible pain in her abdomen. Pulling up her nightshirt, she saw a tumor the size of an almond a few inches above her pubic hair. A few days later, when the almond had become the size of an egg, one of her children developed a mass behind the ear; then the old woman who lived above the afflicted family fell ill with a terrible fever, and the young father who lived below them began to vomit violently; and then a prostitute the father had slept with awoke with pain in her abdomen and then . . .

Moving through the gray, rainy city like a fever, the plague slithered from house to house, street to street, neighborhood to neighborhood. It visited the crowded mercantile quarter on the Right Bank, where the Sienese and Florentine bankers lived; the Grand Rue, where the French cavalry had dashed out to meet the English two years earlier; Les Halles, where local farmers brought their produce on Fridays; St. Jacques-la Boucherie, the butchers' quarter, where the fierce Paris wind made little ripples in the pools of animal blood; and the Right Bank, where crowds assembled each morning to buy goods from the arriving barges.

Crossing the Grand Pont to the Ile de la Cité, the pestilence visited the Hôtel-Dieu, where patients slept three and four to a bed and the clothes of the dead were sold at monthly auctions; the Cathedral of Notre Dame, built on the site of a Roman temple of Jupiter; the rue Nouvelle Notre Dame, begun the same year as the cathedral, 1163, and built wide and straight to accommodate the heavy wagons that carried construction materials to the cathedral; and the exquisite Sainte Chapelle, where Louis IX—saint, anti-Semite, and patron of William of Rubruck—kept his relics, among them the Crown of Thorns and fragments of the True Cross.

On the Left Bank, already a student quarter, the plague found lodgments in the Sorbonne, established a hundred years earlier by the theologian Robert de Sorbon, and destined, a hundred years hence, to become a fierce enemy of Joan of Arc; the College de Navarre, site of the first public theater in Paris; and, of course, the University of Paris, which vied with Bologna for the title of Europe's oldest university (Paris dated its origins to the twelfth-century schools of disputation attached to Notre Dame). Contemporary records indicate that the pestilence took a terrible toll on the university faculty; in 1351 and 1352, some disciplines were so short of teachers, the administration had to relax academic qualifications. Remarkably, however, all the authors of the *Compendium* seem to have survived the pestilence in good health. A 1349 university roll shows that, as in 1347, there were still forty-six masters on the medical faculty. Another notable survivor of the pestilence was the man who commissioned the faculty to write the *Compendium,* King Philip VI.

Corpulent and insecure, Philip was a man of profound contradictions. Though he fought like a lion at Crécy and planned a Viking funeral for himself—the royal heart was to be sent to a church in

Bourgfontaine; the royal entrails to a monastic house in Paris so as to double the number of prayers offered up in repose of the royal soul—Philip fled Paris almost as soon as the pestilence arrived. Over the next year, one catches glimpses of him at Fontainebleau, at Melun, and at the casket of his plague-dead queen, the ill-tempered Jeanne of Burgundy, but Philip does not emerge into full public view again until the early 1350s, when he shocked Paris with a heinous betrayal. The high-minded moralist who hated to hear the Lord's name taken in vain—in Philip's France blasphemers had their upper lip cut off—stole his eldest son's bride-to-be, the beautiful Blanche of Navarre, a few months before the couple were to be married.

Jean Morellet, unlike his king, chose to remain in Paris, and because he did, we have something more fine-grained than the usual chroniclers' estimates with which to measure the city's mortality. Morellet was attached to the parish of St. Germain l'Auxerrois as canon or priest—contemporary records are unclear. Today the parish sits in one of the most congested areas of Paris, surrounded by scores of famous neighbors including the Louvre and the Place de la Concorde, but in 1340, when Morellet became director of St. Germain's building fund, the only landmark of any import nearby was the cobblestoned Grand Rue. From his office at the church, the director would have had an unobstructed view of the windmills, boatmen, and rickety piers along the still unembanked Seine.

Morellet's duties as director of the building fund were not very taxing. When a deceased parishioner left a bequest, he would make a record of it. In the first eight years of his tenure, roughly mid-1340 to mid-1348—the fund received a total of seventy-eight bequests—not enough to keep the director busy. Indeed, the pace of death was so slow in the parish, most of the time Morellet seems to have kept track of the donations in his head. Fund records indicate that he would update the donor list only once or twice a year. However, in the summer of 1348, this pattern changed.

The plague's arrival in St. Germain is announced with a donation of twenty-four sous. At a time when six sous could buy a man a good horse, twenty-four sous was a significant sum. Even more unusual was the purpose of the bequest. Heretofore, all bequests had gone toward maintenance of the parish church and for future building projects. The twenty-four sous were used to purchase burial shrouds for parishioners. Around the same time, the number of bequests suddenly explodes. For most of the 1340s, annual donations to the fund remained in the single digits. In the nine months between June 1348 and April 25, 1349, a total of 445 were received, roughly a forty-five-fold increase.

During the second half of 1349, donations to the fund remained at record levels. In September, fifteen months after the plague's arrival in Paris and seven months after the municipal cemetery, Holy Innocents, had to be closed for lack of burial space, bequests to the building fund reached an all-time high. Morellet, who was now forced to update the fund records monthly because of the volume of donations, notes that the parish received forty-two bequests; in October, the number dipped, but only slightly, to thirty-six bequests.

Director Morellet's figures, which represent a single stream of deaths in a single Paris parish, can't be extrapolated to the entire city. Nonetheless, they show that something unparalleled was happening in Paris, an impression substantiated by chronicler de Venette's account of events at the Hôtel-Dieu. Like most medieval hospitals, including La Pigonette in Avignon, the hôtel was a retirement home for the elderly, and a shelter for the homeless and indigent as well as medical facility. All three roles put it on the front line during the pestilence.

"For a considerable period," says the chronicler, "more than 500 bodies a day were being taken in carts from the Hôtel-Dieu [on the Ile de la Cité] . . . for burial at the cemetery of the Holy Innocents [on the Grand Rue]." Many of the caravans carried the bodies of the *filles blanches,* the young novitiates who nursed the ill. "The saintly sisters of Hôtel-Dieu," wrote de Venette, ". . . worked sweetly and with great humility, setting aside consideration of earthly dignity. A great number of [them] were called to a new life and now rest, it is piously believed, with Christ."

Several historians have called de Venette's mortality figures into question, but they do not seem out of line with the estimates of other contemporaries. The *Grandes Chroniques de France,* kept by the monks of nearby St. Denis, speaks of eight hundred people dying "from one day to the next" in the city. An Italian merchant reports that "on March 13th, [1349] 1573 noblemen were buried not counting petty officials." Another resident claims that "1328 [people] were buried in a single day." During the eighteen months between June 1348 and December 1349, Paris seems to have lost the equivalent of a good-sized village almost every day, and on bad days, a good-sized town. According to Richard the Scot, 50,000 residents died during the plague. "Nothing like it has been heard or seen or read about," wrote a contemporary.

The constancy of the death seems to have dispirited director Morellet. As the year 1349 drew to a close, a certain listlessness becomes apparent in his manner. The director's record keeping becomes intermittent again, and his work shows an uncharacteristic sloppiness; he no longer bothers to write down the names of new donors, just the sums contributed to the fund; it is as if the dead have lost all meaning for him, as if he can no longer envision them as anything but a pile of corpses in one of the little death carts shuttling back and forth in the rain to the cemetery of Holy Innocents. Historian George Deaux believes that this kind of indifference became common later in the mortality as the monotony of death replaced the terror of death. Deaux compares the survivors to "soldiers . . . who have been in the line so long they no longer know or care if their side is winning or losing or even what the terms mean anymore. . . . [W]ar has become an endless course of terror and fatigue, mutated to a sort of boredom that destroys everything but the body's motor functions."

From Paris—and from Normandy, which was also struck in the summer of 1348—the plague spread northward to Rouen, where a new cemetery had to be consecrated to accommodate the dead; to La Graverie, where "bodies . . . decayed in putrefaction on the pallets where they had breathed their last"; to La Leverie, where the family of a noblewoman was unable to find a priest to bury her because the local clergy were all dead and priests from other villages refused to visit one that was flying the black plague flag. At Amiens, burial space was also a problem, until the wandering Philip graciously authorized the mayor to open a new cemetery. In his proclamation, the king declared, "The mortality . . . is so marvelously great that people are dying . . . as quickly as between one evening and the following morning, and often quicker than that."

In the fall of 1348, as the plague approached Tournai on the Flemish border, a local abbot, Gilles li Muisis, recalled a fifty-year-old prophecy and wondered if still worse was to come. "I have been thinking [recently] about . . . Master Jean Haerlebech," wrote the seventy-eight-year-old li Muisis. "When I was a young monk he would often speak to me in secret of things, which afterward came to pass.

". . . He predicted that in 1345 major wars would begin in various places . . . and that in 1346 and 1347 . . . people would not know where to go or where to turn for safety. . . . But he didn't want to tell me anything about 1350 and I was not able to wring anything out of him."

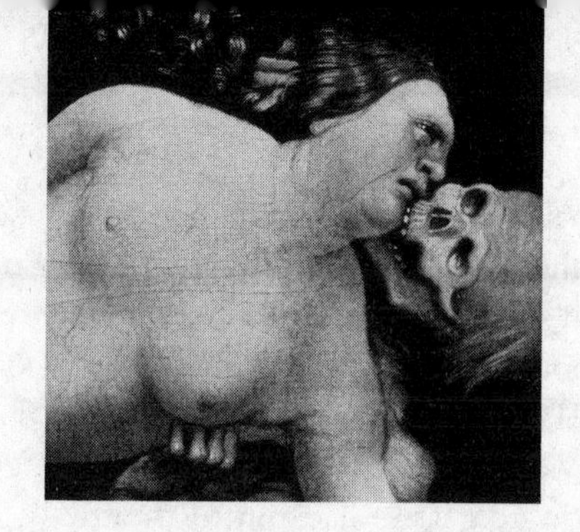

CHAPTER EIGHT

"Days of Death Without Sorrow"

Southwest England, Summer 1348

IN 1348 ENGLAND'S MORALISTS WERE IN A SOUR MOOD. THE monk Ranulf Higden saw pretension everywhere. These days, thundered Higden, "a yeoman arrays himself as a squire, a squire as a knight, a knight as a duke, and a duke as a king." The chronicler of Westminster saw an even more pernicious threat—medieval Spice Girls everywhere! Englishwomen, complained the chronicler, "dress in clothes that are so tight, . . . they [have to wear] a fox tail hanging down inside of their skirts to hide their arses."

However, hardly anyone paid attention to the carping. As mid-century approached, the green and pleasant land of England was in an exuberant mood. The country was flush with military success, awash in French war booty, and best of all, England had a king it could love again. Edward II, the former sovereign, had been a puzzlement to his people. Kings were supposed to like wars, hunting, jousting, and women, but Edward's tastes had run to theatricals, arts and crafts, minstrels—and men. In a guarded reference to the old king's homosexuality, a chronicler wrote that Edward loved the knight Piers Gaveston more than his wife, the beautiful French princess Isabella. In a reign marked by military defeat, famine, and political turmoil, Edward, who gained a reputation for being "chicken-hearted and luckless," lost the support of the English nobility, and, after a coup by Isabella and her lover Roger Mortimer, met death in the vilest imaginable manner. According to legend, Edward's last moments on earth were spent with a hot plumber's iron in his anus.

The only nice thing the author of *The Reign of Edward II* had to say about his subject was that Edward had been remarkably wealthy.

He was also quite handsome, a trait he shared with his son and successor, but otherwise Edward III was everything his father was not: glamorous, romantic, politically deft—and bold. In 1330, at the age of seventeen, Edward seized England's imagination—and avenged his father's memory—by bursting into his mother's bedchamber and arresting the treacherous Mortimer at swordpoint. "Good son, good son, be gentle with dear Mortimer," the queen pleaded, as her lover was led away in chains. However, it was the victory over the French at Crécy in 1346 that turned Edward III into an English demigod. When the chronicler Thomas Walsingham wrote, "[I]n the year of grace 1348, it seemed to the English that, as it were, a new sun was rising over the land," he was thinking of the glorious August morning two years earlier when Edward, in a Shakespearean moment, leaped upon his horse in a field outside Crécy and, with the morning sun at his back, "rode from rank to rank . . . desiring every man to take hede . . . [speaking] so sweetly and with so good countenance and merry cheer, that all such as were discomforted took courage."

In 1348, Walsingham's "new sun" also shone over a nation more politically and socially stable than it had been in more than a generation—and more prosperous than one might expect, given the terrible deprivations of the famine years. In 1348 demand for English wool was such that English sheep outnumbered English people—roughly eight million sheep versus six million humans. And there were the first stirrings of an industrial economy—in the west country and East Anglia, where cloth was being made, and in Wales and Cornwall, where coal and tin were being mined. Meanwhile along the coast, the timber-faced waterfronts of Bristol, Portsmouth, London, and Southampton bustled with high-masted ships from Flanders, Italy, Gascony, and the German towns of the Hanseatic League.

It is impossible to say exactly when in 1348 the plague burst the bubble of English exuberance, but during the early months of the year, the country still seemed in an it-can't-happen-here frame of mind. In January and February, while Avignon was running out of cemeteries, Edward was at Windsor, redecorating the chapel and coyly avoiding the Germans, who were said to want to offer him the emperor's crown; meanwhile, his subjects were shamelessly parading around the English countryside in stolen French war booty. "There was no woman of any standing who had not her share of the spoils of Calais, Caen and other places across the Channel," wrote Walsingham. Strengthening the it-can't-happen-here mood was the fact that the plague was ravaging France, and the insular English considered the French strange even for foreigners; as one contemporary English writer famously noted, your average Frenchman was effeminate, walked funny, and spent too much time fussing with his hair.

It is also difficult to say exactly what changed the national mood. Perhaps it was the onset of the rains in early summer. During the second half of 1348, "scarcely a day went by without rain at some time in the day or night." Maybe, peering into the misty channel that May or June, the English felt a premonitory shudder of dread. Or maybe, more simply, by late spring—with nearly every town on the French side of the channel flying a black plague flag—the danger had become impossible to ignore, even for an island people accustomed to thinking of themselves as a breed apart.

The summer of 1348 was, like the Battle of Britain summer of 1940, a time of stirring backs-to-the-wall rhetoric. "The life of men upon earth is warfare," declared the Bishop of York in July. A month later, the Bishop of Bath and Wells warned his countrymen that the apocalypse was nigh. "The catastrophic pestilence from the east has arrived in a neighboring kingdom [France], and it is very much to be feared that unless we pray devoutly and incessantly, a similar pestilence will stretch its poisonous branches into this realm."

We don't know how many people took the bishop's advice, but however many prayers the English offered to heaven in the rainy summer of 1348, they were not enough. Over the next two years, England would endure the worst catastrophe in its long national life. In the words of the Elizabethan playwright John Ford, between 1348 and 1350:

One news came straight huddling on another
Of death and death and death.

During those two years, perhaps 50 percent of England was consumed in the mortality.*

To Irish monk John Clynn, it felt as if the end of the world had arrived. Writing in the doomsday year of 1349, Clynn declared, "Waiting among the dead for death to come . . . , [I] have committed to writing what I have truly heard and examined . . . in case anyone should still be alive in the future." Given the level of fear, rumor, and confusion current in the summer of 1348, it is probably inevitable that contemporary reports have *Y. pestis* landing at several different points along the English coast, including at Bristol in the west, Southampton

* Estimates of English plague mortality vary. In *A History of Bubonic Plague in the British Isles* (London: Cambridge University Press, 1970, p. 8), J. F. D. Shrewsbury, an English bacteriologist, came up with the highly improbable figure of 5 percent. The consensus figure, as expressed by John Hatcher in *Plague, Population and the English Economy, 1348–1530* (London: Macmillan, 1977, p. 25), is between 30 and 45 percent. But medievalist Christopher Dyer thinks the contemporary evidence supports a mortality rate of close to 50 percent. Dyer has it about right.

and Portsmouth along the channel coast, and even in the north of England.* However, the weight of historical evidence points to Melcombe, a small port on the southwest channel coast, as the most likely initial landing site. Melcombe is mentioned more often than any other port in contemporary writings, including in the chronicle of Malmesbury Abbey, which says that "in 1348, at about the feast of . . . St. Thomas the martyr [July 7], the cruel pestilence, hateful to all future ages arrived from countries across the sea on the south coast of England at the port called Melcombe in Dorset." The *Grey Friar's Chronicle* also says that "In 1348, . . . two ships . . . landed at Melcombe in Dorset before Midsummer. . . . They infected the men of Melcombe who were the first to be infected in England."

Today Melcombe is part of Weymouth, a pleasant channel resort town full of Regency architecture, bright seaside shops, towering cliffs, and history. A D-day memorial in the harbor commemorates the thousands of young Americans who left the town on a rainy June morning in 1944 for the beaches of Normandy; and a plaque near the docks commemorates the departure in 1628 of John Endicott, founder of the Salem colony and first governor of New England. However, the most famous moment in local history may have occurred on the north bank of the Wey, the river that runs through the modern town and gives Weymouth its name. In the Middle Ages, the flat alluvial plain above the north bank was occupied by the town of Melcombe, a name that means "valley where milk was got," or, more simply, "fertile valley," and that apparently originated with the Durotriges, the Celtic tribe who settled the Dorset coast.

In the 1340s Melcombe seems to have been larger and more prosperous than medieval Weymouth, which was squeezed into the narrow strip of land between the south bank of the Wey and the rugged Dorset cliffs—a position that gave it the aspect of being perpetually about to tumble into the English Channel. Melcombe boasted a larger fleet than Weymouth, a crowded trading quarter along Bakeres

* There are also conflicting reports about the day of the plague's arrival. Among the dates cited in the chronicles are June 23, June 24, July 7, August 1, and August 15.

Street—as well as one of the richest men in the county of Dorset, Henry Shoydon, who in 1325 paid an astonishing forty shillings in taxes. A fourteenth-century map shows that in Shoydon's time Melcombe also had a jetty, where visiting ships unloaded goods. Though it is hard to imagine today, in affluent modern Weymouth—surrounded by picture-taking Japanese tourists, T-shirted teenagers, and sleek young couples down from London—on a summer's day in 1348, when the sky was heavy with rain, a vessel or several vessels docked at the jetty carrying "death and death and death."

The vessel—or vessels—was probably returning from Calais, the scene of recent bitter fighting between the French and the English.* After menacing Paris in early August 1346, and defeating a much larger French force at Crécy, Edward swung north and attacked Calais, a walled city of around twenty thousand on what is left of the land bridge that linked Britain to the continent before the end of the last Ice Age. As fighting around the town degenerated into a brutal siege, little Melcombe, not for the last time, found itself drawn into great events across the channel. A surviving royal order from 1346 shows Edward requisitioning twenty ships and 246 mariners from the town to support the English forces in Calais. Some Melcombe men may even have been present a year later to witness one of the most glorious moments in French history. A committee of Calais's leading citizens, the Six Burghers, appeared in the English camp and offered their lives if Edward would agree to spare their fellow citizens. The chronicler Jean le Bel says the English were so moved by the burghers' bravery, "there was not a lord or knight . . . who wasn't weeping out of pity."

Even if Melcombe was not the very first English town to be infected by the pestilence, the links to Calais must have made it a very close second or third. Everything we know about the plague suggests that the French town would have been absolutely toxic with contagion in the summer of 1348. Cramped and walled off to begin with, Calais had just emerged from an eleven-month siege and thus would have

* One contemporary source says the ship that infected Melcombe came from Gascony, but, given the low volume of trade between the two regions, the allegation seems unlikely.

been rich in filth, rats, and malnourished humans when *Y. pestis* arrived in the summer of 1348. On top of that, the English were sending home enormous amounts of "liberated" French war booty. "English ladies were proudly seen going about in French dresses" that summer, and English homes were full of French "furs, pillows, . . . linens, clothes and sheets." Inevitably, some of the "liberated" goods sent home must have carried infected fleas. Additionally, plague-bearing rats most likely slipped into the holds of a few of the Melcombe ships fleeing Calais.

Whichever way the pestilence arrived, by the end of the summer of 1348, the only sounds to be heard in Melcombe were the patter of rain on thatched roofs and the roar of the pounding surf rising above the cliffs. The sole piece of information we have about the fate of the town is inferential. In the early 1350s Edward declared the island of Portland immediately to the south "so depopulated in . . . the late pestilence that the inhabitants remaining are not sufficient to protect it against our foreign enemies." Melcombe's subsequent incorporation into Weymouth suggests that its suffering was even more severe. It is unlikely that the proud townsfolk would have agreed to be incorporated by a smaller and historically less significant Weymouth unless Melcombe had become too weak to sustain itself.

The lack of a documentary record in Melcombe is unusual. As historian Philip Ziegler notes, contemporary English records "offer a fuller picture of the progress of the Black Death than those [of] any other country." On the subject of human suffering, the documents—wills, deeds, manor records, clerical appointments, and property transfers—are not particularly enlightening, but they provide an illuminating guide to *Y. pestis*'s movement and behavior. From the pattern of deaths in the records, for example, it is clear how cunningly the plague exploited England's improving communications network, including the new road networks linking London in the east to Coventry in the midlands and Bristol in the west, and the river networks, which carried a great deal of internal trade. The new ferries and bridges—and the horse, which was replacing the lugubrious ox as a means of haulage—also helped to enhance *Y. pestis*'s mobility.

As Ziegler has noted, in its precision, the plague's initial assault on southwestern England resembled a military campaign. To understand how the assault unfolded, a rudimentary geography lesson is necessary. The west coast of England is shaped like a galloping pig in profile. North Wales forms the pig's floppy ear, central Wales its snout, and the Bristol Channel, below Wales, the space between the pig's jaw and its outstretched, galloping leg. The leg, which thrusts out into the Atlantic, encompasses the southwest of England. On its ocean side, the region extends down the coast from the Bristol Channel to Land's End in Cornwall, the most westerly point in England; and on its channel side up the backside of Cornwall and eastward to the neighboring channel counties of Devon and Dorset, where Melcombe is located.

By the winter of 1348–49, one prong of the pestilence, moving northward from Melcombe, and a second, moving eastward over land from the Bristol Channel (which was infected later in the summer) would intersect, and everything inside the galloping leg would be trapped in a plague pocket.

The Dorset coast is full of brooding Heathcliff vistas—windswept moors, moody gray skies, furiously moving clouds, turbulent reefs, and towering, chalk-faced cliffs, some so sheer they seem to have been cleaved from continental Europe with an ax. Here and there, nestled amid rock and wind and sea, are dozens of little coastal towns like Melcombe and its neighbors: Poole, Bridport, Lyme Regis, Wareham, and West Chickerell. In the Middle Ages, the towns were all sleepy little places where the same two hundred or three hundred sets of genes would chase one another around for centuries, and the rhythm of daily life was as unalterable as sea and sky. In the rainy autumn of 1348, this predictability begins to shatter. The secular records provide brief glimpses of the plague's early progress through the coastal southwest, but the ecclesiastical rolls—especially the meticulously kept clerical mortalities—offer the most sustained view.

In the clerical records, if not in real life, the Black Death begins like a classic English mystery, perhaps an Agatha Christie thriller called

Who Is Killing the Priests of Coastal England? In October—a few months after *Y. pestis* arrives in Melcombe*—West Chickerell to the north and Warmwell and Combe Kaynes to the east suddenly lose priests; in November the clerical death wave spreads to nearby Bridport, East Lulworth, and Tynham, and in December to Settisbury, also nearby, which has to appoint three priests before finding one who won't die on the town. John le Spencer was appointed to a clerical post in Settisbury on December 7, 1348; three days later he was succeeded by Adam de Carelton, who in turn was succeeded on January 4 by Robert de Hoven. In the last two months of 1348, thinly inhabited Dorset had to fill thirty-two clerical vacancies—fifteen in November and seventeen in December, an astonishing replacement rate. And the clerical deaths were just the beginning.

All along the coastal southwest in the black autumn of 1348, the living gathered in the rain to bury the dead. In Poole, to the east of Melcombe, the grave diggers worked to the roar of the pounding surf rolling across the Baiter, a sandy tongue of land converted into a seaside cemetery. In 1348 and early 1349, little Poole seems to have buried the better part of itself in the channel sand. One hundred fifty years later there were still so many abandoned buildings in the town, Henry VIII took time out from his complicated romantic affairs to order their repair. Centuries later, locals were still pointing out the Baiter and regaling visitors with stories of the terrible autumn of 1348, when death and rain howled down upon the town. To the west of Melcombe, in Bridport, which was famous for making the best nautical rope in England, townsfolk buried their dead with a dogged devotion to English legalism. To ensure that every death was properly recorded and processed, in the winter of 1348–49 Bridport doubled its normal complement of bailiffs from two to four. The ecclesiastical records also highlight what may have been the first metastasis of the pestilence. In November and December a sudden burst of clerical deaths in Shaftesbury in central Dorset suggests that

* The season, summer, and the two- to three-month interval between the arrival of *Y. pestis* in July and the first reports of fatalities in October suggests that the plague arrived in England in a bubonic form—"suggests," because the interval could have been caused by a delay in recording deaths.

the plague was now spreading northward through the incessant rain on an inland route that would carry it through the deer-filled forests and sprawling sheep farms of upper Dorset and into the English midlands.

In the seventeenth century, Clevedon, Bridgwater, and the other little towns along the Bristol Channel would become the last bit of green England that African slaves saw on their way to the West Indies. But in the gray, wet August of 1348, sorrow and misery sailed in the opposite direction—not out into the blue Atlantic, but up the channel toward Bristol. On a drizzly morning somewhere between August 1 and 15—chroniclers disagree on the date—an infected ship, or ships, probably from Dorset but possibly from Gascony, docked in Bristol harbor. Very shortly thereafter, the town, the largest seaport in the west of England, literally exploded. "Cruel death took just two days to break out all over town. . . . It was as if sudden death had marked [people] down beforehand," wrote the monk Henry Knighton.

This was the second prong of *Y. pestis*'s assault, and it quickly turned Bristol into a charnel house. "The plague raged to such a degree," says one writer, "that the living were scarcely able to bury the dead. . . . At this period," he adds, "the grass grew several inches high in High Street and Broad Street."

According to the *Little Red Book,* a list of local government officials, fifteen of Bristol's fifty-two city councilors died in 1349, a death rate of almost 30 percent. However, surviving ecclesiastical records suggest that Bristol's clergy suffered far more grievously; one series of records shows that ten of eighteen local clerical posts had to be filled in 1349—a death rate of about 55 percent. The town historian estimates that 35 to 40 percent of Bristol died in the mortality.

Below the seaport, in the Bristol Channel town of Bridgwater, the Christmas of 1348 was desolate beyond imagination. The local water mill had stood idle since November, when William Hammond, the miller, died, and the autumn rains had turned the low-lying ground around the town into a swamp, ruining crops and making movement difficult. Then, over Christmas, while Bridgwater prayed for deliverance, the pestilence went on a rampage and killed at least twenty residents at a nearby manor.

To the north of Bristol, in Gloucester, anxious local officials ordered the town gate closed and all residents of the seaport banned from the town. But later in the fall, when the pestilence came up the road from Bristol—tired, hungry, and eager to get out of the rain—the quarantine proved as useless as it had been in Catania a year earlier. In 1350 someone scrawled Gloucester's epithet on a church wall: "Miserable, wild, distracted. The dregs of the people alone survive."

Sometime between November 1348 and January 1349, the Bristol prong of the pestilence—now thrusting inland on a road network that led eastward across the shires of south-central England to London—and the coastal prong—advancing northward from Dorset into the midlands—crossed paths.

As the plague pocket snapped shut, the blustery Bishop Ralph of Shrewsbury issued what would become one of the most famous pronouncements of the mortality. It was not quite Churchill's we-will-fight-them-on-the-beaches speech, but Bishop Ralph, whose diocese of Bath and Wells lay trapped at the eastern edge of the plague pocket, was a master rhetorician himself. In January, when even the hopeful had lost hope, the bishop lifted hearts across the diocese with a startling announcement.

"Wishing, as is our duty, to provide for the salvation of souls and to bring back from their paths of error those who have wandered . . . [w]e declare that those who are now sick or should fall sick in the future . . . if they . . . cannot secure the services of a priest, then they should make a confession to each other as is permitted in the teachings of the apostles . . ."

A few paragraphs later, Ralph conferred another even more extraordinary right on the laity—or rather, on some of its members. The bishop said that if no priest could be found, "The Sacrament of the Eucharist [Holy Communion] . . . may be administered by a deacon [a layman who assists in ecclesiastical functions]."

The bishop's extraordinary decision to give the faithful the right to administer the holy sacraments was born of desperation. The diocese of Bath and Wells would lose almost half its normal complement of priests in the mortality, and as Ralph's proclamation noted, the dangers of ministering to the sick had become so great, no surviving priest could be found who was "willing whether out of zeal or devotion or for a stipend to take on the [duties of] pastoral care."

Alas, the bishop's personal courage did not quite measure up to his muscular rhetoric. In late January, as the plague danced through the streets of Bath, he retreated to the relative safety of his manor in rural Wiveliscombe. In fairness, it was Ralph's normal practice to winter at the manor. Moreover, in 1349 he was hardly the only august personage to pay a prolonged visit to rural England. Edward III himself spent the early months of 1349 in the rural southeast and around Windsor—very fretfully, apparently, since he sent up to London for holy relics. Still, Ralph's decision to leave Bath probably contributed to an ugly incident later in the year.

In December, as the plague was waning, he decided to visit the little town of Yeovil to celebrate a Mass of thanksgiving. However, after a year of "death and death and death," the townspeople apparently became upset at the sight of the pink, plump Ralph riding into town at the head of a rather large and resplendent entourage. After the bishop disappeared inside the local church, an angry mob gathered in a nearby square. Voices were raised, fists waved, weapons brandished, denunciations made—then suddenly the crowd was running toward the church. Bishop Ralph's *Register,* a kind of official diary, describes the rest of the story. "Certain sons of perdition, [armed with] a multitude of bows, arrows, bars, stones and other kinds of arms" burst into the church, "fiercely wound[ing] very many . . . servants of God," then "incarcerated [us] . . . in the rectory of said church until on the day following [the attack] the neighbors, devout sons of the church . . . delivered us . . . from our prison."

On returning to Wiveliscombe, a very angry Ralph ordered Yeovil citizens Walter Shoubuggare, Richard Weston, Roger Le Taillour, and John Clerk—as well as other "sons of perdition"—excommunicated.

The men were commanded to "go around the parish church . . . on Sundays and feast days . . . bare head[ed] and foot[ed] . . ." in a penitential manner. In addition, at High Mass, they had to hold a candle "of one pound of wax" until the hot wax had melted onto their hands. Perhaps the memory of the cemetery at Yeovil—"polluted by violent effusions of blood"—began to haunt the bishop, or perhaps he started to feel guilty about wintering in Wiveliscombe. Whatever the reason, soon after issuing the excommunication order, Ralph revoked it. "Lest . . . the teaching [of] Christ be diminished and the devotion . . . of God be weakened," he wrote the vicar at Yeovil, "we suspend the said interdict."

In Oxfordshire, the county to the east of Ralph's diocese, the plague caused such devastation, the documents that survived have an end-of-the-world feel to them. In 1359 we hear that taxes can no longer be collected in the little hamlet of Tilgarsley because it has been deserted since 1350; in nearby Woodeaton manor, that after "the mortality of men . . . scarce two tenants remained . . . and they would have departed had not Brother Nicholas of Upton . . . made an agreement with them." In Oxford, which lost three mayors to the plague, the few surviving documents include a petition from a university official and a death estimate. The official complains "that the university is ruined and enfeebled by the pestilence . . . so that its estate can hardly be maintained or protected." The death estimate comes from a former chancellor, the fierce Richard Fitzralph, Bishop of Armagh. In 1357 the bishop wrote that there used to be "in ye University of Oxenford . . . thrilty thousand scolers [scholars] . . . and now [in 1357] there beth unneth [under] six thousand." Since the town of Oxford itself, let alone the university, could not have held thirty thousand souls, the bishop was surely exaggerating. Nonetheless, what was true was that a great many people had died. The most trustworthy account of what happened in Oxford may come from an eighteenth-century scholar who says that when the plague arrived in the autumn of 1348, "those that had places and houses in the country retired [though overtaken there also] and those that were left behind were almost totally swept away. The school doors were shut, colleges and halls relinquished and none scarce left to keep possession or . . . to bury the dead."

What such postmortems leave out are the details of daily life that autumn: the never-ending tattoo of rain on thatched roofs, the muffed thud of shovels in church graveyards, the lamentations of parentless children and childless parents, and, in the soggy fields beyond the little villages of southern England, the rotting carcasses of thousands of sheep and cows. Mass animal die-offs were common in the mortality, but, judging from contemporary accounts, in England they achieved a special intensity. "In the same year [as the pestilence]," wrote a chronicler, "there was a great sheep murrain [epidemic] throughout the realm, so much so that in one place five thousand sheep died in a single pasture, and the bodies were so corrupt that no animal or bird would touch them."

The English animal die-offs seem to have been caused by rinderpest and liverfluke, herd diseases that flourish in damp weather. In 1348 and 1349, their spread was probably further abetted by a lack of shepherds to tend threatened flocks. However, in other places the die-offs may have been part of the pestilence. In Florence dogs, cats, chickens, and oxen died in the plague, and, like people, many had buboes. The most haunting report of an animal die-off comes from a medieval Arab historian, who says that in Uzbekistan lions, camels, wild boar, and hare "all lay dead in the fields afflicted with the boil."

In England, as in other countries, one of the few things that offered a measure of protection against the pestilence was privilege. The stone houses of the wealthy were less vulnerable to rat infestation,* and the aristocracy and the gentry tended to enjoy better health in general. Indeed, most modern fifty-year-olds would envy the superb physicality of Bartholomew Burghersh, a knight and diplomat of Edward III's time. When Burghersh died in late middle age, he still had a lean, muscular build, broad shoulders, a complete set of teeth, and

* This is a point of controversy. Evidence from the Third Pandemic suggests that stone houses can be penetrated by rats. But even if such homes are permeable, they were less permeable than wattle-and-daub peasant huts.

no signs of osteoarthritis, a common feature of medieval skeletons. According to one estimate, only 27 percent of aristocratic and wealthy England died in the mortality, as opposed to 42 to 45 percent of the country's parish priests, and 40 to 70 percent of the peasantry.

However, as a poem of the time noted, no one, not even the highborn, was immune from the universal pestilence.

Scepter and Crown
Must tumble down
And in the dust be equal made
With the poor crooked scythe and spade.

Southeast England, Early Fall 1348

Across the long quays of medieval Bordeaux flowed bales of wool, packets of produce, barrels of burgundy wine, and, on an early August day in 1348, a golden-haired princess of England, Joan Plantagenet, youngest child of Edward III. With her heart-shaped face and air of regal certitude, Joan must have seemed like a miracle to the weary French stevedores working the quays. For weeks they had known nothing but pestilential death; now, suddenly, here in their midst was a royal fairy tale, complete with four brightly pennoned English ships, a silky Spanish minstrel, a gift from Joan's betrothed, Prince Pedro of Castile, a hundred regally dressed archers, and two of Edward's most lofty servants, Andrew Ullford, leathery veteran of the French wars, and Robert Bourchier, lawyer and diplomat.

Little is known about the princess's visit to Bordeaux except that she stopped there en route to Spain, where she was to marry Prince Pedro in the fall, and that the morning of her arrival the mayor, Raymond de Bisquale, was waiting on a quay to warn her and the rest of her wedding party about the presence of the plague. We also know that the English brushed aside the warning, though we do not know why. The princess's foolishness can probably be attributed to her age. Fifteen-year-old royals must be even more inclined than ordinary fifteen-year-olds to think themselves immortal. However, the recklessness of the two senior figures in the royal party, Bourchier and Ullford, is harder to fathom. Perhaps after surviving Crécy, where the English had been outnumbered four or five to one, Ullford had developed his own delusions of immortality. Or maybe Mayor de Bisquale struck him and his fellow mandarin, Bourchier, as one of those inconsequential little Frenchmen with fussy hair and a funny walk.

The two royal officials should have known better.

On August 20 Ullford died a hard plague death, leavened only by a good view. The old soldier's final hours were spent in the Château de l'Ombriere, a sumptuous Plantagenet castle overlooking Bordeaux harbor. Several others in the wedding party also died, including, on September 2, Princess Joan, who left behind the memory of a girlish laugh and an unworn wedding gown made from four hundred and fifty feet of rakematiz: a thick, rich silk embroidered with gold. What the princess did not leave behind was a body. In October Edward offered the Bishop of Carlisle a huge sum to go to pestilential Bordeaux and retrieve the princess's corpse, but whether the prelate actually visited the town is unclear. In any case, Joan's body was never recovered. Historian Norman Cantor thinks this is because her corpse was burned in October, when Mayor de Bisquale ordered the harbor set aflame. Intended to check the spread of the pestilence, the fire blazed out of control and destroyed several nearby buildings, including, notes Professor Cantor, Château de l'Ombriere, where Princess Joan died.

On September 15, when Edward III wrote to King Alfonso, Pedro's father, to inform him of Joan's death, the English demigod sounded like every parent who has ever tried to make sense of something as senseless as a child's death. "No fellow human being could be surprised if we were inwardly desolated by the sting of grief, for we are human too. But we who have placed our trust in God . . . give thanks to Him that one of our own family, free of all stain, whom we have loved with pure love, has been sent ahead to heaven to reign among the choirs of virgins where she can gladly intercede for our offences before God."

As Edward wrote, a new metastasis was developing in the south of England. During the fall, the plague appeared in Wiltshire, the county immediately to the east of Dorset, then, almost simultaneously, in Wiltshire's eastern neighbors, Hampshire and Surrey. A circumstantial case can be made for Southampton, on the Wiltshire coast, as the source of the new outbreaks. Ships from France, including from Bordeaux, visited the port almost daily. However, since there were no telltale bursts of ecclesiastical death in Southampton until December, it also seems possible that this new assault by the plague originated in Dorset—perhaps from Melcombe, *Y. pestis* spread east as well as north.

All that can be said with certainty seven hundred years later is that in the fall of 1348, people in the counties to the east of Dorset knew that death was about to burst upon them. On October 24 Bishop William Edendon, whose Hampshire diocese of Winchester stood directly in harm's way, issued an ominous warning. Evoking the lamentations of Rachel in Matthew 2:18,* the bishop declared, "A voice has been heard in Rama. . . . We report with anguish the serious news . . . that this cruel plague has begun a . . . savage attack on the coastal area of England. . . . We are struck with terror lest . . . this brutal disease should rage in any part of our city or diocese."

On November 17, with the plague now howling at Hampshire's borders, Bishop Edendon took the occasion to remind the faithful of "the radiant eternal light which glows . . . in the dark core of human suffering." In a second proclamation, the bishop declared that while "sickness and premature death often come from sin . . . by the healing of souls this kind of sickness [the plague] is known to cease." There is no record of how many of the faithful drew hope from the bishop's words.

Like Tolstoy's unhappy families, over the first winter of the pestilence, the little towns of southern England began to die, each in its own way. The following May, as a spring sun warmed the gray slabs of Stonehenge, the only sound to be heard on the nearby Wiltshire estate of Carleton Manor was birdsong. Carleton's water mills stood quiet, its

* "A voice in Rama was heard, lamentation and great mourning; Rachel bewailing her children and would not be comforted because they are not."

farmlands untilled, and twelve of its thatched cottages empty. The years 1348 and 1349 may have been the quietest years in the English countryside since the land was first inhabited by man. At Ivychurch priory, near the Wiltshire-Hampshire border, twelve of the priory's thirteen canons lay dead. One can scarcely imagine the feelings of the survivor, James de Grundwell, but by March 1349 he must have found it odd not to wake to the sound of rain pounding against the roof and the moans of dying monks echoing through the dank monastery halls. Edward thought de Grundwell's good fortune worthy of both a promotion and a mention in a royal dispatch. "Know ye that since . . . all the other canons of the same house, in which hitherto there had been a community of thirteen canons regular, have died . . . , we appoint James de Grundwell custodian of the possessions, the Bishop testifying that he is a fit and proper."

In Winchester, ancient capital of England, an all-too-familiar problem produced a bitter division in January 1349. Concerned about unburied corpses "infecting" the air—and hence spreading the pestilence—the laity wanted to dig a plague pit outside the city; but the clerical establishment, led by Bishop Edendon, resisted. The plague pit would be on unconsecrated ground, and people buried on such ground might be overlooked on Resurrection Day. On January 19 Bishop Edendon tried to mollify popular discontent about the Church's stance on burials with another proclamation. There was good news, the bishop declared. "[T]he Supreme pontiff . . . had . . . on account of the imminent great mortality, granted to all people of the diocese . . . a plenary indulgence at the hour of death if they departed in good faith." However, with piles of unburied bodies everywhere, secular Winchester had become impatient with religious bromides. A few days after the bishop's announcement, anger about the plague pit boiled over into violence. A group of townsfolk attacked a monk while he was saying a funeral Mass.

After the attack, the Church bowed to popular will and ordered the town's existing cemeteries expanded and new burial sites dug in the countryside. However, Bishop Edendon, who had one of those minds that have kept the Catholic Church in business for the last two thousand years, was not about to let the townspeople have the last word in the dispute. As part of the cemetery expansion, he announced that a plot of diocese land used by local merchants as a market and fair-grounds for over a century would be converted into a burial site. For good measure, the bishop also had the town of Winchester fined forty pounds for encroaching on diocese property.

Over the winter of 1348–49, as "every joy . . . ceased . . . and every note of gladness . . . hushed," perhaps half of Winchester died. The diocese, which included the neighboring county of Surrey, had one of the highest clerical mortality rates in England; 48.8 percent of the beneficed—or salaried—clergy in the diocese perished. Figures for the town of Winchester are less precise, but by 1377 a preplague population of eight thousand to ten thousand had shrunk to a little more than two thousand. Not all the missing eight thousand townspeople were plague fatalities, but one historian thinks a death figure of four thousand for Winchester is "conservative." Other parts of Hampshire county also became "abodes of horror and a very wilderness" over the first winter of the pestilence. In Crawley the plague carried off so many people, the village did not attain its preplague population of 400 again until 1851, five centuries later.

At Southampton, where the Italians came to buy English wool and the French to deliver wine, contemporary records indicate that as much as 66 percent of the beneficed clergy may have died in the first winter of the plague. Hayling Island, near Portsmouth, also suffered grievously. "Since the greater part of the . . . population died whilst the plague was raging," Edward III declared in 1352, ". . . the inhabitants are oppressed and daily are falling most miserably into greater poverty."

While some villages vanished entirely during the Black Death, one of the great English legends—that hundreds of villages were obliterated in the pestilence—has turned out to be partially myth. Recent research indicates that many of the "lost" villages actually succumbed to economic atherosclerosis; others, though given a final nudge into oblivion by the mortality, were already so weak economically, death was inevitable. However, the legend of the lost plague villages is not entirely untrue. Here and there in the green and pleasant English countryside, one comes across the odd ruin—the crumbling wall or overgrown path—that still echoes of the time when, everywhere upon the land, there was "death without sorrow, marriage without affection, want without poverty and flight without escape."

Undoubtedly, the average Englishman found the mortality as frightening as the average Florentine or Parisian, but a phlegmatic, self-contained streak in the English character kept outbursts like those at Winchester and Yeovil and quarantines like those at Gloucester relatively infrequent. For every English "son of perdition" overcome by fear, there were a dozen John Ronewyks: solid, undemonstrative men who ignored the danger and quietly went about their work. As one English historian has observed, "With his friends and relations dying in droves, . . . with every kind of human intercourse rendered perilous by the possibility of infection, the medieval Englishman obstinately carried on in his wonted way."

Ronewyk was reeve—or manager—at the Hundred of Farnham, one of more than forty manors owned by Bishop Edendon's diocese of Winchester. Little is known about John's personal life, but from contemporary documents we can piece together an idea of how he sounded and looked. Like modern German or Spanish speakers, medieval Englishmen pronounced their long vowels, so when John told the dairymaid at Farnham, "I have a liking for the moon," he would have said, "I hava leaking for the moan." John also would have been a pretty unprepossessing dresser; his wardrobe probably consisted of one copy of four basic items—breeches (underwear), hose, a shirt, and a kirtle, an all-purpose overgarment. On nights when he was planning a visit to the dairymaid, John's mother would have ironed his single shirt with a "sleck stone," a heavy, flat-sided object that she would warm by the fire first. Like many of his contemporaries, John probably slept naked. This common medieval habit must have made the work of *X. cheopis,* the rat flea, and *P. irritans,* the human flea, a good deal easier.

Farnham, the estate that John managed, was located in Britain's Champion country. The region, which stretches from southern Scotland to southern England, is one of most fertile areas in the world. Only the Ukraine and parts of western Canada and the United States have the same combination of good soil and clement weather. Today much of Champion country is buried under high streets and minimarts, but in John's time, the region was a vast golden sea of wheat and barley, interspersed with neat rows of trees and picturesque little villages of thirty to forty families. Each house was built almost exactly like its neighbor—thatched roof, timber frame, and walls made from wattle and daub—and each village was its own little island world in the vast, swaying grain sea.

In the summer of 1347—the summer before the plague—life in Farnham was better than it had been in decades. Bishop Edendon might have been an irascible character, but in the fourteenth century his well-managed estates were among the few places in England where crop yields were approaching thirteenth-century levels again. There was also more fresh meat and ale on village tables, and many of the vexatious old feudal obligations were beginning to disappear. In a way, 1066 was even being avenged. As mid-century approached, vernacular English, the language of ordinary folk like John, was replacing French, the language spoken by the Norman conquerors of 1066, and the official language of the English aristocracy and government. In the fourteenth century the scholar John Trevisa remarked that, these days, English aristocrats, many of Norman blood, knew "no more French than their left heel."

With several thousand acres and three to four thousand souls under his care, John Ronewyk was, in effect, managing a good-sized agribusiness. An eleventh-century book called *Gerefa* lists a reeve's principal duties as agricultural management and estate maintenance, but a man like John was also expected to know how everything worked on the manor and to help collect the rents, taxes, and fees due the lord. However, *Gerefa* had nothing to say about the problems that John began to encounter in the autumn of 1348, when the pestilence arrived at Farnham.

The timing of *Y. pestis*'s appearance is interesting. Farnham lies to the east of Dorset and Wiltshire, near the border between Hampshire and Surrey, yet the manor recorded its first plague deaths around the same time as these more westerly counties. The timing raises two possibilities; either Surrey was infected independently—possibly by sea—or else contemporaries like Bishop Edendon underestimated how quickly the plague spread along the coast. On October 24, while the bishop was issuing his "Rama" warning in Winchester, about twenty-five to thirty miles to the *east,* Farnham already had or was about to have its first two plague cases. The manor rolls for 1348 show that two tenants died in October. In November there were three more deaths, and in December, eight. In January the death toll fell to three, and in February to one. The combination of rain, cold weather, and the rapid dash along the wintry English coast had left *Y. pestis* winded. In June the mortality rate was still holding at one death per month, but then, in midsummer, when the fields were gold with wheat and the manor barns echoed with buzzing flies, *Y. pestis* revived. As July turned into August, "People who one day had been full of happiness, were the next found dead."

From the fall of 1348 to the fall of 1349, the first year of the pestilence, 740 people died at Farnham—a mortality rate of about 20 percent. As the second winter approached, people no doubt hoped for another reprieve, but this time, energized by the fresh summer air and country sun, the pestilence continued its killing ways through the cold weather. From the fall of 1349 to the fall of 1350, an additional 400-plus residents died. By the time *Y. pestis* finally left Farnham in early 1351, the mortality had claimed almost fourteen hundred people at the manor. Since we have only a rough estimate of Farnham's population, it is difficult to arrive at a percentage figure, but one scholar who examined the rolls estimates that more than a third and perhaps as much as a half of the manor died.

In many other parts of Europe, death on so vast a scale led to great chaos and social dislocation. But except for the wailing that echoed through the manor villages on soft summer nights and the number of people who wore black to church on Sundays, outwardly life at Farnham seemed to change very little. During the first year of the pestilence, the crops were harvested on time and in the usual amounts—Bishop Edendon's castle received its annual set of repairs, and the ponds got their annual refurbishment. John's friend, the dairymaid, even made her usual six cloves of butter (a clove equals seven to eight pounds); and after the harvest, the fortunate haymakers received a bonus of four bushels and four quarters of oats. At Christmastime the staff at the bishop's castle were feted with their traditional three holiday dinners.

Over the winter of 1348–49 Farnham experienced its first plague-related economic dislocations. As manpower shortages developed, labor costs exploded, while the price of farm animals plummeted. "A man could have a horse previously valued at forty shillings for half a mark, a good fat ox for four shillings, and a cow for 12 quid," wrote a contemporary. The sharp drop in animal prices was rooted in a provision of feudal law that came back to haunt estate owners during the pestilence. When a tenant died, the lord of the manor was entitled to the dead man's best beast of burden as a death tax. Usually the heriot, as the tax was called, was a boon for the lords, but so many peasants were dying in the winter of 1348–49, estates had more animals than they knew what to do with. John, a shrewd manager, decided to hold on to most of the livestock that had come into Bishop Edendon's possesion until prices firmed up, but he was the exception. At most manors, "heriot" animals were dumped onto the market, depressing prices.

Thanks to John's prudence and firm leadership, Farnham prospered during the first year of the plague. Receipts amounted to 305 pounds, while working expenses were only about 43 pounds.

The second year was harder. Death had become so pervasive, whole families were being obliterated now. Forty times that second year, the name of a deceased tenant was read aloud in manor court, and forty times no blood relation came forward to claim his vacant holding. As the "harshness of the days stiffen[ed] men's malice," the surviving tenants, John included, began to work two farms, their own and a dead neighbor's. In a time of death without end, marriage all but vanished. In 1349 there were only four weddings at Farnham.

In 1350 John Ronewyk had less of everything—money, good weather, and labor, which had now become ruinously expensive and very surly. "No workman or laborer was prepared to take orders from anyone whether equal, inferior or superior," wrote a chronicler. Despite obstacles unimaginable to the author of *Gerefa,* in 1350, as in the previous years, John got the crops in on time. In the third year of the pestilence, the harvest was less than it had been in 1349, but not substantially less. In 1350 John's friend, the dairymaid, even made her usual number of cheeses—26 in winter and 142 during the summer—and butter: eight cloves, or roughly fifty pounds.

Bishop Edendon must have marveled at John's ingenuity. In the midst of one of the bleakest years in all of English history, he organized a small army of tillers, masons, plumbers, carpenters, sawyers, and quarrymen. In 1350 the bishop's manor at Farnham got more than its usual annual refurbishment. Somehow John found the princely sum of twenty-two shillings and five quid to pay the workmen's inflated wages.

Social cohesion is a complex phenomenon, but applied gently—with a respect for the vast differences in time and place—the Broken Windows theory of human behavior may speak to the relatively low level of upheaval in Black Death England.

The theory, which informs much modern police work, holds that the physical environment buttresses the psychological environment the way a beam buttresses a roof. Why? Broken windows, dirty streets, abandoned cars, boarded-up storefronts, empty grass- and refuse-covered lots send the message: "No one is in charge here." And when authority and leadership break down, people become more prone to lawlessness, violence, and despair, in the same way that a defeated army becomes more prone to panic if the officers fail to provide resolute leadership.

England in 1348 and 1349 was hardly free of physical or emotional chaos, but enough John Ronewyks stepped forward—to harvest the crops, maintain the land and buildings, keep the records, man the courts—to convey the sense that the country was not slipping into anarchy, that authority was being maintained. Their steady leadership may have helped to sustain order, self-discipline, and lawfulness at a very difficult moment.

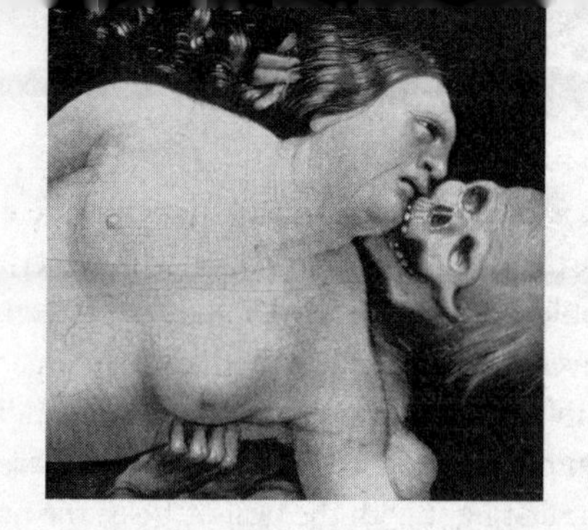

CHAPTER NINE

Heads to the West, Feet to the East

London, Early Fall 1348

A NIGHTTIME WALK ACROSS MEDIEVAL LONDON WOULD PROBAbly take only twenty minutes or so, but traversing the daytime city was a different matter. Crammed into the narrow mile between the malodorous Fleet River (on the western border) and the Tower of London (on the eastern border) were sixty thousand to a hundred thousand turbulent souls, and at least an equal number of noisy chickens, pigs, cows, dogs, cats, oxen, geese, and horses, as well as innumerable tumbrels and carriages. All this mayhem was compressed into lanes barely wide enough for a fat man to turn around in. The chronicler who described London as "among the noble cities of the world" may have been thinking about the capital's lovely walled gardens and church squares, but he was more likely simply deluded, since even in these islands of tranquillity the insistent din of medieval urban life was only yards away. The city sounds began at first light with the peal of bells, the mournful cries of beasts going to slaughter, and the creak of country carts rolling southward in the chilly dawn air toward Cheapside, London's premier commercial district. As the morning sun rose above St. Paul's, the walled city yawned, flung open its gates, and the bounty of the English countryside flowed into the capital from Cow Lane and Chicken Lane and Cock Lane in the northwest suburbs—a place of "pleasant flat meadows, intersected by running waters, which turn revolving mill-wheels with a merry din."

At the Shambles and at Butchers Row, immediately inside the Newgate entrance to the London wall, goods were triaged: butchers in bloodstained aprons took the large animals in hand for slaughter; other merchandise flowed directly onto Cheapside, a few hundred yards to the south. Imagine a shopping mall where everyone shouts, no one washes, front teeth are uncommon, and the shopping music is provided by the slaughterhouse up the road, and you have Cheapside, the busiest, bawdiest, loudest patch of humanity in medieval England. The street was home to more than four thousand individual market stalls, hundreds of musicians and beggars, countless rogues and scalawags, and innumerable gannocks, tapsters, and tranters—roving ale ladies. It may be true, as the chronicler claimed, that "the citizens of London were renowned beyond all others, for 'their fine manners,'" but none of those well-mannered Londoners lived in Cheapside.

As London's principal commercial center, Cheapside was also where people came to see and be seen. In the spring of 1348, a visitor might encounter eight-year-old Geoffrey Chaucer or Sir Walter Manny, one of Edward's great captains, taking a constitutional, or John Rykener, who would later achieve notoriety as a prostitute named Eleanor. According to a Corporation of London report, one night after the plague, Rykener "was detected in women's clothing . . . in a stall near Soper's Lane committing detestable, unmentionable and ignominious acts with a John Britby." Britby apparently mistook Rykener for a woman and when he discovered his mistake, concluded that no one was perfect.

If Cheapside pulsated with the bloody energies of the slaughterhouse, the smoky back streets of the capital throbbed to the gritty rhythms of industrial life. "They drive me to death with the din of their dints," a resident complained of the blacksmiths, who—along with the tanners and dyers, goldsmiths and silversmiths—produced much of the capital's manufacture. However, the much maligned smithies did make one important contribution to London life: the charcoal, wood, and newfangled sea coal they used in their work were aromatic, and foul London air was in desperate need of aromatic odors. Even by medieval standards, sanitary conditions in the city were appalling. Periodically, the Fleet River, the principal municipal sanitation dump, would be rendered impassable by refuse from the dozens of garderobes, or private outhouses, that lined the river banks like incontinent sentinels, and London's cesspools were so full, an unfortunate citizen, Richard the Raker, actually drowned in one. Edward III complained that "[t]he air in the city is very much corrupted . . . and most filthy stinks." However, the principal danger from London's refuse was not that it smelled, but that it attracted disease-bearing rats.

Next to Cheapside, the busiest place in London was the Thames riverfront. "To this city, from every nation . . . under Heaven, merchants rejoice to bring their trade in ships"—for once, the chronicler did not exaggerate. From a distance, the rows of ostrich-necked cranes and high-masted ships thrusting into the gray London sky above the harbor looked like a primeval forest; closer up, the wood and canvas thicket revealed itself to be inhabited by a sweaty tribe of reedy, foul-mouthed stevedores who swarmed across the docks unloading spices from Italy, wine from Gascony, silks from Spain, linen from France, and timber, fur, iron, and wax from Scandinavia. At night the harbor assumed yet another guise—it became the Kingdom of the Rat. While London slept, thousands of hungry rodents, their wet noses twitching in the cool night air, would follow the fetid London odors out of the stilled ships, across the docks, and into the darkened city beyond.

In 1348, once a southbound traveler crossed London Bridge, the only bridge over the Thames, he was out of the city and in Southwark, a squalid little suburb of narrow streets, small workshops, petty criminals, and alleyway sex. When London banned prostitution, the capital's sex workers relocated to Southwark, where they became known as "Winchester geese," in honor of the town's one architectural grace note, the Bishop of Winchester Park.

London's other major suburb, Westminster, lay about a mile to a mile and a half to the west of the bridge, and it was famous for its great abbey, and for Westminster Palace (the king's residence), as well as for being a sanctuary for wrongdoers. Since the eleventh century, when the village became the seat of the English Crown, Westminster had witnessed many dramatic moments, but none of equal gravity to that of September 1348, when the pestilence was rushing toward London inland from Bristol and Oxford and along the coast from Wiltshire and Hampshire. That September, one imagines scenes of high drama at Westminster Palace: Edward III and his ministers anxiously studying maps; clerks furiously scribbling orders; messengers scurrying from office to office; and arriving horsemen shouting out the latest news from the fronts at Hampshire and Bath and Winchester.

During the Great Mortality, England continued to be governed vigorously. The royal courts and the Exchequer (Treasury) remained open, tax collectors collected taxes, and the diligent king kept an eye on everything from the French to rising wages, which he froze in 1349 and again in 1351. However, Edward's initial response to the mortality lacked his characteristic boldness. September 1348 found the king in a moody, brooding silence. Probably the loss of Princess Joan, who had died earlier that month, weighed heavily on him, but one English historian, Professor William Ormrod, thinks that initially Edward—and his ministers—underestimated the dangers of the pestilence. During the fall, says Professor Ormrod, the government seems to have gone from one extreme to the other: from apathy and indifference to something akin to panic. In December Edward retreated to the countryside; shortly thereafter he sent up to London for his relics and ordered the parliamentary session scheduled for January 1349 canceled.

According to most contemporary accounts, the plague arrived in the capital on a rainy early November morning, but from whence it came remains unclear. Geoffrey le Baker, an Oxfordshire clerk, suggests the city was infected via the Bristol prong of the pestilence; in le Baker's account, the epidemic spreads eastward across the shires of south-central England to "Oxford and London." An invasion by way of Kent, the coastal county to the south of London, is also a possibility. However, since London seems to have been infected before the surrounding countryside, the most probable source of infection is the sea. A visiting ship may have deposited the plague bacillus on the Thames docks, and from there *Y. pestis* launched an all-points attack.

Surprisingly for the city of Shakespeare and Dickens, London produced no great plague chroniclers on the order of Agnolo di Tura or Giovanni Boccaccio. But Thomas Vincent's evocative description of London during a later outbreak of plague suggests what the city must have been like in the terrible winter and spring of 1349. "Now, there is a dismal solitude," wrote Vincent. "Shops are shut . . . people rare, very few walk about . . . and there is a deep silence in almost every place. If any voice can be heard, it is the groans of the dying, and the funeral knell of them that are ready to be carried to their graves." Daniel Defoe, who survived the Great Plague of London (1665) as a child, conjures up an even more terrible picture of daily life in the city. In some people, wrote Defoe, "the plague swellings . . . grew so painful . . . not able to bear the torment, they . . . threw themselves out of windows. Others, unable to contain themselves, vented their pain by incessant roarings. Such loud and lamentable cries were to be heard as we walked along the streets that would pierce the very heart to think of."

However, the only people who know firsthand what happened in Black Death London are the dead, and not long ago they were interviewed by a group of British archaeologists. In the mid-1980s, as rush-hour traffic whizzed by overhead, the archaeologists descended into a plague pit dozens of feet below the modern city. If a measure of a civilized society is the ability to bury its dead with dignity, then evidence from the plague pit suggests that civilization held in London.

The mixture of caskets, shrouds, individual graves, and trenches at the site indicates that on days when the dying was light, an effort was made to observe traditional burial rites; people got individual graves and some kind of funeral. Even on days when the death carts came back full and there was no time for ritual, bodies were not simply tossed willy-nilly into a pit. Some of the plague dead in the trenches were buried in caskets and shrouds, and everyone was laid out the same way: side by side, heads to the west, feet to the east. An effort may even have been made to segregate plague victims by age and gender. When archaeologists excavated the middle section of one trench, dozens of London children gazed up into the English sky for the first time in seven hundred years.

The charcoal and ash found in many coffins and shrouds also speaks to the order and organization of civilization. Since both ash and charcoal can help slow the putrefaction process, it may be that on days when the dying was heavy, rather than just toss corpses into a sea of elbows and knees and upturned buttocks, the grave diggers stockpiled plague fatalities for proper burial the following day. Another possibility is that corpses were preserved because a triage system was in operation. This occurred during the Great Plague of 1665, when the dead were transported across the city to cemeteries.

To determine how many Londoners perished, it is also necessary to interview the dead at another, more famous mortality burial site. In 1348 Ralph Stratford, the Bishop of London, "bought a piece of land called No Man's Land" northwest of the city, amid the "pleasant flat meadows" of West Smithfield. A year later, Sir Walter Manny, a famous veteran of the French wars, expanded the site by purchasing "thirteen acres and a rod adjoining . . . the said 'No Man's Land.'" The Smithfield site is, by far, the largest Black Death cemetery in London, but how large has been a matter of controversy for centuries.

Robert of Avesbury, who clerked for the Archbishop of Canterbury, claims that the pestilence "grew so powerful [in London] that between Candlemas [February 2, 1349] and Easter [April 12], more than two hundred corpses were buried almost every day in the new burial ground made next to Smithfield." A sixteenth-century historian named John Stow claims that, in his time, the cemetery bore the inscription, "A great plague raging in the year 1349 A.D., this churchyard was consecrated; wherein . . . were buried more than fifty thousand bodies of the dead." Smithfield cemetery has long since disappeared under urban sprawl, but, even assuming Stowe's memory was accurate, the figure of fifty thousand burials sounds astonishingly high. Assuming medieval London had a population of a hundred thousand—the high end of current estimates, once the plague dead in the city's hundred-plus regular cemeteries are included, London's overall mortality rate would have to have been in the 65 to 80 percent range—highly unlikely. Assume London had a population of sixty to seventy thousand—the low end of current estimates—and the city would have been virtually depopulated by August 1349.

Since medieval statisticians were given to terrific flights of fancy,* probably what the plague's author meant to say is that a great many people were buried at Smithfield. A recent estimate puts the cemetery's population at seventeen thousand or eighteen thousand and London's overall mortality rate at twenty thousand to thirty thousand, with thirty thousand being the more probable figure. If medieval London had seventy thousand souls, a reasonable estimate, that would mean a death rate of close to 50 percent.

Some historians believe the plague in London may have followed the pattern in Avignon—pneumonic in winter, bubonic in the spring and summer—though firm evidence on this point is lacking. Contemporary sources are more helpful on the question of who died. In London *Y. pestis* seems to have killed with egalitarian abandon, claiming no fewer than two archbishops of Canterbury, John Offord and his successor, Thomas Bradwardine, and numerous members of the royal

*Besides overly vivid imaginations, one other reason medieval statisticians were so inaccurate is that calculations were often done with Roman numerals. Try multiplying CCXLIV by MCIX, and you get the idea. Only after Arabic numerals gained favor did it become possible to add, subtract, divide, and multiply with ease and precision. (George Gordon Coulton, *The Black Death* [London, 1929], p. 29.)

household, including the king's physician, Roger de Heyton, and the wayward guardian of Princess Joan, Robert Bourchier, who had escaped the plague in Bordeaux only to die in London. In a fit of antiunionist frenzy, the pestilence also struck down the leaders of many of the city's powerful trade guilds, including eight wardens from the Company of Cutters, six wardens from the Hatters Company, and four wardens from Goldsmith Company.

The plague also claimed twenty-seven monks at Westminster Abbey, and the number would have been twenty-eight had not the hot-tempered, disagreeable Abbot Simon de Bircheston fled to his estate in Hampshire, to no avail. During its sweep through coastal England, the plague stopped in Hampshire and killed him.

In the waning months of the pestilence, Cheapside emptied and the Shambles fell quiet as farmers began to boycott the capital for fear of infection. There were now so few people, even with the boycott, the second Horseman of the Apocalypse, Famine, could gain no purchase. A 1377 poll tax assessment put the population of the postplague capital at thirty-five thousand.

Had the assessors examined London's moral state, they would have found that it, too, had fallen precipitously. John of Reading, a Westminster monk, observed that in the years following the pestilence, priests, "forgetful of their profession and rule, . . . lusted after things of the world and of the flesh." The cackling Henry Knighton noted that many highborn women "wasted their goods and abused their bodies." A similar moral decline was evident elsewhere in postplague Europe and the Middle East. "Civilization," noted the Muslim scribe Ibn Khaldun, "both in the East and West was visited by a destructive plague. . . . It swallowed up many of the good things . . . and wiped them out. . . . Civilization decreased with the decrease of mankind. . . . The entire world changed."

In *On Thermonuclear War,* one of the most exhaustive studies ever conducted on the effects of nuclear war, strategist Herman Kahn states, "Objective studies indicate that even though the amount of human tragedy would be greatly increased in the postwar world, the increase would not preclude normal and happy lives for the majority of survivors." The aftermath of the plague suggests that Dr. Kahn's assessment of postapocalyptic life may be half right. Survivors of the mortality did indeed rebuild their lives and their societies, but as the poem "The Black Death of Bergen" observes, the memory of what they had endured never left them:

Sights that haunt the soul forever
Poisoning life til life is done.

Writing in the aftermath of World War I, James Westfall Thompson, a University of Chicago psychiatrist, noted several parallels between the Lost Generation of the Great War and the generation that lived through the Black Death. "The superficial yet fevered gaiety, the proneness to debauchery, the wild wave of extravagance, the gluttony—all these phenomena [are] readily explicable in terms of the shock and trauma of the Great War," declared Dr. Thompson, and all have parallels in the behavior of the plague generation.

East Anglia, Spring 1349

On a map, the eastern coast of England forms a reasonably straight line down from Yorkshire to the Wash, a large bay on what the Victorians used to call the German Ocean (otherwise known as the North Sea). Below the Wash, the coast suddenly boils out like a head bursting though a wall; that cartographical illusion is East Anglia.

Perhaps because the ocean and sky always seem to beckon, the region has long been a departure point for the restless. In the seventeenth century, colonists from Norfolk and Suffolk, East Anglia's two counties, helped to settle New England, bringing with them not only many local place-names, including Yarmouth, Ipswich, Lynn (Massachusetts), Norwich, and Norfolk (Connecticut), but the speech patterns that gave rise to the Boston accent. However, well before East Anglia discovered the New World, it discovered how to make an unpromising environment blossom. In the years before the plague, the peasants of fertile Champion Country watched in awe as their counterparts who lived along the German Ocean transformed a land of light sandy soils, moody skies, and small, uneconomic farms into the most densely cultivated region of England.

In the fourteenth century, East Anglians who did not farm made cloth, the major industry in the region. In hundreds of towns and villages across Norfolk and Suffolk, the *thwack!* of the fuller and his "stocks" echoed from dawn to dusk. Charged with cleaning and thickening wool before it was spun, the fuller borrowed his techniques from wine making and the Inquisition. A fuller spent half the day in a trough of water, jumping up and down on a pile of wool; the other half, beating the wool senseless with a wooden bar—the "stocks"—until it surrendered the necessary degree of spotlessness and thickness.*

By the eve of the plague, East Anglia had grown into the most populous area of England, and its leading commercial center, Norwich—the suffix "wic," as in Nor*wic*h and Ips*wic*h, is an ancient designation for trading place—had become the second city of the realm, with a population of perhaps twenty thousand, many of them descendants of peoples who lived on the other side of the German Ocean. In Roman times, the fierce Saxons invaded so often, legionnaires called the region the Saxon Shore, and the Saxons were followed in the ninth and tenth centuries by the even fiercer Vikings. But the fiercest conqueror in East Anglia history came not by longboat, but by tumbrel or cart or saddlebag. Sometime around the Ides of March 1349, *Y. pestis* came up the road from London, and by the time it left, East Anglia, like Florence and Siena and Avignon, had experienced the equivalent of a thermonuclear event.

Though many regions of England suffered grievously in the mortality, it is hard to dispel the notion that, in the east of England in general and in East Anglia in particular, the suffering achieved a horrible new intensity. It was almost as if, unhinged by the bloodletting in the narrow, fetid lanes of Bristol and London and Winchester, *Y. pestis* had forgotten the first rule of survival for an infectious disease: leave some survivors

* By the time of the mortality, the water mill was beginning to do much of the fuller's work.

behind to carry on the chain of infection. In most of England, plague mortalities seem to have ranged from 30 to 45 percent; in the counties along the German Ocean, the average may have been nearer 50 percent, and in some places along the coast, higher. Dr. Augustus Jessop, the Victorian historian, author of what still remains the most comprehensive examination of East Anglia, wrote that in the "year ending 1350, more than half the population . . . was swept away. . . . [And] if any one should suggest that *many more* [italics added] than half died, I should not be disposed to quarrel with them."

By 1377 the population of Norwich had shrunk from a preplague high of about twenty thousand to under six thousand. As in Winchester and London, not all the missing residents died of plague, but the holocaust in the city was so great, for centuries after, the Black Death haunted civic memory. In 1806 a historian wrote that in 1349 Norwich "was in the most flourishing state she ever saw and more populous than she hath been ever since." Great Yarmouth, the leading seaport of East Anglia, also bore the plague's scars for centuries. You can almost hear the wind whistling through the empty streets in a sixteenth-century report prepared for Henry VII. "Most . . . of the dwelling places . . . of [Yarmouth]," wrote the authors, "stood desolate and fell into utter ruin and decay."

The fact that the valley of the Stour in lower Suffolk was among the first places in East Anglia to be struck supports the notion of London as the source of infection. The capital is only forty to forty-five miles to the south. While those miles were a good deal longer in 1349 than they are today, it still seems odd to find the peasants of Conrad Pava, a medium-sized estate in the valley, arguing about land and dowries—as manor court records show—as the year opened. People must have been thinking about pestilence. Indeed, with London so close, they were probably thinking of little else. Perhaps that bleak January, residents found comfort in arguing about the traditional issues of manor life.

By the time the manor court convened once again in March, the mortality had become impossible to ignore. The names of nine plague victims—six men and three women—are entered in court records.

Such a large number of deaths in so short a time must have given rise to hopes that the worst was over, but the worst had not even begun. On May 1, when the manor court met for its third session of the year, fifteen new deaths were recorded—thirteen men and two women. Seven of the deceased left no heirs; at Conrad, as at Farnham, whole families were obliterated. In the summer of 1349, while London buried the last of its dead, the mortality peaked on the manor, producing more victims. On November 3, at the last court of the year, thirty-six new deaths were recorded; this time thirteen of the deceased left no heirs. In six months, twenty-one families on a manor of perhaps fifty families had been wiped out.

In April, in the little village of Heacham in Norfolk, near the western coast of the German Ocean, the pestilence intruded on Emma Goscelin's life with the randomness of a stray bullet. A month earlier, as *Y. pestis* was traveling north under the remains of a late winter sky, Emma and her husband, Reginald Goscelin, were engaged in a bitter dispute over Emma's dowry. The court rolls are unclear about the cause of the dispute—perhaps Reginald was a wastrel who squandered Emma's money on the local ale ladies. Whatever he did, Emma was mad enough about it to take him to court. The case of Goscelin v. Goscelin was scheduled to be heard at manor court in Heacham on April 23, 1349, and Emma was not planning to attend court alone. Records indicate that several witnesses had agreed to testify on her behalf. That spring, if Reginald probably thought his life could not possibly get any worse, he was wrong. On April 23, Emma had to tell the court that the errant Reginald was dead, as were all her witnesses.

In Norwich, the epicenter of the storm, the dead quickly began to outnumber the living. Think of the survivors, writes Dr. Jessop, with only a touch of Victorian hyperbole, "threading [through] the filthy alleys, . . . stepping back into doorways to give the death carts passage, . . . [being] jostled by lepers and outcasts." Think, he continues, of the city's cemeteries: "Tumbrels discharging their load of corpses all day long, tilting them in huge pits made ready to receive them; the stench of putrefaction palpitating through the air . . . [people]

stumbling over the rotting carcasses . . . breathing all the while the tainted breath of corruption."

As elsewhere in England, in East Anglia the social order held, although there was enough postapocalypse lawlessness to give point to G. B. Neibuhr's assertion that in "times of plague . . . the bestial and diabolical side of human nature gains the upper hand." Probably no one better illustrates Neibuhr's point than William the One-Day Priest, an errant cleric who robbed six days a week and celebrated Mass on the seventh. Among those to run afoul of One-Day William was Matilda de Godychester, who was relieved of her purse and a ring in Epping forest. Later, Matilda told a court she was happy to escape with her life. Also active in the postplague era was the con artist Henry Anneys, whose specialty, tax avoidance schemes, would make him right at home today. One day in the early 1350s, Henry showed up at the door of Alice Bakeman, no paragon of virtue herself. Hearing that Alice wanted to avoid paying a heriot, or death tax, on some inherited property, the silky Henry proposed a trade—one of his best tax schemes for one of Alice's best milking cows. Henry got his cow, and Alice her scheme, but, alas—and probably predictably—the tax authorities saw through the scheme and Alice had to pay the heriot.

William Sigge was as low as Henry Anneys was cunning. William's crimes included stripping lead from one dead neighbor's roof, stealing pots and pans from another dead neighbor's cottage, and altering the boundary of a third dead neighbor's farm, so as to extend his own property. By all rights, Catherine Bugsey, who also preyed on the dead, should have been dead herself, at least ten times over; Catherine's specialty was stealing clothing from plague victims. But when arrested in her latest acquisition, a leather jerkin, Catherine was the picture of health.

After the sixth-century Plague of Justinian, the historian Procopius observed, "Whether by chance or Providential design, [the pestilence] strictly spared the most wicked."

In East Anglia, history seems to have repeated itself.

* * *

Few aspects of mortality scholarship are more fraught with controversy than the question of whether priests died in greater or lesser numbers than the general population. Some historians believe that the clerical mortality was higher because priests as a group were older, and if they performed their duties conscientiously, more likely to be exposed to risk. Other scholars believe that since clerics were better fed and housed, they may have died in slightly smaller numbers than the general population. Even if we take the second position, the ecclesiastic death rates in the county of Lincoln, to the north of East Anglia, are so high, a general death rate of 55 percent for the county seems probable. In the city of Lincoln alone, 60 percent of the beneficed—or salaried—clergy died; in the village of Candleshoe, 59 percent; in Gartree, 56 percent; while another village, Manlake, had one of the highest ecclesiastical death rates in England: an astonishing 61 percent.

Despite the losses sustained by the clergy, the plague weakened the authority and prestige of the institutional Church. To some degree, this was a by-product of disillusionment. For a thousand years, the Church had presented itself as God's representative on earth. Yet the universal pestilence had shown it to be as powerless, as far from God's favor, as every other institution in medieval society.

Leading clerics attempted to rationalize away the Church's impotence by depicting the Black Death as an opportunity for salvation. "Almighty God uses thunder, lightning and other blows . . . to scourge the sons he wishes to redeem," declared the ever blustery Bishop Ralph of Shrewsbury. Another common rationalization was to depict the plague as a necessary and just punishment for a wicked humanity. "O ye of little faith . . . Ye have not repented of your sins . . . therefore I have sent against you the Saracens and heathen people, earthquake, famine, beasts . . . etc. etc." warned the Heavenly Letter, one of the most widely circulated public documents of the Black Death. But none of the rationalizations was entirely successful. Europe emerged from the plague still a believing society, but after a four-year journey through the heart of darkness, people did not believe in quite the same way they had before.

Neither was the standing of the Church helped by a penchant for blaming the victim, a habit particularly pronouced among the English clergy. "Let us look at what is happening now," declared the Bishop of Rochester. "We [English] are not stable in faith. We are not honorable in the eyes of the world—on the contrary, we are, of all men, the falsest and in consequence, not loved by God." Henry Knighton could not have agreed more, though, in Friar Knighton's view, it was tournament groupies that brought down God's wrath against the English. The plague, he wrote, was a consequence of the bands of beautiful young women who corrupted public morals by attending tournaments in provocative dress. Admittedly, Knighton was a great crank—upon observing that the pestilence had killed 140 Franciscans in Marseille, he could not resist adding, "And a good job, too!" But even the normally sober John of Reading became a little unhinged on the subject of English tomfoolery. "And no wonder," John declared, in a passage describing the plague's arrival, "given the empty headedness of the English, who remained wedded to a crazy range of outlandish clothing without realizing the evil that would come of it."

Another important factor in the Church's decline was the post-plague state of the clergy, and, again, this trend was particularly pronounced in England. After the Black Death, there were far fewer priests to comfort or minister to the laity, and since many talented clerics had died, ecclesiastical leadership deteriorated. "At that time," wrote Knighton, "there was such a dearth of priests that many churches were left without the divine offices, Mass, Matins, Vespers, sacraments and sacramentals." Compounding the problem of thinned clerical ranks was the greed of many survivors. "One could hardly get a chaplain to serve a church for less than ten pounds or ten marks," says Knighton. ". . . [W]hereas before the pestilence, when there were plenty of priests, anyone could get a chaplain for five or even four marks." The low quality of the new clerical recruits also produced disillusionment. Many of the replacements were either very young and ill trained, as in Norwich, where sixty clerks, "though only shavelings" were pushed into vacant rectories, or equally ill trained middle-aged men, many widowers who had no real vocation.

In many cases, too, the clergy became swept up in the "Lost Generation" spirit. In the postplague years, ecclesiastical discipline slackened and holiness declined. One Franciscan chronicler complained that "the monastic orders, and in particular the mendicants, began to grow both tepid and negligent, both in that piety and that learning in which they had up to this time flourished."

However, it was the behavior of priests during the plague, not after it, that may have had the most negative effect on the Church. This may seem odd, given the 42 to 45 percent mortality rate among English parish priests, but there may be a ready explanation. While most clerics remained at their posts, many clerics performed their duties in a less than heroic manner. "The picture one forms," writes Philip Ziegler, ". . . is that of a clergy doing its daily work but with reluctance and some timidity, thereby incurring the worst of the danger but forfeiting the respect it should have earned. Add to this a few notorious examples of priests deserting their flocks . . . and some idea can be formed of why the established Church emerged from the Black Death with such diminished credit."

North of England, Spring 1349

Above East Anglia, England constricts, as if concentrating its energies for a collision with the rough, borderland Scots. In theory, this narrowness, which keeps the sea very near in the north, should have made the region vulnerable to a waterborne assault, but, initially at least, *Y. pestis* seems to have traveled north overland, perhaps with a group of refugees from London, perhaps in a tumbrel of grain. All that is known for certain is that on a dull May morning in 1349, as a spring sun rose above the cathedral, York, the leading city of the north, with a population of almost eleven thousand, began to die.

The northern counties, Lancashire (on the west coast), Yorkshire (on the east), and above them, Cumberland and Durham, had a long time to contemplate their fate. It took the pestilence ten months to reach the region. In the meantime, residents had little to do but plow the fields, listen to rumors from the south, and contemplate the words of William Zouche, Bishop of York. "Almighty God," thundered the bishop, "sometimes allows those whom he loves to be chastened so that their strength can remain complete by the outpouring of grace in a time of spiritual infirmity."

Over the winter and spring of 1349, as the rest of England writhed in agony, nature, as if expecting an important new houseguest, was busy preparing the north for the arrival of *Y. pestis*. On the last day of 1348, a winter flood submerged the western parishes of York; then, a few days before Passion Sunday, an earthquake rocked the Abbey of Meaux, also in Yorkshire. If modern experience is any guide, both events may have facilitated *Y. pestis*'s work by disrupting rodent habitats and sending local rats fleeing toward human settlements.

Compared to the urban areas of Lincoln and Norfolk, the city of York escaped relatively lightly. Clerical losses amounted to 32 percent, 10 percent or more below the national average for parish priests and almost 30 percent below those in Lincoln. The Abbey of Meaux was not so fortunate. The plague claimed forty of its fifty monks and lay brothers, including six victims—among them, the abbot—on the single terrible day of August 12, 1349. While abbeys, with their large complement of unwashed bodies, undiscarded food, and dank corridors, were magnets for rats and fleas, the chronicler at Meaux seemed inclined to think that evil portents were also at work at the monastery. In particular, he mentions the recent death of a pair of Siamese twins, who were "divided from the navel upward . . . and sang together very sweetly" in nearby Kingston-on-Hull. According to the chronicler, "a short time before the pestilence" the twins had died the saddest of imaginable deaths. When one passed on, "the survivor held it in its arms for three days." And given what happened in Hull, perhaps the twins' deaths were a portent.

Edward III was abnormally unforgiving about taxes—in one case, he sent seven tax collectors to a post, until finally he found one the plague couldn't kill—but the pestilence left the twins' native Hull in such a state of depredation, even the king was moved to pity. "Considering the waste and destruction which our town of Kingston-on-Hull has suffered," he decided to remit certain taxes back to Hull.

Despite York's relatively low clerical mortality rate, overall the county of Yorkshire's losses were in line with the 40 to 50 percent national average. To the west, *Y. pestis* may have been more lethal. In the fourteenth century Lancashire, which borders the Irish Sea, was one of the most thinly populated regions of England; but a postplague survey of ten local parishes came up with a mortality rate of more than thirteen thousand. In Derbyshire to the south, the most eloquent set of mortality statistics are in a small parish church where a plaque commemorates the Wakebridge family's brush with annihilation in the summer of 1349.

18 May, Nicholas, brother of William
16 July, Robert, brother of William
5 August, Peter, father of William and Joan, sister of William
10 August, Joan, wife of William and Margaret, sister of William

William himself survived the pestilence.

Another William, the enterprising William of Liverpool, seeing opportunity where others saw only misery, concluded that 1349 was the perfect year to start a funeral business. Documents from medieval Lancashire say that William "caused one third of the inhabitants of Everton [a Lancashire town] to be brought to his house after death," presumably to be buried at a price.

Durham and Cumberland, the two most northerly counties in England, were accustomed to random death. For more generations than anyone could remember, they had served as the frontline in a series of predatory wars against the Scots, whom the expansionist English could defeat but never quite conquer. In 1352 war-weary and plague-devastated Carlisle, the principal city of Cumberland, was forgiven its taxes, because, according to a royal decree, the town was "more than is usual depressed by the pestilence." In Durham, a wave of unrest seems to have swept through the county in the summer of 1349. There are reports of peasants refusing to pay fines and refusing to take on the holdings of deceased tenants, but it is unclear whether the refusals were isolated acts or part of some larger, organized movement. Lacking reliable data, we have to deduce morale in the borderlands by an image. It is of a mad, lone peasant who, in the years after the plague, wandered the villages and lanes of the region, calling out for his plague-dead wife and children. The man is said to have greatly upset the populace.

The Scots, still laboring under the impression that the plague was an English phenomenon, were enjoying themselves immensely in the summer of 1349. "Laughing at their enemies . . . [and] swearing 'by the foul death of England,'" in March 1350 they amassed a large army in the forest of Selkirk near the English border, "with the intention of invading the whole realm." But before the attack could be launched, "the revenging hand of God" reached across the border and scattered the gathering Scots with "sudden and savage death."

Henry Knighton enjoyed Selkirk even more than the 140 dead Franciscans in Marseille. "Within a short space of time," he wrote, "around five thousand [Scots] died and the rest, weak and strong alike, decided to retreat to their own country. But the English following, surprised them and killed many." The routed Scots took the plague home with them, but *Y. pestis* did not take as well to the rough terrain and cooler climate of the Highlands. A third of Scotland may have died, perhaps less. Whatever the exact mortality rate, it was lower than England's.

For the Welsh, who were already under the English boot, the pestilence brought an end to hope, but not to poetry. In the desperate spring of 1349, as *Y. pestis* lapped at the Welsh borderlands north of the Bristol Channel, the poet Jeuan Gethin wrote: "We see death coming into our midst like black smoke, a plague which cuts off the young, a rootless phantom which has no mercy for fair countenance. Woe is me of the shilling [bubo] in the arm pit; it is seething, terrible, where ever it may come, a head that gives pain and causes a loud cry, a burden carried under the arms, a painful angry knob, a white lump . . ."

Beyond the poetry, very little is known about the mortality in Wales, except that it affected both the English colonists—or "Englishry"—in the lowlands, and the natives—the "Welshry"—who lived in the mist-shrouded hills. We also know that, as the plague gathered strength, the Welsh countryside filled with hard men, like Madoc Ap Ririd and his brother Kenwric, who "came by night in the pestilence to the house of Aylmar after the death of the wife of Aylmar and took from the same house one water pitcher and basin, value one shilling . . . [and who also] stole three oxen from John le Parker and three cows, value six shillings."

In Ireland, which may have been infected in the late summer of 1348 by way of Bristol, contemporary accounts indicate that the pestilence did discriminate between foreigner and native. Geoffrey le Baker observed that in Ireland the plague "cut down the English inhabitants in great numbers, but the native Irish living in the mountains and uplands were scarcely touched." This observation is echoed in a 1360 report prepared for Edward III; its authors note that the pestilence "was so great and so hideous among the English . . . [but] not among the Irish." A seventeenth-century Irish historian, perhaps with the atrocities of the Cromwellians still fresh in mind, wrote that while the pestilence "made great havoc among the Englishmen . . . for those who were true Irish-men born and dwelling in hilly countries, it scarce just saluted them."

The Anglo-Irish tendency to cluster in seacoast towns probably did make them more vulnerable. The plague seems to have landed first on the east coast at Howth or Dalkey, two towns to the north and south of Dublin, then spread to the city itself. *Y. pestis* could not have selected better landing sites. Using the cluster of little villages and towns around Dublin as a human bridge, the pestilence quickly vaulted into the more sparsely populated hinterland. By December 1348, Kildare, Leath, and Mouth—three counties around the capital—were infected, and by late summer 1349 the plague was in Clare and Cork on the west coast.

One historian puts the death rate among the Anglo-Irish at 35 to 45 percent; the native Irish probably suffered less, though how much less is unclear. Whatever their death rate, there was more than enough suffering to go around in Ireland in 1348 and 1349.

The final sentence in the manuscript of John Clynn, the Irish monk who wrote "waiting among the dead for death to come," was written by another monk. It reads: "And Here it seems the author died."

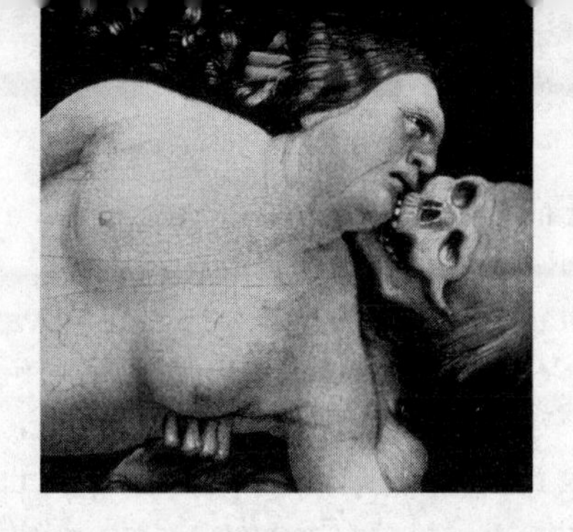

CHAPTER TEN

God's First Love

Lake Geneva, September 1348

On a September morning in 1348, as the little villages of southern England died in an autumn rain, a small vessel glided across the silvery blue surface of Lake Geneva. In the morning light, the sweeping expanses of sky, sea, and mountain around the boat looked like a backdrop for the Ride of the Valkyries, but there were no great-bodied Norse goddesses on the lake today, just some sleepy local burghers fortifying themselves against the morning chill with a flask of wine—and a surgeon named

Balavigny, who sat alone at the prow of the boat in a funnel-shaped *Judenhut*, a Jew's hat.

The Great Mortality occasioned one of the most vicious outbreaks of anti-Semitic violence in European history. The first pogroms in southern France in April 1348 had been traditional acts of Holy Week violence, but during the summer, as the pestilence swept across Europe with unbroken might, the attacks changed character. With fear and unreason everywhere, the Jews, accused of every other sin, were now accused of fomenting the pestilence. By mid-September, when surgeon Balavigny was arrested in the lakeside town of Chillon, the well-poisoning accusation had grown into a conviction—and the conviction into an international Jewish conspiracy, complete with a mastermind, a sinister Spanish rabbi named Jacob; an army of secret agents; and a goal so evil, it filled every Christian heart with fear and trepidation. The Jews were contaminating the wells because they sought world domination.

Over the summer, the conjunction of anti-Semitism, paranoia, and medieval detective work had produced a detailed description of the plot and of the Jewish poison distributed by Rabbi Jacob's agents, including its packaging and the way it worked. According to one plotter, "if anyone suffering the effects of the poison comes into contact with someone else, especially while sweating, the other person will be infected." In Chillon, where Balavigny was interrogated after his arrest, local authorities had also obtained information about the agents distributing the poison and the letter Rabbi Jacob sent to coconspirators. According to another plotter, the letter commanded the recipient, "on pain of excommunication and by the obedience he owed the Jewish law, to put the poison in the larger public wells . . ."

When the particulars of the plot were described to him at his first interrogation on September 15, surgeon Balavigny must have felt like Alice upon stepping through the looking glass, though Alice was never "put to the question." The phrase was a medieval euphemism for torture, and the interrogators at Chillon seemed to have regarded their work with Balavigny as a particularly outstanding example of the torturer's art. A note on the surgeon's transcript boasts that after only being "briefly put to the question" on the fifteenth, the surgeon confessed freely and fulsomely to complicity in the well poisonings, and that at a subsequent interrogation on the nineteenth, Balavigny disclosed the names of his coconspirators without being "put to the question" at all.

It is unclear how long after his interrogation the surgeon was taken across Lake Geneva to Clarens, the destination of this morning's boat ride, but not more than a week could have elapsed. Clearer is the purpose of the trip: earlier in the summer Pope Clement VI had vigorously denounced the persecution of the Jews. "Recently," he declared, ". . . it has been brought to our attention by public fame—or, more accurately, infamy—that numerous Christians are blaming the plague . . . on poisonings carried out by the Jews at the instigation of the devil, and that out of their own hotheadedness they have impiously slain many Jews, making no exception for age and sex." In such an atmosphere, Balavigny's jailers probably felt it prudent to acquire physical evidence of the surgeon's guilt. So thus it came to pass that on this brilliant September morning, while rainy London awaited the mortality, Friar Morellet counted the dead in Paris, and Matteo Villani wept bitter tears for his plague-dead brother Giovanni in Florence, surgeon Balavigny and his sleepy burgher guards set sail for Clarens in search of imaginary evidence for an imaginary crime.

One can only guess at Balavigny's thoughts as he sat huddled at the prow of the boat, watching the sun lick away the last of the morning mist like frosting from a cake, but the surgeon's state of mind cannot have been very different from Primo Levi's on arriving at Auschwitz on a desolate Polish morning seven hundred years later. "No human condition is more miserable than this," wrote Levi. ". . . They have taken away our shoes, our clothes, even our hair; if we speak they will not listen . . . [and] if they listen they will not understand." In the camps, Levi discovered that when a man lost everything, he often ended up "losing himself." If a measure of losing yourself is embracing the dementia of your tormentors, by the time surgeon Balavigny disembarked at Clarens, he had stepped through the looking glass. When asked if a village spring looked familiar, Balavigny replied, yes, "this is the spring where I put the poison." And when one of the burghers, a sharp-eyed notary named Henri Gerard, found a rag near the spring, the surgeon "confirmed that it was the . . . cloth in which the poison had been enclosed."

Three weeks after losing himself, the surgeon lost his life. In early October, Balavigny was burned at the stake.

We go
Do not ask: where?
We go
We have been told to go
From the days of fathers' fathers
Abram went, Jacob went,
They all had to go,
Go to a land, go from a land,
All of them bent
Over the full path of the farer . . .

The millenia-long wanderings of the Jews began as they ended, in holocaust. Between A.D. 66, when the Jews of Palestine rose against Rome in the Great Revolt, and A.D. 70, when the victorious imperial standard was planted atop the ruins of the Temple Mount, 1,197,000 Jews were slain or sold into slavery, according to Tacitus. Indeed, for a time after the Revolt, it is said that it was cheaper to buy a Jew than a horse in Rome. In A.D. 128 "nearly all of Judea was laid waste" again. For the second uprising, the historian Dio Cassius puts the butcher's bill at 985 villages and towns and 50 forts destroyed, 580,000 Jews killed in battle, and "countless numbers of others destroyed by starvation, fire and sword." Dio and Tacitus probably exaggerated Jewish losses, but not egregiously. The sixty years between A.D. 70 and A.D. 130 tore the heart out of Jewish Palestine. Visiting Jerusalem in the fourth century, St. Jerome found that the memory of those years still hung heavy upon the land. Of the Jewish remnant, Jerome wrote, "a sad people . . . decrepit little women and old men encumbered with rags and years, showing both in their bodies and dress, the wrath of the Lord."

Seven hundred years later, Benjamin of Tudela, a peripatetic gem merchant with a taste for adventure, picked up the story of the Diaspora. In 1183, Benjamin began a three-year odyssey through the Jewish communities of Europe and the Near East. He visited Constantinople, where he found that no Jew, no matter how wealthy, was allowed to ride a horse, "except for Rabbi Solomon, the Egyptian, who is the King's doctor." In Spain, Benjamin found Jews not only rode horses, but rode them with knightly panache, dressed like emirs in exquisite silks and many-gemmed turbans, served as ambassadors and administrators, and became renowned physicians, scholars, and philosophers. At Crisa on Mount Parnassus, the gem merchant came across a colony of straw-hatted Jewish farmers sweating under the Mediterranean sun; in Aleppo, Jewish glass blowers with exaggerated cheek muscles; in Brindisi, dyers with stained hands; and in Constantinople, tanners who polluted the streets of the Jewish quarter with the effluvia of their work. But on the whole, the Diaspora communities that Benjamin described in his *Book of Travels* were mercantile in character and often quite small.

In the general population collapse of the early Middle Ages, the Jews suffered disproportionately. From eight million in the first century—roughly 10 percent of the Roman Empire—their numbers fell to a million and a half by Benjamin's time.* Spain, home to the largest and most prosperous Jewish community in medieval Europe, may have had between a hundred and a hundred and fifty thousand Jews; but the devout and fierce Ashkenazi community of Germany barely numbered twenty-five thousand. However, wherever they lived, the Jews of the Diaspora were usually more prosperous and better educated than their Christian neighbors.

In *World on Fire,* a recent book on globalization, Yale scholar Amy Chua notes that in many modern third world countries, a small, skilled, nonnational elite often acts as a go-between to the global economy. In

* Like most statistics on the Middle Ages, this one comes in several variations. While a million and a half Jews is the consensus figure, estimates of the medieval Jewish population range as low as four hundred thousand (by the Italian scholar Anna Foa) and as high as two and a half million (by the historian Norman Cantor). (Anna Foa, *The Jews of Europe After the Black Death,* trans. Andrea Grover [Berkeley: University of California Press, 1992], p. 87; Norman F. Cantor, *In the Wake of the Plague: The Black Death and the World It Made* [New York: Free Press, 2001], p. 150.)

modern Southeast Asia, the overseas Chinese play this role; in modern sub-Saharan Africa, the Lebanese. In the early Middle Ages, when Christian literacy and numeracy rates were close to zero, the comparatively well educated Jews played an analogous role in Europe. "By virtue of their experience in commerce and superior knowledge in commodities, markets and monetary transactions, their versatility in languages and the dispersion of their coreligionists, Jews occupied a preeminent position in international trade," write scholars Mordechai Breuer and Michael Graetz. Indeed, so preeminent was the Jewish commercial role in the first half of the Middle Ages that many writs of privilege and ordinances contained the formulation *judaei et ceteri mercatores*—"Jews and other merchants."

In the ninth and tenth centuries, Jewish merchants could be found in the pepper markets of India, the silk markets of Samarkand and Baghdad, the slave markets of Egypt (where they sold pagan slaves called "Canaties"), and on the vast, empty expanses of the Silk Road, atop camels laden with jewels and spices. In a letter to his famous philosopher brother Moses, the merchant David Maimonides described the death of one these intrepid Jewish trader-travelers. "The greatest misfortune that has befallen me during my entire life, worse than anything else, was the death of [my colleague]," wrote David. "May his memory be blest; [he] drowned on the Indian Sea carrying much money belonging to me, to him and to others. . . . He was well versed in the Talmud, the Bible and knew [Hebrew] grammar well and my joy in life was to look at him."

If a trader's life could be hazardous, it could also be quite lucrative. In the first millennium, Jewish living standards were so far above the European norm, "the term 'dark early Middle Ages' . . . has as little applicability for the Jewish medieval period as for the Byzantine Empire," note scholars Breuer and Graetz. Commerce made some Jews not merely wealthy but fabulously rich. When Aaron of Lincoln died, a special branch of the English Exchequer (Treasury) had to be established to tabulate his fortune, and a French Jew named Elias of Vesoul, anticipating the Rothschilds, built a far-flung banking and commercial empire as early as the eleventh century. However, even men like Aaron and Elias lived the anxious, uncertain existence of outsiders. Upon Aaron's death, the English Crown seized most of his fortune; from Abraham of Bristol, another wealthy English Jew who was imprisoned in 1268, the Crown extracted one tooth a day until Abraham agreed to deposit ten thousand silver marks in the royal coffers.

The anti-Semitism that cost Abraham his teeth was grounded, first of all, in theology. The early Church fathers held the Jews culpable of so many sins that between the third and eighth century, a new literary genre, *Adversus Judaeos,* was created to describe them all. An early example of the genre, *An Answer to the Jews,* accused the Jewish people of forsaking God and worshipping false images; another early example, *Rhythm Against the Jews*, of trading God the Father for a calf and God the Son for a thief. Other works in the *Adversus Judaeos* tradition included *On the Sabbath, Against the Jews,* which accused the Jews of grossness and materialism; *Eight Orations Against the Jews,* which likens the Jew to a stubborn animal spoiled by kindness and overindulgence; and *Demonstration Against the Jews,* which equates Jerusalem with Sodom and Gomorrah. *Homilies Against the Jews* declares uncircumcised Gentiles the new Chosen People, a point also made by *Books of Testimonies Against the Jews,* though the latter work makes the point with more literary grace. Using the parable of Jacob and his two wives, *Testimonies* presents the elder wife Leah, with her defective eyes, as the embodiment of the synagogue, and the beautiful young wife Rachel as the symbol of a Church Triumphant. *Contra Judaeos* also uses biblical imagery, although in this case to get at the core of the theological argument with the Jews. *Contra* presents Cain as the symbol of the Jewish people, and Abel, the brother he slew, as the symbol of Christ.

In its mildest form, early medieval Christianity expressed its theological grievances against the Jews as a complaint: the Jewish people rejected Christ, the Light and the Way. In a sterner formulation, the grievances acquired the menace of accusation: although recognizing Christ's divinity, the Jews rejected Him because He was poor and humble. And in their most vitriolic form, the grievances became a bill of indictment for murder: Jews were Christ killers.

Political and social factors have also helped to fuel anti-Semitism through the ages. Thus, in the decades after Christ's death, Christian Jews, eager to separate their new religion from its temple roots, launched an attack on their orthodox counterparts, an attack that grew increasingly more expansive as the decades passed. Thus, in Mark, the earliest gospel, written around A.D. 68, Satan is associated with the scribes. In Luke, written ten years later, the "evil one" becomes affiliated with a wider segment of Jewish society, but the target is still individual groups like "the chief priests and the captains of the temple." But by the time John writes, around the year A.D. 100, Satan's allies have become simply "the Jews." The phrase "the Jews" appears seventy-one times in John, compared to a total of sixteen times in Matthew, Mark, and Luke.

The Jewish faithful often responded to Christian Jews in kind. "May the *minim* [the heretics] perish in an instant; may they be effaced from the Book of Life and not be counted among the just," goes the *Shemoneh Esrei* prayer. The orthodox establishment also disparaged Christ as the illegitimate child of a Roman solider named Panthera, denounced His miracles as tricks and the Resurrection as a hoax.

As Christianity became non-Jewish, religious rivalry replaced intragroup conflict as an engine of anti-Semitism. During the early Middle Ages especially, Church authorities became alarmed at the number of Christians who were attracted to Jewish teachings. John Chrysostom, hammer of the "Judaizers"—Christians attracted to Jewish teachings—declared, "I know that many people hold a high regard for the Jews and consider their way of life worthy of respect at the present time. This is why I am hurrying to pull up this fatal notion by the roots. . . . A place where a whore stands on display is a whorehouse. What is more, the synagogue is not only a whorehouse and a theatre; it is also a den of thieves and a haunt of wild animals." Another seminal anti-Semite, the ninth-century bishop Agobard of Lyon, believed Christians who broke bread with Jews risked spiritual seduction. Agobard lived just long enough to see one his most paranoid fantasies realized. On a trip to Rome in the 820s, Bodo, father-confessor to Louis the Pious, Charlemagne's son and successor, fled to Spain, converted to Judaism, and married a Jewess.

Of the impious Bodo, Agobard's successor at Lyon, the dyspeptic Archbishop Amulo, thundered, "Now he lives in Spain, . . . his bearded figure squats in the synagogues of Satan and joins with other Jews in blaspheming Christ in His Church."

In the centuries leading up to the Black Death, anti-Semitism also became a useful tool for financiers and nation builders. In 1289 English-controlled Gascony expelled its Jews and seized their property. The following year, 1290, the English Crown turned on native Jews. Edward I, grandfather of Edward III, ordered the Jews of England expelled and their goods confiscated, although, having long been a favorite target of the English Exchequer, the Jews did not have much left to confiscate. In the mid-thirteenth century, when the treasury shook down Aaron of York, it got more than 30,000 silver marks; by the expulsion of 1290, together the Jews of eleven leading English towns could barely raise a third that sum.

In France, where feelings against Jews traditionally ran high, the monarchy used a policy of expulsion to win popular support and to enrich itself. Thrown out in 1306, the Jews were readmitted in 1315, expelled again in 1322, brought back again in 1359, and expelled again in 1394.

One day in the late fourth century, a woman stood on a pier in Carthage and, "wild with grief," watched as a ship slipped over the horizon, taking with it everything she loved and ever would love. The woman's name was Monica, and there was more than a touch of the domineering matriarch about her. It would be an exaggeration to say that St. Augustine would never have become St. Augustine without the overbearing, suffocating Monica, but had she not been so inescapable and controlling, the dissolute young pagan might have wallowed in the fleshpots of Milan for another decade before embracing Christianity in 387.

Like Churchill, another man with a difficult mother, Augustine was also a nonstop talker; his words, recorded by an ever-present staff of copyists, grew to fill nearly a hundred books, including two works of historic importance. These are the autobiographical *Confessions,* where the sound of a personal voice can be heard for one of the few times in the Middle Ages—"I closed her eyes and a great sorrow surged through me," Augustine wrote of his mother's death—and *City of God,* which helped to define Christian policy toward the Jews for nearly a millennium. When the eighteenth-century philosopher Moses Mendelssohn exclaimed that, without Augustine's "lovely brainwave, we [Jews] would have been exterminated long ago," he was referring to *City of God*.

While *City,* and Augustine's other "Jewish" writings, rehearse all the familiar arguments of Christian anti-Semitism, including the Jews' unwillingness to recognize the divinity of Christ, Augustine's novelty was to add a "but" at the end of the traditional indictment. In Augustine's vision, the Jews had a divinely appointed role; God intended them to "bear witness" to a Christianity Triumphant.* And since the Jews had to remain Jews to fulfill the role, the Augustinian "but" amounted to a ticket to survival—the only such ticket early Christianity issued to a dissident minority. As Jacob Neusner has noted, "Judaism endured in the West for two reasons. First, Christianity wanted it to endure and, second, Israel, the Jewish people, wanted it to. The fate of paganism in the fourth century shows the importance of the first of the two factors."

During the nearly seven-hundred-year ascendancy of the Augustinian "but," the virus of anti-Semitism remained in an attenuated form. Even vicious anti-Semites like Agobard of Lyon rarely talked of mass conversion, mass expulsion, or mass extermination. The ninth and tenth centuries were a period of relative peace and prosperity for European Jews, particularly in Spain and Germany, where immigrants from northern Italy established the first Ashkenazi settlements. Indeed, Louis the Pious, leader of the Carolingian Empire, the largest empire of its day, was renowned as a friend of the Jews, as was his father Charlemagne. However, the human mind can only hold two contradictory thoughts for so long. Consequently, after the turn of

* Augustine imagined that, seeing the prophecies of the Old Testament fulfilled, at End of Days, the Jews would embrace Jesus on their own.

the millennium, the complex Augustinian formulation, "Hate the Jews, respect the Jews" gave way to the simpler formulation, "Hate the Jews." In 1007 there were persecutions in France, and in 1012 forced conversions in Germany; then, in 1096, an apocalypse. For centuries afterward the names of the Jews slaughtered during the Crusader pogroms* of 1096 would be read aloud in European synagogues on Saturday mornings.

The pogroms originated in Rouen. Shouting, "We depart to wage war against the enemies of God, while here in our very midst dwell the . . . murderers of our Redeemer," a group of Crusaders ran through the streets of the town, slaying Jews. In Speyer and Cologne, resolute action by a local bishop managed to avert wholesale slaughter, but in Mainz, which had a weak bishop and an unsympathetic citizenry, the carnage was terrible. As a Crusader force breached the town wall, the local Jewish community gathered in the courtyard of the bishop's palace. Rabbi Solomon bar Simson describes what happened next. "In a great voice, they all cried, 'We need tarry no longer, the enemy is already upon us. Let us hasten to offer ourselves as a sacrifice before God. . . . The women girded their loins with strength and slew their own sons and daughters and then themselves. Many men also mustered their strength and slaughtered their wives and children and infants. The most gentle and tender of women slaughtered the child of her delight. [Then] they all arose, men and women alike, and slew one another. . . . Let the ears hearing this and its like be seared, for who has heard or seen the likes of it . . ." In Worms, the Jewish community recited the ancient Shema prayer: "Hear, O Israel, the Lord, our God, the Lord is One," as they fell beneath the Crusader swords. Afterward the dead were stripped naked and dragged away.

In Trier, where the Torah was defaced and trampled, a pretty young Jewess taunted the marauding Crusaders. "Anyone who wishes to cut off my head for fear of the Rock, let him come and do so." To emphasize the point, the young woman stuck out her neck in defiance.

* It is perhaps not entirely coincidental that a favorite motif of Nazi propaganda art was Hitler dressed in gleaming Crusader armor.

Earlier, two local Jewish leaders had been killed for a similar taunt. But, according to a Jewish chronicler, the young woman "was comely and charming . . . and the uncircumcised ones did not wish to touch her." The taunter was told she would be spared if she agreed to convert, but, fierce in her faith, like the Jews of Mainz, she chose a suicidal martyrdom.

The pogroms in Mainz, Worms, and Trier were an early expression of a new, more militant Christianity. The *Civitas Dei*—or God State—grew out of the wave of intense pietism that swept through Europe in the Central Middle Ages. The new state's controlling metaphor was the body: just as its various limbs fit together into an organic whole, so, too, does—or should—Christian society. Inspired by this corporatist vision, the angry sword of orthodoxy struck out at dissident minorities, such as the Albigensian heretics of southern France and the Jews. Many aspects of modern anti-Semitism date from the period of the *Civitas Dei*.

For example, the menacing figure of the hook-nosed Jew, whom Chaucer described as "hateful to Christ and all his company," first appears in twelfth-century paintings of the Crucifixion. The blood libel accusation, another famous anti-Semitic canard, is also a twelfth-century creation. Two days before Passover 1144, the body of a skinner's apprentice named William was found horribly mutilated in a wood outside Norwich in East Anglia. Upon hearing that William, whose head had been shaved and "punctuated with countless stabs," was last seen alive entering the house of a Jew, his mother, Elvira, accused the local Jewish community of the murder. Two village girls who worked for local Jewish families then stepped forward to offer corroborating evidence. The girls said a group of Jews had seized William after synagogue, gagged him, pierced his head with thorns, and bound him to a cross.

Slowly, a legend began to grow up around the unfortunate William, who was quickly sainted for his services to Christianity. First in East Anglia, next in England, then throughout Christendom, stories circulated about the ritual murder of Christian children during Passover. In most versions of the rumor, the killings were said to be a reenactment of

Christ's crucifixion, but in one particularly strange iteration the Jews were alleged to kill Christian children to relieve hemorrhoidal suffering. Supposedly, all Jews had suffered from hemorrhoids since they called out to Pilate, "His blood be upon us and our Children." And according to Jewish sages, the only known relief for the condition was Christian blood.

A few years after William's death, an apostate Jew named Theobald of Cambridge added a new charge to the blood libel accusation, one that would resonate though the pogroms of the Black Death and beyond. William, said Theobald, was the victim of an international Jewish conspiracy. "It was laid down by [the Jews] in ancient times, that every year they must sacrifice a Christian in some part of the world . . . in scorn and contempt of Christ." Anticipating another aspect of the Black Death pogroms, Theobald also placed the powerful Spanish rabbinate at the heart of the conspiracy. "Wherefore the leaders and rabbis of the Jews who dwell in Spain assemble together [each year] . . . they cast lots for all the countries which the Jews inhabit . . . and the place whose lot is drawn has to fulfill the duty [of killing a Christian child]." In 1934 the Nazi publication *Der Stürmer* was still recycling the blood libel accusation. The magazine devoted an entire issue to the ritual murder of Christian children.

The early thirteenth century saw another important anti-Semitic landmark. At the Fourth Lateran Council in 1215, Church authorities decreed that "Jews and Saracens of both sexes in every Christian province and at all times shall be marked off . . . from other people through the character of their dress." From this measure emerged the yellow badge of the French Crown, which became the yellow star of the Nazi state; the Judenhut of surgeon Balavigny, which looked like an upside-down saucer; the pointed green hat of Polish Jews; and the tablet-shaped cloth strips that English Jews wore across the chest.*

As anti-Semitism intensified, petty degradations became a daily occurence for many Jews. In Turin, Jews caught on the street during first

* Jews, however, were not the only people to be singled out by a special dress code. Lepers and several other groups also were required to wear distinctive garments.

snowfall were pelted with snowballs unless they paid a ransom of twenty-five ducats. In Pisa, students celebrated the Feast of St. Catherine by capturing the fattest Jew they could find and making the local Jewish community pay his weight in sweets.

Historian Norman Cantor thinks incompetent internal leadership may have added to the woes of the Jews. "That the Jews were victims is clear," says Professor Cantor. "That the leadership of their intellectual elite might have made things worse has been underinvestigated." Rabbi Solomon ben Abraham could serve as exhibit A of Professor Cantor's point about incompetent leadership.

Rabbi Solomon's particular bête noire was Maimonides, the greatest Jewish thinker of the Middle Ages. In the rabbi's view, Maimonides' *Guide for the Perplexed* and *Mishneh Torah* (a code of Jewish law) were full of Aristotelian notions, and Aristotle, the rabbi fervently believed, was bad for the Jews. Agreeing, the conservative Ashkenazi rabbinate of northern France supported Solomon's denunciation of Maimonides. However, in Provence and Spain, regions with traditions of toleration and cosmopolitanism, the rabbinate sided with Maimonides.

According to contemporary accounts, Rabbi Solomon was so aggrieved by the stance of the Mediterranean liberals that he turned to leaders of the Inquisition, the Church arm that enforced Christian orthodoxy, for assistance. Beyond this, the story grows murky. One Mediterranean liberal claims that, feeling aggrieved, Solomon handed over Maimonides' works to the Inquisitors for inspection. "Behold," the rabbi is supposed to have told the Inquisitors, "most of our people are unbelievers and heretics, for they were led astray by the words of Rabbi Moses of Egypt [Maimonides] who wrote heretical books. Now, while you are exterminating the heretics among you, exterminate our heresies as well.'"

However, the liberal was probably trying to discredit Rabbi Solomon. The rabbi may have had an "uncircumcised heart," as one critic charged, but he was not stupid. There is no evidence that he handed over Maimonides' works to a hostile Church body. Nevertheless, there is an element of truth in the "spirit" of the liberal's accusation. Intrigued by the rabbi's complaints about Maimonides, the

Inquisitors began to peruse other Jewish religious works. Predictably, it did not take them long to find an offensive one.

In 1240, eight years after the Maimonidean controversy, a second confrontation occurred; this time, however, it was over a central text of Judaism, the Talmud, and the confrontation was between Christian and Jew. The two central figures in the affair were Nicholas Donin, a converted Jew turned Franciscan,* who brought the overlooked Talmud to the attention of the Vatican, and Rabbi Yehiel ben Joseph, who defended the book in a famous disputation with Donin in 1240.

In most accounts of the event, Rabbi Yehiel emerges as a skilled and wily advocate. In one exchange, Donin asks, Is it not true that the Talmud insults Jesus? Yes, replies Rabbi Yehiel. The Talmud disparages *a* Jesus, but then, referring to the reigning French monarch, Louis IX, he adds, "not every Louis born in France is the king of France. Has it not happened that two men were born in the same city, had the same name and died in the same manner? There are many such cases."

Rabbi Yehiel also had to concede Donin's point that the Talmud forbade Jews to mix with Christians. But, again, he was able to outmaneuver his opponent. He reminded the officials overseeing the disputation that Christian law also discouraged intercourse between the Christians and Jews. Moreover, added the rabbi, despite such injunctions, in daily life the two groups often intermingled freely. "We [Jews] sell cattle to Christians, we have partnerships with Christians, we allow ourselves to be alone with them, we give our children to Christian wet nurses." Although the Talmud could not have had a more clever advocate, two years after the disputation, in 1242, the work was convicted of heresy and publicly burned in a Paris square.

During the Central Middle Ages, as knowledge about Judaism grew, Christian attitudes began to harden. Now the Jews were not

* In medieval Spain, Christian-Jewish disputations also covered nonreligious subjects. Topics included such questions as: "Why did Christians tend to be fair-skinned and good-looking while Jews tended to be black and ugly?" The Jewish answer to the paradox had two parts. One, Christian women had sex during menstruation, and hence were more prone to pass on a redness of blood to their offspring and, two, Gentiles often "had sex surrounded by beautiful paintings and [thus tended to] give birth to their likeness." (Paul Johnson, *A History of the Jews* [New York: Harper & Row, 1987] p. 218.)

merely "obstinate in their perfidy," an old charge; they also threatened "injury to the Christian Faith." This was a new charge, and with its suggestion of subversion it opened the door to policies the Augustinian "but" had helped to keep in check.

Over the next half century, there were mass expulsions of the Jews in England and France, forced conversions, and mass exterminations.

One would be accusing God of cruelty if one thought that the Jews' steadfast bearing of suffering could remain unrewarded. . . . The Jews are oppressed with the heaviest taxes, as if each day they had to buy anew the right to live . . . if they want to travel, they have to pay sums to gain . . . protection . . . [and they] cannot own fields or vineyards . . . So the only profession open to them is that of usury, which only increases the Christians' hatred of them.

—PETER ABELARD

The Middle Ages, the birthplace of the stereotypical Jewish nose, was also the birthplace of the stereotypical Jewish moneylender. In *The Treasure and the Law,* a story about the signing of the Magna Carta, one of the central documents of the Middle Ages, Rudyard Kipling manages to squeeze almost every medieval cliché about the moneylender into a single sentence. "Doors shut, candles lit," Jewish moneylenders put off their rags and cringing and decide the fate of the world with their secret knowledge of that "mighty underground river," gold. However, the desciption of Peter Abelard, father of Scholasticism and lover of Heloise, is far closer to the mark. Medieval Jews became moneylenders out of desperation, not out of a desire to run their hands through "mighty underground rivers" of gold.

During the economic boom of the twelfth and thirteenth centuries, the Jews' near-monopoly on mercantile and financial skills began to dissolve, and with it, their dominance in traditional "Jewish" professions. Increasingly, international trade wore an Italian face, especially an avaricious Venetian and Genoese one, while at home commerce and finance became the province of Flemings, Florentines, Germans, and Lombards, whose reputation for unscrupulousness was legendary. For a people in need of a new profession, moneylending offered many attractions. It required neither travel nor landownership—activities restricted for Jews. And money, being a highly mobile commodity, could easily be transported in case of expulsion. Most important of all, in the matter of interest, medieval law actually favored the Jew. Though Christians often subverted the ban, it was against canon law to loan money for a profit, but not against Judaic law. Usury was permitted as long as the client was a non-Jew.

For a people under economic pressure, moneylending also promised the relief of handsome profits. In Burgundy a lender could charge up to 87 percent interest; in other parts of France, more than 170 percent. Thus, a loan of 140 florins taken out in 1334 by Guillaume, Lord of Drace, earned his moneylender 1,800 florins by the time it was paid off. A few Jews saw high interest rates as a way to strike back at a hated oppressor as well as to make money. "One should not benefit an idolater . . . [but] cause him as much damage as possible without deviating from the path of righteousness," declared Levi ben Gershom. However, most of the men who became moneylenders did so out of the need to make a living. "If we are condemned to live in the midst of nations, and cannot earn our living in any other manner except by money dealings with them; therefore the taking of interest is not prohibited," declared one medieval Jewish scholar.

Moneylending personalized anti-Semitism in a way Church doctrine never could; it brought hatred of the Jew into the home and made it intimate, personal, and vicious.* The average peasant farmer or rustic knight knew little about Agobard of Lyon, but he knew about 90 and 100 percent interest rates and what a lien on his cattle would do to him; he also knew from rumors that if he missed a loan payment the moneylender would sell his wife into prostitution. Though many of the things said about moneylenders were clearly slanderous, loan collection is not an activity designed to bring out the best in any people. According to historian Norman Cohn, "Jewish moneylenders

* There is some question about how common moneylending was among medieval Jews. But Rabbi Joseph Colon, who lived a century after the plague, says the Jews of Italy and France scarcely knew another profession.

often reacted to insecurity and persecution by deploying a ruthlessness of their own."

As the Christian world grew increasingly hostile, the Jews turned to princes, kings, bishops, and city councils for protection, but these alliances had a Faustian element. Frequently a ruler, loath to raise taxes, would use the local Jewish community to "sponge" the populace. Jewish moneylenders would be allowed to charge a high interest rate and to use royal courts to collect nonperforming loans, but then the "protector" would confiscate the profits, leaving the Jews to face the resentment of the populace. Often, the alliances also had the unhappy effect of making the Jews a surrogate for the local authority. Thus, while some anti-Semitic attacks were motivated by anger over high interest rates, others were expressions of anger at a local bishop or prince who was too powerful to attack directly. As populist anti-Semitism grew, physical violence became a daily occurrence. In Speyer a mob attacked a Jewish woman named Minna, cutting off her lips and thumbs; in eastern France Jacob Tam was wounded five times on the head to atone for the wounds the Jews had inflicted on Christ.

Pogroms also became more common. There were major outbreaks of anti-Semitic violence in 1146, 1189, 1204, 1217, 1288, 1298, and 1321. The last of these, the pogrom of 1321, was notable for being a kind of dress rehearsal for the anti-Semitic violence of the Black Death. Many of the characteristics that marked the pogroms of 1348 and 1349 were also present in 1321. There were the same whispers about an anti-Christian international conspiracy and the same torch-bearing Holy Week mobs. The two pogroms also began similarly. In both cases they were ignited by accusations of well poisoning, and in both cases at first the accusations were directed not at the Jews, but at another fringe element in medieval society—lepers, criminals, and vagrants, even the English.

For one French chronicler, the year 1321 was noteworthy mostly for remarkable meteorological events. There was a great snowfall in February, another before Lent, and later in the spring, a great rainfall. Almost parenthetically, as if such things were part of the natural order, the chronicler added that between the first and the second snowfall, the lepers of France were exterminated.

An account by a Dominican Inquisitor, Bernard Gui, is more forthcoming. The exterminations were provoked by the discovery of a lepers' plot to overthrow the French Crown. "You see how the healthy Christians despise us sick people," a coup leader is alleged to have said when the plotters met secretly in Toulon to elect a new king of France and appoint a new set of barons and counts. It is not entirely clear how the plot first came to light, but by Holy Week 1321 nearly everywhere in southern France one heard the same story; the lepers, "diseased in mind and body," were poisoning local wells and springs. Alarmed, Philip V, "the Long One," ordered mass arrests. Lepers who confessed complicity in the plot were to be burned at the stake immediately; those who professed innocence, tortured until they confessed, then burned at the stake. Pregnant lepers were allowed to come to term before being burned, but no such stays were offered to lepers with children. In Limoges a chronicler saw leprous women tearing newborns from their cribs and marching into a fire, infants in arm.

Almost immediately, the populace concluded that the Jews were also involved in the plot. This popular verdict was based on guilt by association. Like the lepers, who wore a gray or black cloak and carried a wooden rattle, Jews were required to dress distinctively. Additionally, both groups were considered deceitful. As an inscription at Holy Innocents cemetery in Paris reminded the unwary: "Beware the friendship of a lunatic, of a Jew, of a leper." The two groups were also hated, although after the recent Great Famine the Jews were probably hated more because of their moneylending. There was another important connection, though no bill of indictment mentioned it: wealth. The Jews, who, despite their vulnerable economic position, still sat on a substantial amount of private capital, and the leper asylums, whose treasuries were flush with contributions and endowments, made lucrative targets. For the mobs, it seemed like that rare opportunity to do well while also doing good. In early June, even before the mass arrests of the lepers began, the populace struck out at the Jews. One chronicler reports that on a summer morning, a group of 160 Jews marched into a fire near Toulon, singing "as if marching to a wedding feast." Near Vitry-le-François forty Jews slit their own throats rather than fall into Christian hands. In Paris the local Jewish community had to pay 150,000 livres in protection money; some Parisian Jews were murdered anyway.

The French Crown was brought into the pogroms later in the summer by the alarming "discovery" of a secret covenant between the Jews, the Muslims, and the lepers. The compact first came to light at the end of June, during a solar eclipse in Anjou and Touraine. For a period of four hours on the twenty-sixth, the afternoon sun appeared swollen and horribly engorged, as if bursting with blood; then, during the night, hideous black spots dimpled the moon, as if the craters on its acned face had turned inside out. Certain that the world was coming to end, the next morning the populace attacked the Jews. During the rampage, a copy of the secret covenant was discovered inside a casket in the home of a Jew named Bananias. Written in Hebrew and adorned with a gold seal weighing the equivalent of nineteen florins, the document was decorated with a carving of a Jew—though the figure could have been a Muslim—defecating into the face of the crucified Christ.

On reading a translated copy of the covenant, Philip V was horrified. The Muslim ruler of Jerusalem, through his emissary, the viceroy of Islamic Granada, was extending to the Jewish people the hand of eternal peace and friendship. The gesture was occasioned by the recent discovery of the lost ark of the Old Testament and the stone tablets upon which God had etched the Law with His finger. Both were found in perfect condition in a ditch in the Sinai Desert and had awoken in the Muslims, who discovered them, a desire to be circumcised, convert to Judaism, and return the Holy Land to the Jews. However, since this would leave millions of Palestinian Muslims homeless, the King of Jerusalem wanted the Jews to give him France in return. The guilty homeowner Bananias told French authorities that after the Muslim offer, the Jews of France concocted the well-poisoning plot and hired the lepers to carry it out.

After reading the translation and several corroborating documents, including a highly incriminating letter from the Muslim King of Tunisia, Philip ordered all Jews in France arrested for "complicity . . . to bring about the death of the people and the subjects of the kingdom." Two years later, any Jewish survivors of the royal terror were exiled from the country.

The pogroms of 1348 were also fed by rumors of well poisoning and secret cabals, and, as in 1321, it took some time for the rumors to attach themselves to the Jews.

The first pogrom of the Black Death, on April 13, 1348, was a traditional act of Holy Week violence aggravated by the pestilence. Medieval Europeans knew that whenever bad things happened to Christians, the Jews were to blame. On the night of the thirteenth, several dozen Jews were dragged from their homes in Toulon* and murdered amid the glare of torchlight and the sound of heavy-footed tramping through the town. The next morning, as the mutilated bodies of the dead lay drying in the spring sun, rumors about plague poisons were already circulating through southern France, but as yet, the rumors had not attached themselves to the Jews. On April 17, in a letter to Spanish officials, who had written requesting information about the pestilence, Andre Benezeit, the vicar of Narbonne, asserted that the plague had two causes: unfavorable planetary alignments and poisonings. Around Narbonne, the vicar said, beggars and vagrants and the "enemies of the Kingdom of France"—in other words, the English—were helping to spread the plague with secret potions.

A week earlier, French officials had given Pedro the Ceremonious, King of Aragon, similar information. The officials claimed that the plague, which had not yet reached Spain, was spread by a poison sprinkled in water, food, and "the benches on which men sit and put up their feet." In this iteration of the rumor, the poisoners were described as men posing as pilgrims and friars, not beggars and vagrants. Given the panicky atmosphere in southwestern Europe that spring, it is not surprising that musician Louis Heyligen had heard similar rumors in Avignon. At the end of the month, Heyligen wrote to friends in

* Toulon was in Provence, and Provence was unaffected by the 1322 exile order, since the region lay outside the direct control of the French Crown. Its ruler was Queen Joanna, a beacon of tolerance. After the pogroms in April and May of 1348, Joanna reduced taxes on the Jews.

Flanders that "some wretched men were found in possession of certain powders and [whether justly or unjustly, God knows,] were accused of poisoning the wells with the result that anxious men now refuse to drink the water. Many are burnt for this and are being burnt daily."

Later in the spring, when *Y. pestis* entered Spain, a new round of pogroms erupted. In Cervera, eighteen people were killed, and in Tarrega, "on the tenth day of the month of Av," a Christian mob yelling, "Death to the traitors!" murdered three hundred Jews. On May 17, two months after the plague arrived in Barcelona, twenty Jews died in a bizarre street brawl. After some thatch from the roof of a Jewish building fell on a passing funeral, the angry mourners attacked the building and killed several of the occupants. Despite fifteen thousand plague dead in Barcelona, in Spain as in southern France, Jews were killed for the sin of being Jews, not for contaminating wells. "Without any reason they [the Christians] injure, harass, and even kill the Jews," concluded a 1354 report on the pogroms in Aragon.

But north of the Pyrenees, the whispers continued.

. . . rivers and fountains
That were clear and clean.
They poisoned in many places.

During the spring and early summer, several groups were auditioned for the role of poisoner, but history had already ordained who would get the role. Despite the allure of lepers, the charm of beggars, and the novelty of the English and the pilgrims, some tropism in the European soul always brought it back to the Jews.

In July the pestilence, the Jews, and the well-poisoning accusations finally joined hands in Vizille, a little market town just beyond the eastern borders of medieval France. Early in the month, nine Jews, possibly the sons and daughters of refugees who had fled to Vizille after the French Crown banned its Jews in 1322, were tried for contaminating local wells. The fate of the defendants is unknown, but that summer several other Jews in eastern France were burned at the stake for poisoning wells.

On July 6, in a papal bull, Clement VI pointed out that "it cannot be true that the Jews . . . are the cause . . . of the plague . . . for [it] afflicts the Jews themselves." However, with death coming up every road and through every door, few people were in a mood to listen to reason. Europe desperately needed a villain—someone it could snatch by the throat and throttle in retaliation for all its weeping mothers and dead children, for the squalid, rain-soaked plague pits and the tortured cities. From Vizille the pogroms spilled northeastward across the somber French countryside toward Switzerland. In many places, rumors of the well poisonings arrived months before the plague, but that did nothing to dim their power. Marrying the Jews to the well poisonings gave people a sense of empowerment. Increasingly, in villages and forest clearings, men and women told one another: Perhaps if we kill all the Jews, the plague won't come to our village. Even if the plague did come, with the Jews dead, at least debts to Jewish moneylenders would be canceled. Later, after the violence subsided, one chronicler would write that the "poison which killed the Jews was their wealth."

A few resolute leaders defended their Jewish communities, but others, fearing the population would turn on them, stepped aside and allowed the crowd to vent its fear and rage.

Amadeus VI, ruler of Savoy, the region around Lake Geneva, took a middle course. Amadeus did not want angry mobs tossing Jews down wells in Savoy, as the mobs were doing in eastern France; on the other hand, he did not want to appear unsympathetic to popular feeling. Amadeus resolved his dilemma with a traditional bureaucratic maneuver—he ordered an investigation. In the late summer of 1348, eleven local Jews, including the surgeon Balavigny and a woman called Belieta, were arrested and interrogated in the Lake Geneva town of Chillon.

A transcript of Belieta's interrogation—or, rather, interrogations, since there were two—still exists. On October 8, when she was first questioned, Belieta admitted a knowledge of, but not complicity in, the plot. At "midsummer last," she told her interrogators, a conspirator had given her a packet of poison, but she had disobeyed his order "to put the poison into the springs," giving it instead to a "Mamson and his wife for them to do it."

At her second interrogation on October 18, Belieta was more forthcoming. This time she confessed that she had indeed done "as she was told"; she put poison in the "springs so that people [who] used the water would fall sick and die." Like Bona Dies, a Jew from Lausanne, who was "racked" for four nights and four days, Belieta may have been tortured beyond endurance. The transcript of her first interrogation says she was only "briefly put to the question," but the transcript of her second contains no such qualifiers. It is also possible that Belieta was trying to protect her son, Aquetus, another alleged "conspirator," who was not standing up well under questioning. She may have hoped that if she confessed, the authorities would spare Aquetus; they did not.

Broken in mind and body, after being "put to the question to a moderate degree," a few days later a distraught Aquetus told his interrogators "that by his soul, the Jews richly deserved to die, and that, indeed, he had no wish to live, for he too richly deserved to die."

The interrogations at Chillon were an important turning point in the pogroms. Though "documentary evidence" about the well poisonings had already begun to circulate in Germany and Switzerland, both largely plague-free in the early fall of 1348, a sizable segment of educated opinion remained skeptical. Echoing Clement's papal bull, the doubters asked, "If the Jews are poisoning wells and springs, why do they die of the plague like everyone else?" Whoever prepared the transcripts of the Chillon prisoners' "confessions"—and the confessions were transcribed, and the transcripts widely circulated—was a master propagandist. Crisp, cogent, and richly detailed, they contained the kind of human moments and details likely to convince a sophisticated medieval reader. For example, after poisoning a spring in his home village of Thonon, surgeon Balavigny supposedly comes home and "expressly forbids his wife and children from using the spring without telling them why": just the kind of behavior one would expect of a conscientious husband and father.

Describing a recent visit to Venice, another plotter from the Lake Geneva region, a silk merchant named Agimetus, recalls the quality of the water where "he scattered some poison." It was "a well or cistern of fresh water near the house of the . . . Germans." Agimetus was also given the busy schedule of an international conspirator. After leaving Venice, he hurries south to Calabria and Apulia to poison wells, then on to Toulouse for more poisonings.

The notion of conspiracy is underlined by the repetition of certain names and places in the transcripts. There are, for example, several references to meetings outside the "upper gate in Villeneuve," a town near Montreux, where "the leading members of the Jewish community always discuss matters," and to a bullying secret agent named Provenzal, who tells one timid conspirator, "You're going to put the poison . . . into that spring or it'll be the worse for you." Another recurring character is the gentle Rabbi Peyret, who tells Agimetus, as he is about to leave for Italy, "It has come to our attention that you are going to Venice to buy merchandise. Here is a satchel of poison. . . . Put a little in the wells."

The plot's mysterious mastermind, Rabbi Jacob of Toledo, remains a shadowy presence in the transcripts, but, thanks to a thousand years of Christian teaching, every reader already knew what he looked like. Hooked-nosed, bent-shouldered, black-bearded, when the rabbi spoke of his plan for Jewish world domination, the "mighty underground river" of gold echoed in his voice.

In German-speaking Europe, reaction to the Chillon transcripts—and other incriminating documents—was swift and furious. "Within the revolution of one year, that is from All Saints Day [November 1] 1348 until Michaelmas [September 29] 1349, all the Jews between Cologne and Austria were burnt and killed," wrote Heinrich Truchess, Canon of Constance.

In November, barely a month after Balavigny, Belieta, her son Aquetus, and Agimetus were executed, the first pogroms broke out in Germany. The towns of Solden, Zofingen, and Stuttgart killed their Jews in November; and Reutlingen, Haigerloch, and Lindau killed theirs in December. As January 1349 dawned cold and bright along the Rhine, it was Speyer's turn. The Jews who did not immolate themselves in their homes were hunted down in the winter streets and bludgeoned to death with pikes, axes, and scythes. This happened so frequently, unburied corpses became a public health problem. "The people of Speyer . . . ," wrote a chronicler, "fearing the air would be infected by the bodies in the streets . . . shut them into empty wine casks and launched them into the Rhine." Farther down river in Basel, the city council made a halfhearted attempt to protect the local Jewish community, but when a mob protested the exile of several anti-Semitic nobles, the council lost its nerve. Basel spent the Christmas season of 1348 constructing a wooden death house on an island in the Rhine. On January 9, 1349, the local Jewish community was herded inside. Everyone was there, except the children who had accepted baptism and those in hiding. After the last victim had been shoved into the building and the door bolted, it was set afire. As flames leaped into the cobalt blue sky, the screams and prayers of the dying drifted across the river and into the gray streets of Basel.

In February, when the pogroms reached Strassburg, a bitter winter wind was blowing off the Rhine. The mayor, a tough patrician named Peter Swaber, was a man of conscience and resolve. If the Jews are poisoning wells, he told an angry crowd, bring me proof. The city council supported the mayor, and officials in Cologne sent a letter of encouragement, but in the end, all Swaber had to offer the people of Strassburg was the opportunity to act righteously, while his opponents could promise relief from Jewish debt and access to Jewish property. On February 9, a government more in tune with the popular will unseated Swaber and his supporters. Five days later, on February 14, under a dull winter sun, the Jews of Strassburg were "stripped almost naked by the crowd" as they were marched "to their own cemetery into a house prepared for burning." At the cemetery gates, "the youth and beauty of several females excited some commiseration; and they were snatched from death against their will." But the young and beautiful and the converts were the only Jews to see the sun set in Strassburg that Valentine's Day. Marchers who tried to escape were chased down in the streets and murdered. By one estimate, half of Strassburg's Jewish population—900 out of 1,884—were exterminated at the cemetery.

A few weeks later the Jews of Constance were "led into the fields at sunset. . . . [S]ome proceeded to the flames dancing, others singing, the rest weeping." In Brandenburg, the Jews were burned on a grill like meat. "These obstinate Jews . . . heard the sentence with laughing mouths and greeted its execution with hymns of praise," recalled an eyewitness, who says that "not only did they sing and laugh on the grill, but, for the most part, jumped and uttered cries of joy, and, thus . . . suffered death with great firmness." In Erfurt, where the pogroms were carried out with less resoluteness, a leader named Hugo the Tall had to admonish slackers, "Why are you standing around? Go . . . look for the Jews and beat them nicely." In Nordhausen, Landgrave Frederick of Thuringia-Meisen also had to steel the weak. "For the praise and honor of God and benefit of Christianity," the landgrave admonished a wavering city council, burn the Jews immediately.

According to Canon Truchess, "once started, the burning of the Jews went on increasingly. . . . They were burnt on 21 January [1349] in Messkirch and Waldkirch . . . and on 30 January in Ulm, on 11 February in Uberlingen . . . in the town of Baden on 18 March, and on 30 May, in Radolfzell. In Mainz and Cologne, they were burnt on 23 August . . .

"And, thus, within one year," wrote the Canon, "as I have said, all the Jews between Cologne and Austria were burnt. And in Austria they await the same fate for they are accursed of God. . . . I could believe that the end of the Hebrews had come if the time prophesied by Elias and Enoch were now complete, but since it is not complete, it is necessary that some be reserved."

The Canon was more optimistic than Jizchak Katzenelson, who, before his murder in Auschwitz on April 29, 1944, wrote a poem called "Song of the Last Jew":

Not a single one was spared. Was that
just ye heavens? And were it just, for whom?
For whom?

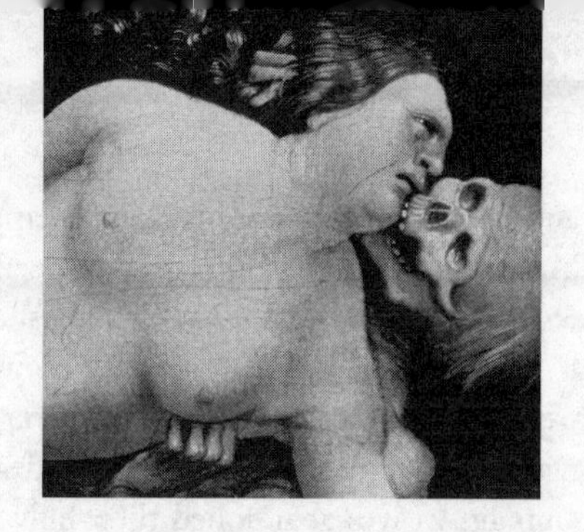

CHAPTER ELEVEN

"O Ye of Little Faith"

LONG BEFORE THE PESTILENCE REACHED THE RHINE, BARRELS with dead Jews inside were floating downstream to the river's headwaters above Lake Constance. The plague did not arrive in force in Central Europe until the winter of 1348–49, eight months after the well-poisoning rumors began and six months after the first Jews were executed in Vizille. The pestilence seems to have penetrated Central Europe through the Venetian-dominated Balkans. In the Middle Ages, the Adriatic coast of Croatia was home to tens of thousands of citizen-exiles of the ruler of "half and a quarter and of the Roman empire." Split—or Spalato, as the Venetians called it—seems to have been the first city in the region struck. On Christmas Day 1347 or thereabouts, a Venetian galley "storm driven from the east" and equipped with enough destructive power to obliterate the Adriatic coast from end to end visited Split. The following April, attracted by the scent of death in the spring air, packs of mountain wolves abandoned their hilly redoubts above the city and descended on Split to attack survivors.

On January 13, 1348, a second major Venetian colony, Dubrovnik—or Ragusa, as it was known—became infected. In a later outbreak of the plague, the city would gain fame as the creator of the quarantine. In the spring of 1348, Dubrovnik established another remarkable, if less imitated, custom. With total extinction looming, in early June municipal authorities ordered every citizen to make out a will. So great was the agony in the city that word of its calamity spread across the Adriatic and up the Alpine passes to Germany, where, later in the year, officials would send a letter of condolence to survivors, expressing sympathy for the "terrible mortality by which the population has been greatly diminished." Over the summer Istria, farther north on the Balkan coast, was stricken. Then in August, after death had done all that death could do, *Y. pestis* bade farewell to the Balkans and moved northward into Hungary, Austria, and Germany.

Traditionally, stories about the Black Death in Germany begin on a warm June day in 1348, when a surprise visit by *Y. pestis* kills "1,400 of the better class of inhabitants" in the Bavarian town of Muhldorf. Everything about the story is correct except perhaps the final digit of the date. Apparently intending to write a nine, the chronicler wrote an eight instead. On June 29, 1348, the alleged day of the Muhldorf massacre, the available evidence suggests that *Y. pestis* was astride almost every approach to Germany, but yet not in the country. One plague prong was racing eastward across France, another northward through Switzerland, and a third westward from the Balkans via Austria. For a time that fall, it seemed as if the Austrian prong would breach medieval Germany before the end of 1348. In October, after crossing the Brenner Pass in the Alps, the prong arrived just west of Innsbrück; Bavaria was only a few dozen kilometers to the north, but instead of persevering, *Y. pestis* unexpectedly stopped and made camp for the winter.

Over the next few months, as the Black Death marshaled its forces along the German borders, from the Baltic to Bavaria to the Rhine, an anxious populace waited and watched. No doubt, most expected the assault, when it finally came in the spring, to come from Austria. In April or May, a rested, reinvigorated plague would decamp from Innsbrück and thrust the final forty-five kilometers north into Bavaria. But according to historian Ole Benedictow, who has meticulously reconstructed the Black Death's movements across Europe, the first breach of German territory occurred in the west. In May 1349, a transport ship or traveler carried *Y. pestis* up the Rhine from Basel, where despite the winter's anti-Semitic measures, the plague was raging, to the little German town of Lichtenau.

Some historians have alleged that Germany suffered relatively lightly during the Black Death. But the available statistics—and admittedly, they are not nearly as detailed as those for England—indicate that the national mortality rate was on a par with Germany's neighbors. Strassburg and Mainz experienced horrible mortalities, as did Hamburg, where as much as two-thirds of the population may have died, and Bremen, where the death toll reportedly reached 70 percent. In Erfurt, 12,000 are said to have perished, in Mainz, 11,000. By comparison, Frankfurt, which was struck later than other German cities, escaped relatively lightly; two thousand deaths in seventy-two days. No one knows the death toll in the Baltic town of Lübeck, where the plague, sweeping down from Scandinavia, opened a new northern front against Germany in the summer of 1349. But the following year, will-making in Lübeck increased by 2,000 percent.

In the spring of 1350 the German countryside bore the wild, hopeless look not seen in Central Europe again until the spring of 1945. A chronicler speaks of "men and women, driven to despair, wander[ing] around as if mad . . . [of] cattle left to stray unattended in the fields, [of] "wolves . . . [coming] down from the mountains to attack sheep . . . [and] acting in a way which never had been heard of before." Unlike their Balkan counterparts, the German wolves, "as if alarmed by some invisible warning, turned and fled back into the wilderness."

In Vienna, which was infected in the spring of 1349, and where one out of every three faces would vanish by year's end, there arose the legend of Pest Jungfrau, a malignant plague goddess who emerged from the mouths of the dead in the form of a glowing jet of blue flame, and who had only to raise her hand to smite the living. There are few eyewitness accounts of Black Death Vienna, but the report of one Abraham Santa Clara, who lived through a recurrence of the plague in the city, gives us some sense of what life was like in the Austrian capital in May and June 1349. Santa Clara writes of "small children [who] were found clinging to the breast of their dead mothers," and of one fierce little girl, who, as her "dead mother was placed on the cart, . . . tried to accompany her by force, and with a lisping tongue continued to cry, 'Mammy, Mammy,' bringing water to the eyes of the rough, hardened corpse bearers."

Sweeping though Central Europe, the Black Death seems to have stirred some netherworld deep within the turbulent Teutonic soul. Having already given birth to the pogroms, the region now gave birth to another bizarre phenomenon, the Flagellants.

"A race without a head," the Dominican friar Henrici de Hervordia wrote of the Flagellants, who spread across Central Europe like an exotic jungle growth in late 1348 and 1349, bringing a thrilling eroticism and the specter of salvation to a populace grown weary of death and fear. Describing a self-flagellation, one practitioner wrote of how he had "stripp[ed] himself naked" and beat his "body and arms and legs til blood poured off him." Then in an ecstasy of pain and joy, he fell to his knees in his cold monk's cell and "naked and covered in blood," and shivering in the frosty air "pray[ed] to God to wipe out his sins."

The Flagellants' genius was to transform this erotically charged private self-abuse into public theater. During the mortality, troupes of fifty to five hundred Flagellants tramped across the landscape of central and northern Europe, bringing their passion play of blood, pain, and redemption to the multitudes. At each new town or village, a troupe would announce its arrival with a lusty chorus of deep-throated singing. As the sound of a "sweet melody" rose above the tree line like a "Heavenbound Angel," church bells would peal, windows open, people rush into the streets. Quickly a crowd would gather in the town square. As the singing grew nearer, the townsfolk would grasp hands and begin to sway back and forth in rhythm. Then, just as villagers' eardrums were about to burst from the roar of approaching voices, a wall of brilliant purple and gold banners would appear at the far end of the village square. At the sight of Flagellants—shoeless, hooded, and dressed in white cloaks, with a red cross on the front and back—shouts of "Save us!" would rise from the crowd. Some spectators wept; swooning women clasped their hands to their breasts; a few people laid their dead in the square for a blessing. If the town had any Jews, they went into hiding; the Flagellants were violently anti-Semitic. As the troupe marched to the local church, their brilliant pennants billowing and snapping in the wind, the members would sing:

Your hands above your heads uplift
That God the plague may from us shift,
And now raise up your arms withal
That God's mercy on us fall.

Inside the church, the marchers would strip to the waist in preparation for the first part of the Flagellant ceremony, a penitential walkabout. As members walked in a circle around the churchyard two abreast, each man would lash himself violently on his naked torso until it became "swollen and blue."

"Ahhh!" the crowd would exclaim, when the troupe suddenly fell down, "as if struck by a bolt of lightning." On the ground, each man would assume the position of the particular sin he was most culpable of: adulterers lay on their stomachs, murderers on their backs, perjurers on their sides with three fingers extended above the head. This part of the performance concluded with the Flagellant master passing among the fallen, lashing the sea of bleeding, sweating flesh beneath him with a scourge one contemporary described as "a kind of stick from which three tails with large knots hung down." Each knot contained "iron spikes as sharp as needles, and each spike was "about the length of a grain of wheat." As the master flogged the prone men, occasionally a spike would drive "so deeply into the flesh, [it could] only be pulled out by a second wrench."

The centerpiece of each performance was the collective flagellation. On the master's command, the troupe would form a circle around three members designated as cheerleaders. As the men began beating themselves rhythmically on the back and chest, the cheerleaders would exhort individual Flagellants to apply the lash more sternly, igniting furious contests of self-abuse. Gradually, the Ancient Hymn of the Flagellants would rise from the crowd:

Come here for penance good and well,
Thus, we escape from burning hell
Lucifer's a wicked wight
His prey he sets with pitch alight

At intervals, the troupe would fall and rise; after each interruption, the flogging would grow more intense, eventually attaining a frenzied, tom-tom-like rhythm. After the circle of men had fallen to the ground for a last time, the townsfolk would move among the bloody, sobbing figures, dipping handkerchiefs into raw, oozing wounds. Then, as the spectators stroked their cheeks with Flagellant blood, they would listen to the master read the Heavenly Letter.

Written by God and deposited on the altar at the Church of the Holy Sepulchre in Jerusalem in 1343,* the letter was a stern warning to a wicked humanity. "O ye children of men, ye of little faith, . . . Ye have not repented of your sins nor kept My holy Sunday . . . There-

* The original Heavenly Letter was discovered in the thirteenth century. In 1343 an angel left a revised, updated version of the letter at the same Jerusalem church where the original was found.

fore, I sent against you the Saracens and heathen people . . . earthquake, famine, beasts; serpents, mice, and locusts; hail, lightning and thunder . . . water and floods." A few paragraphs later, the letter warned of even greater evils to come, failing repentance. "Thus, I had thought to exterminate you and all living things from the earth; but for the sake of my Holy mother, and for that of the holy cherubim and seraphim [angels] who supplicate for you both day and night, I have granted a delay. But I swear to you . . . , if ye keep not My Sunday, I will send upon you wild beasts such as have never been seen before, I will convert the light of the sun into darkness . . . and I will smother your souls in smoke."

Long before the Middle Ages, the Indians of Brazil, who whipped themselves on the genitals, and the Spartans, who whipped themselves everywhere, were enthusiastic self-floggers. But whereas the Spartans associated the practice with fertility, in the medieval world flagellation was designed to appease divine wrath. Eleventh-century Italian monks, among the first medieval practitioners, used flagellation to atone for personal sin: by punishing himself for his misdeeds, a sinner could stay God's vengeful hand.

Flagellation as atonement for collective sin began in 1260, when Italy was visited by a terrible series of epidemics, wars, and crop failures. Convinced that a wrathful God was punishing a sinful mankind, troupes of scarred, sunburned Flagellants began tramping through the devastated Italian countryside. Within a year, the German countryside was also filled with roving bands of men, whipping themselves as they marched. North of the Alps, the movement acquired an organization, rituals, and songs.

The Italian branch of the Flagellants eventually fell under the control of the church, but the German wing, which had a deep anarchic strain, resisted clerical authority and was banned in 1262. However, whenever a major disaster struck, bands of hymn-singing German Flagellants would suddenly appear, Lazarus-like, out of nowhere, as in 1296, when a terrible famine struck the Rhineland, and again in 1348, when the plague swept into Bavaria.

According to legend, the Flagellants of the Black Death arose from a planetary alignment on "the third hour after midnight on March 12, 1349." Reportedly, a few weeks later—3 a.m. on March 29, to be precise—"gigantic women from Hungary came to Germany, . . . divested themselves publicly of their clothing and to the singing of all kinds of curious songs beat themselves with rods and sharp scourges."

Whatever its true origins, the Flagellant movement of the Black Death, like its thirteenth-century predecessor, found its true spiritual home in Germany. After building a base in the southern and central part of the country, the movement quickly spread outward: first, to the rest of German-speaking Europe; then to France, Flanders, Holland, and finally, in 1350, to an unreceptive London. The phlegmatic, insular English seemed puzzled by the sight of half-naked men, publicly "lash[ing] themselves viciously on their . . . bodies . . . now laughing now weeping." What would the foreigners think of next? You can almost see the chronicler Thomas Walsingham shake his head as he wrote, "They do these things ill-advisedly."

The movement, which was called the Brotherhood of the Flagellants and the Brethren of the Cross, displayed a fair degree of organization. Before joining, an applicant had to obtain the permission of a spouse and agree to make a full confession of all sins committed since the age of seven. New recruits also had to pledge to whip themselves three times daily for thirty-three days and eight hours, the length of time each Flagellant spent on pilgrimage. The pilgrimages were really marches, and their time frame was keyed to Christ's earthly life—thirty three and a third years. However, since pilgrims were forbidden to bathe, shave, or change clothing, the Flagellant columns that tramped from town to town often became plague vectors. Along with washing, marchers were also forbidden to sleep in a bed or to have sex. Indeed, if a Flagellant so much as spoke to a woman on a march, he was flogged by the master of the troupe, who would conclude the whipping with the words, "Arise by the honor of pure martyrdom, and, henceforth, guard yourself against sin."

Initially the Flagellants operated with a measure of self-restraint. However, the movement had begun with an implicit anticlerical message—priests are unnecessary for salvation—and as the mortality worsened, and with it, disillusionment with the Church, that implicit message became explicit. Increasingly, the Flagellants began to view themselves not as a selfless sufferers doing penance for a wicked humanity but as a mighty host of glorious saints with divinely endowed powers, including the ability to drive out devils, heal the sick, and raise the dead. Members boasted of supping with Christ and conversing with Mary; there was even talk of extending the pilgrimages to thirty-three and a third years. Fiercely anticlerical, members disrupted Masses, drove priests from churches, looted ecclesiastical property, defamed the Holy Eucharist, and denounced the hierarchy.

A changing demographic may have hastened the turn toward radicalism. As more conservative members of the movement died or were expelled, the Flagellants became increasingly younger, poorer, more criminal, ignorant, anticlerical, and anti-Semitic. For a moment in March 1349 the pogroms seemed about to vanish, but during the spring, as the Flagellants spread out across Germany, the anti-Semitic violence reignited. Flagellant columns killed Jews wherever they found them, including in Frankfurt, where the arrival of a troupe of marchers inspired one of the bloodiest pogroms of the mortality. The local Jewish quarter was sacked, its inhabitants killed, and their goods stolen.

The public seemed unable to get enough of the Flagellants. In 1349 Strassburg played host to a new pilgrimage every week for six months, and Tournai saw a pilgrimage begin every few days. From mid-August to mid-October 1349, fifty-three hundred Flagellants reportedly passed through the town. However, official Europe, sensing the movement's revolutionary potential, was far less enthusiastic. The magistrates of Erfurt closed the city to Flagellant bands, while Philip VI of France declared everything west of Troyes a Flagellant-free zone, and Manfred of Sicily threatened to kill any member of the movement who so much as set foot in his domain. Despite its members' brazen anticlericalism, Clement VI seemed prepared to tolerate the movement—initially, at least. It was probably only a public relations gesture, but when a troupe of Flagellants passed through Avignon in the spring of 1348, Louis Heyligen reports that the "Pope took part in some of these processions."

The turning point came early in the fall of 1349, when a report prepared by a Sorbonne scholar named Jean de Fayt arrived on Clement's desk. Alarmed at its contents, he issued a stinging denunciation of the movement. "Already the Flagellants under the pretense of piety have spilled the blood of the Jews . . . and frequently also the blood of Christians. . . . We therefore command our archbishops and suffragans . . . as well as the laity, to stand aloof from the sect and never again to enter into relations with them."

A year later, the Flagellants had "vanished as suddenly as they had come, like night phantoms or mocking ghosts."

The Flagellants and the pogroms never expanded much beyond the European heartland, but the plague would penetrate to almost every corner of the continent, Christian or otherwise. In the summer of 1349, it arrived in Poland, where King Casimir, influenced by his beautiful Jewish mistress, Esther, offered asylum to Jews fleeing persecution in Central Europe. The communities that the refugees established would last unbroken until the Second World War.

On the other side of Europe, the plague expanded as far west as the Atlantic beaches of the Iberian peninsula. The island of Mallorca, a hundred or so miles off the Mediterranean coast, was a major distribution center for the spread of the plague in Spain. One report has a ship out of Marseille bringing the plague to the island in the bitter December of 1347; if true, the account raises the possibility that the island was infected by a member of the little death fleet, which musician Louis Heyligen says left a trail of contamination along the southern coast of Europe whose "horror can scarcely be believed, let alone described." From Mallorca, the busy coastal shipping lanes quickly delivered the plague to the shores of continental Spain. By March, Barcelona and Valencia were infected, though judging from contemporary accounts, citizens of both cities remained unaware of the infection until early May.

In the same month, May, a ship from Mallorca heading southward toward the Gibraltar Gap carried the Black Death to Almeria, a principal city of Granada on the southern tip of the Iberian peninsula and the last Muslim stronghold in Spain. Islamic law held that since God alone decided who lived and died, in the case of plague, there was nothing to do but wait for Him to render a judgment. However, the citizens of Almeria apparently felt there was nothing un-Islamic in giving God a little help. Active prophylactic measures kept the plague from breaking out of the city until autumn. As the pestilence swept along the golden Spanish coast toward Gibraltar, King Alfonso of Castile, who was besieging the Muslim stronghold, was urged to flee to safety. However, the king, who had lost a future daughter-in-law to the plague two years earlier—Princess Joan of England—insisted on remaining with his army. On March 26, 1350, a Good Friday, Alfonso became the only reigning European monarch to die of the pestilence. In 1350 Spain's other royal house, Aragon (roughly speaking, Aragon occupied what is today Mediterranean Spain, Castile, and Atlantic Spain) also experienced several deaths. In May King Pedro lost a daughter and a niece to the plague—and in October, a wife.

There is some evidence that the hyperlethal septicemic form of plague was active in Spain. There are several descriptions of what sound like the symptoms of the disease, and a number of Spanish plague stories revolve around almost instant death, which is a characteristic of septicemic plague. One such tale can be found in the chronicle of the old Abbot Gilles li Muisis.

The story concerns a French cleric who visited pestilential Spain on a pilgrimage. One evening he stopped at a little country inn, where he supped with the widowed innkeeper and his two daughters, then booked a room for the night. Awakening the next morning, the Frenchman found the inn empty. Puzzled, he called out to the innkeeper. Failing to get a reply, he tried first one and then the other of his daughters. Again, no answer. Finally, the Frenchman called for the family servant. Silence for a fourth time. Now, deeply perplexed, he began to search the inn.

Encountering another guest, the Frenchman asked after the innkeeper, his family, and the servant. All dead, sir, replied the other guest. The four were taken with the plague during the night and died almost immediately.

The account of Granada physician Ibn Khatimah suggests that pneumonic plague was also active in Spain. For the Muslim physician, the disease's two outstanding characteristics were contagiousness and a bloody cough.

The Atlantic coast of Portugal marked the westernmost boundary of the plague's advance, and by the time the disease arrived on the sandy beaches of the region, it was running out of steam. Except for the city of Coimbra, Portugal was lightly affected.

Three other areas of Europe are also supposed to have largely escaped the ravages of the Black Death: Poland, the Kingdom of Bohemia (roughly speaking, modern-day Czech Republic), and an odd-man-out region composed of Flanders and the southern Netherlands. However, new research suggests that, like the perfect guest, *Y. pestis* considered no place in Europe too small or unimportant to visit.

Poland, like Germany, was squeezed by an octopus-type envelopment. In the July 1349, the first wave of plague entered the country near one of history's favorite playgrounds, the Polish city of Danzig, where World War II (and later the Solidarity Movement) began. In a series of follow-up assaults, the Black Death took the country from the south via a disease prong advancing northward through Hungary from the Venetian-dominated Balkan coast, and from the east via a thrust out of Russia. Then, in 1351, just as the survivors were telling one another the worst was over, *Y. pestis* sent a fourth plague prong across the River Oder from Frankfurt to conduct a mopping-up operation. There are no reliable death figures available for Poland but, tellingly, as in England and France, wages in the country soared after the plague, due to a tremendous manpower shortage.

Bohemia's supposed immunity also has come under question recently. The kingdom was long assumed to have been spared the worst ravages of the Black Death because of its remoteness from the trade routes that carried the plague through the European heartland, but that view has been challenged of late. According to Professor Benedictow, in the decades before the Black Death, Bohemia, with its lucrative mining industry; gleaming capital, Prague, and energetic population of a million and a half, was one of the most prosperous and bustling regions of Europe, exporting tin and silver and importing salt (for the preservation of meat) and iron (for making farm implements).

As with Poland, no reliable mortality figures are available for the region. But as proof that the Black Death ravaged the kingdom, which was infected sometime in 1349 or 1350, Professor Benedictow points to a story in the *Chronicle of Prague*. It concerns the visit a group of Bohemian students studying in Bologna made to the kingdom in the waning days of the plague. According to the account, the "students . . . saw that in most cities and castles . . . few remained alive, and in some all were dead. In many houses also those who had escaped with their lives were so weakened by sickness that one could not give the other a draught of water, nor help him in any way and so passed the time in great affliction and distress. . . . In many places, too, the air was more infected and more deadly than poisoned food, from the corruption of the corpses, since there was no one left to bury them."

Recently, there has also been a critical challenge to the third "spared" region, the southern part of the Netherlands and Flanders, but here the revisionists have had less success. A wealth of local data indicates that compared to their neighbors, the two areas experienced relatively low mortality rates. Even Professor Benedictow, a profound skeptic on the subject of spared regions, concedes that "the Netherlands did not suffer such great losses in the Black Death as Italy or England."

There may be an explanation for this "miracle." Flanders, which had a relatively low Black Death mortality of 15 to 25 percent, and the southern Netherlands both lost large numbers of children during the Great Famine, and that may have left the two regions with fewer vulnerable adults when the plague struck—that is, adults with deficient immune systems due to early exposure to famine.

However, as a general proposition, it is fair to say that almost no area of Europe entirely escaped the Black Death. In addition to blanketing the continent from east to west, by 1350, the plague would also cover it from north to south.

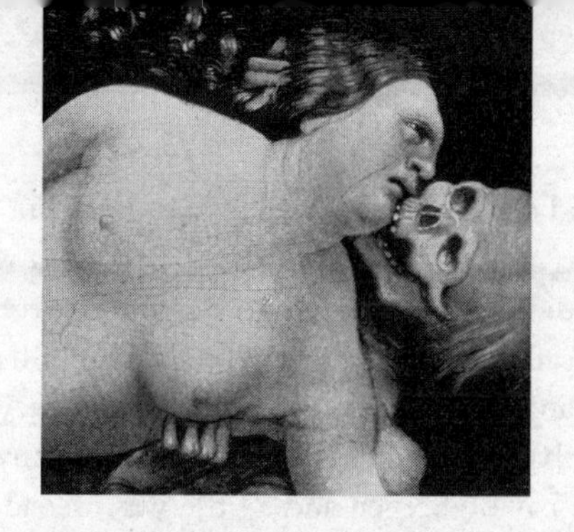

CHAPTER TWELVE

"Only the End of the Beginning"

IN MAY 1349, AS WILL-MAKING REACHED A PEAK IN PESTILENTIAL London, and the Jews of Strassburg sat shivah for the dead, out on the windswept latitudes of the "German Ocean" a frothy spring sea was carrying the plague northward toward Scandinavia, where it would complete its circumnavigation of Europe with a final thrust through Norway and Sweden before vanishing back into the Russian wilderness. Legend has it that Norway was infected by an English merchantman, which left London sometime in late April and was next seen a month later beached on a spit of land near Bergen with all hands dead. But Oslo, Norway's other major city, may have been infected first. And while an English merchantman did carry the plague to Bergen, the crew was still alive when the ship sailed into port, though they did not have much longer to live. According to the *Lawman's Annual,* a medieval Scandinavian chronicle, "At that time a ship left England with many people aboard. It put into the bay of Bergen and a little was unloaded. Then all the people on the ship died. As soon as goods from the ship were brought into the town, the townsmen began to die. Thereafter, the pestilence swept all over Norway."

Scandinavia posed formidable challenges for *Y. pestis*. Thinly populated, in the Middle Ages the region was the back of the beyond; it had little to offer in the way of crowded streets or dense cities. In the desolate north, the plague would have to become a feral scavenger surviving on the occasional farm family, the odd rustic knight and his dog, and the little fishing village perched above a fjord. Even more formidable than the thin population was the hostile climate. In Scandinavia summers are a blink of the eye and winters eternal and bitter. Accordingly, historians have long assumed that pneumonic plague predominated in the region. And, indeed, several medieval Nordic sources describe what sounds like pneumonic plague; says the *Lawman's Annual,* "People did not live more than a day or two with sharp pangs of pain. After that they began to vomit blood." However, Professor Benedictow thinks that the seasonality of outbreaks, the pattern of dissemination, and degree of lethality all point to bubonic plague as the dominant form of the disease in Scandinavia. He argues that what sources like the *Annual* are describing is pneumonic plague *secondary* to bubonic plague—that is, cases of the disease where the plague bacilli metastasize from the lymph nodes to the lungs.

The answer to the Scandinavian conundrum may lie in the Russian theory of marmot plague, with its tropism for the lungs. In this regard, it is worth comparing two recent outbreaks of *Y. pestis*. In 1991, when plague broke out near marmot foci in China, almost half the victims went on to develop pneumonic plague. By contrast, in Vietnam, where the rat predominates, the wartime epidemic of the 1960s was almost exclusively bubonic—a case rate of 98 percent.

Whether pneumonic or bubonic, the pestilence spread across Scandinavia with its customary ferocity. A few months after James de Grundwell, sole survivor of the plague's visit to Ivychurch priory in England, was elevated to the position of head abbot, his Norwegian counterpart, a priest who was the sole surviving cleric in the diocese of Drontheim, was elevated to the position of archbishop. After taking a monstrous toll in Bergen, the plague all but obliterated the remote mountain village of Tusededal. Months after the pestilence burned itself out, a rescue party reaching the village found only one survivor, a little girl who had been made so wild by her solitary existence, the rescuers named her Rype—wild bird.

From Norway, the plague thrust eastward across the interior of Scandinavia to Sweden, where, in 1350, the blustery King Magnus II issued a thunderous, albeit tardy, warning. "God," declared Magnus, "for the sins of men, has struck the world with this great punishment of sudden death. By it, most of the people in the land to the west of our country [i.e., Norway] are dead. [The plague] is now . . . approaching our Kingdom of Sweden." To repel the threat and to appease a wrathful God, Magnus ordered foodless Fridays (except for bread and water) and shoeless Sundays (Swedes were ordered to walk to church barefoot). But just as it killed Italians, Englishmen, and Frenchmen who avoided windows with a southern exposure and inhaled fragrant scents, *Y. pestis* killed Swedes with and without shoes, on foodless Fridays and on gluttonous Saturdays. Among the dead were the king's two brothers, Knut and Hacon.

In its travels across Eurasia, the pestilence had encountered every manner of ecological phenomenon. It had seen mountains collapse into lakes (China), plumes of volcanic ash swallow the noon sun (Italy and China), torrential floods gulp up villages (China, France, Germany), swarms of locusts three German miles long (Poland and China), tidal waves as high as a cathedral spire (Cyprus), and skies that poured rain for six months (England). But approaching the coast of Greenland, *Y. pestis* encountered a new natural wonder. Rising out of the frigid, white-capped sea, it gazed up at monstrous cliffs of silvery ice shimmering in the brilliant, bitter sunlight of a new Little Ice Age.

From Scandinavia, one prong of the plague crossed the Baltic and reentered Russia. Striking Novgorod, *Y. pestis* traveled south, clinging to the trade routes like a blind man feeling his way along a narrow corridor, until, at last, the golden onion-shaped domes of Moscow rose above the Russian plain. As the crow flies, the Russian capital, which was devastated by a terrible epidemic in 1352, lies only about seven hundred miles to the north of Caffa, where *Y. pestis* had set sail for Sicily several years earlier. Having closed the noose, the hangman rested.

On a glorious morning, Christendom awoke to find the plague gone. Life and joy, denied for so long, demanded their due. Survivors drank intoxicatingly, fornicated wildly, spent lavishly, ate gluttonously, dressed extravagantly. In England craftsmen took to wearing silk cloth and belts with silver buckles, and ignored a royal ordinance forbidding the lower orders from eating meat and fish at more than one meal a day. In Orvieto, where almost half the town lay buried in local plague pits, couples copulated on the freshly laid grass above the pits. In France "men became more miserly and grasping." And everywhere survivors luxuriated in the sudden abundance of a commodity that only a few months earlier had seemed so fragile, so perishable—time: wonderful, glorious, infinite time. Time for family, for work. Time to gaze into an evening sky. Time to eat and drink and make love. "There are three things a man may say properly belong to him," declares a character in a work by the Florentine humanist Leon Battista Alberti. When a companion asks what they are, Alberti's character replies: a man's fortune, his body—"and a very precious thing, indeed."

"Incredible, what is it?" asks the companion.

"Time, my dear Lionardo," Alberti's character replies.

The burst of postplague debauchery disappointed, though did not surprise, moralists like Matteo Villani, the dour brother of the plague-dead Giovanni. It was further proof—as if Matteo would ever need further proof—of the innate wickedness of man. "It was thought," he wrote after the pestilence, "that people whom God by his grace in life had preserved . . . would become better, humble, virtuous and catholic, avoiding inequities and sins and overflowing with love and charity for one another. But . . . the opposite happened. Men . . . gave themselves over to the most disordered and sordid behavior. . . . As they wallowed in idleness, their dissolution led them into the sin of gluttony, into banquets, taverns, delicate foods and gambling. They rushed headlong into lust." Agnolo di Tura, who lived in Siena, where they were still counting the dead, offered a more succinct description of Europe's post–Black Death mood. "No one could restrain himself from doing anything."

The hysterical gaiety was the very thin veneer of a profound and abiding grief and sense of dislocation. In 1349, as the plague lifted from Italy, a mournful Petrarch wrote to his friend Louis Heyligen: "The life we lead is a sleep; whatever we do, dreams. Only death breaks the sleep and wakes us. I wish I could have woken before this."

Petrarch would get his wish. Before the mortality was finally over, there would be tens of millions more dead to mourn, but by that time, Europe would be in the shadow of the Enlightenment, and the poet long dead.

The plague of Moscow in 1352 was, to borrow a phrase from Churchill, not "the end [of the plague] or even the beginning of the end but only the end of the beginning."

One can scarcely imagine with what heavy heart an English chronicler wrote the following words: "In 1361, a grave pestilence* and mortality of men began throughout the whole world." Barely eleven summers passed between the Black Death and the *pestis secunda,* as the second outbreak of plague was called. The new epidemic, which began in 1361, marked the beginning of a long wave of plague death that would roll on through more than three centuries. Had it not occurred in the

* Significantly, John of Reading says the reappearance of the plague in England in the spring of 1361 was accompanied by ecological upheaval. According to John, there was "a damaging drought . . . and because of the lack of rain there was a great shortage of food and hay." The shortages must have driven infected rodents into homes and barns in search of food. (John of Reading, excerpted in *The Black Death, Manchester Medieval Sources,* trans. and ed. by Rosemary Horrox [Manchester: University of Manchester Press, 1994], p. 87.)

immediate shadow of the Black Death, today the second pestilence would be spoken of as an epic tragedy in its own right. In rural Normandy 20 percent of the population died; in Florence, already shrunken to a remnant by the Black Death, the mortality also reached 20 percent. In England, losses among the landed gentry almost matched those of 1348–49, roughly 25 percent plus. However, to contemporaries, it was less the scope of the *pestis secunda* than its victims that left the greatest impression.

To the people who lived through it, the *pestis secunda* seemed to strike down the young in disproportionate numbers. Indeed, many contemporaries referred to the 1361 outbreak not as the *pestis secunda* but as the "Children's Plague," or "les mortalite des enfauntz." Surgeon Guy de Chauliac, who was still practicing medicine in 1361 and had one of the keenest clinical eyes of the Middle Ages, says that "a multitude of boys and few women were attacked." Modern scientific opinion holds that no population group has a special vulnerability to plague. But, like the tarabagan born in a surge year, the children born after the Black Death may not have had an opportunity to acquire the temporary immunity that comes to survivors after exposure to *Y. pestis*.

The *pestis secunda* was followed by the *pestis tertia* of 1369. Thereafter, for the next several centuries, Europe would scarcely know a decade without plague somewhere on the continent. In the Netherlands alone, there were epidemics in 1360–62, 1362–64, 1368–69, 1371–72, 1382–84, 1409, 1420–21, 1438–39, 1450–54, 1456–59, 1466–72, 1481–82, 1487–90, and 1492–94.

However, the Renaissance plague, as the post–Black Death wave of pestilence is sometimes called, differed from its predecessor in several crucial respects. Although there were occasional catastrophic exceptions, like the Great Plague of London in 1665, over the centuries *Y. pestis* steadily abated in ferocity. Outbreaks became local in nature, while, on average, mortality rates shrank to 10 to 15 percent. The pestilences of the fifteenth and sixteenth centuries also were different in other respects. If pneumonic symptoms like blood spitting persisted, no one wrote or talked about them anymore; the later plagues appear to have been largely bubonic in character, and, as in the Third

Pandemic, seasonal in nature; they struck in summer rather than all year round and moved at a slow Third Pandemic pace. Instead of leaping from city to city, they crept from neighborhood to neighborhood. Contagion remained a prominent feature, but instead of flying randomly from person to person, in its later iterations, the plague struck specific clusters of people—say, householders living in a particular alley or lane, or members of a family who slept in the same bed or wore the same clothes.

Anthropologist Wendy Orent has an interesting theory about the reasons for the change. Dr. Orent shares the Russian view that plague strains become species-specific over time; that is, the lethality, rate of dissemination, and other characteristics of a particular strain of plague are shaped by its interaction with a particular host species—hence, marmot plague is different in certain respects from rat plague, because *Y. pestis* has had a different history with the two species. Dr. Orent hypothesizes that sometime in the 1320s and 1330s, after marmot plague jumped into humans, *Y. pestis* reinvented itself as a human ailment. "The Black Death became, in a limited, short-term sense, a human disease," she says, "much of it spread lung to lung . . . [although] perhaps sometimes rats and fleas passed on the disease as well."

However, since the human version of plague represented a biological dead end—it was so lethal, it risked obliterating its host population—Dr. Orent thinks that after the Black Death, *Y. pestis* returned to its roots as a rat disease, and as it did many of the plague symptoms that have puzzled historians and scientists for centuries—such as the malodorous smell of plague victims, gangrenous inflammation of the throat and lungs, and the vomiting and spitting of blood—disappeared with it. "There is no doubt that [the rat and flea] played the principal role in converting the plague into a constant, if somewhat less virulent menace, over the next several centuries," says Dr. Orent. The nature of Europe's rodent population offers some support for her thesis. Europe does not have the right kind of wild rodent population to support permanent plague foci; *Y. pestis* requires the warmth and humidity of a burrow to survive; and European rodents do not dig the kind of burrows that the pathogen needs to sustain itself. In the post–Black Death era, the burden of sustaining the chain of infection fell to the black rat and its cousin, the Norwegian rat, neither of which is ideally suited to provide plague foci. Indeed, the truly catastrophic outbreaks of Renaissance plague, like the epidemic that struck Marseille in 1720, may not have been the work of European rodents, but rather have come from a form of plague transported to Europe from the eastern Mediterranean or the Middle East.

In the words of *Piers Plowman,* one of the most famous English literary works of the late Middle Ages, the century after the Black Death was a time of

> *feveres and fluxes*
> *Coughs and caricles, crampes and toothaches*
> *. . . Byles and bocches and brennyng agues*
> *. . . pokes and pestilences*

However, the long stream of illness that afflicted post–Black Death Europe was not just *Y. pestis*'s doing. All across the continent, the effects of chronic persistent plague were compounded by recurring waves of smallpox, influenza, dysentery, typhus, and perhaps anthrax. Sometimes several diseases would strike at once. In England, France, and Italy, for example, the *pestis secunda* was accompanied by a major outbreak of smallpox. Sometimes an illness would appear alone. In the 1440s, a major wave of smallpox—or red plague, as the disease was then called, swept through northern France, claiming even more lives than had a recent outbreak of bubonic plague. Two decades later a smallpox epidemic killed 20 percent of an English town. Influenza, another late medieval killer, produced large mortalities. In 1426–27 a major flu epidemic swept through France, the Low Countries, Spain, and eastern England, where it may have killed as much as 7 percent of the population. Another major disease of the era, the sweating sickness—or the Picardy sweat—appeared six times between 1485 and 1551, mostly in the region around the English Channel; often by the time the "sweat" had burned itself out, 10 percent of the population was dead. Poor sanitation also produced a wave of waterborne enteric fevers, especially intestinal dysentery, or the "bloody flux," and infantile diarrhea, which historian Robert Gottfried believes may have been an important contributor to the perhaps 50 percent infant mortality rate of the Middle Ages. In 1473 East Anglia, already ravaged by plague and influenza, lost 15 to 20 percent of its adult population to dysentery.

The fifteenth century also saw the emergence of "modern diseases" such as typhus, which originated in India, and syphilis, whose origins remain a source of debate. Gonorrhea, or the "French pox," long the classic venereal disease, continued to debilitate armies and upset kings. Complained Edward IV of England, after a campaign in France, "Many a man . . . fell to the lust of women and were burned by them, and their penises rotted away and fell off and they died."

In the century between 1347 (when the plague first arrived in Sicily) and 1450, estimates of Europe's population loss range from 30 to 40 percent to as high as 60 to 75 percent—a demographic collapse on the scale of the Dark Ages. Florence shrank by two-thirds—from 120,000 in 1330 to 37,000—and England by perhaps as much. Eastern Normandy may have suffered even more grievously; between the last quarter of the thirteenth century and the last quarter of the fourteenth century, the region's population shrank by 70 to 80 percent.*

As often happens after a major demographic catastrophe, immediately after the Black Death the birth rate surged. Like many of his contemporaries, the French monk Jean de Venette was astonished at the number of pregnant women on the streets. "Everywhere," he says, "women conceived more readily than usual. None proved barren; on the contrary, there were pregnant women wherever you looked. Several gave birth to twins, and some to living triplets."† Indeed, historian

* It is worth noting that in roughly the same period, 1200 to 1400, the population of China fell by half, from approximately 120 million to 60 million.

† De Venette says another oddity of the baby boom is that "when the children born after the plague began cutting their teeth, they commonly turned out to have only 20 or 22 [teeth], instead of the 32 usual before the plague."

John Hatcher argues that, had the post–Black Death demographic recovery proceeded unimpeded, by the 1380s Europe would have replaced its twenty-five million to thirty million plague dead.

The reasons for the collapse of the demographic recovery are complex. The most obvious is the torrent of disease; indeed, there was so much infectious illness in the century after the Black Death, the period is sometimes called the "Golden Age of Bacteria." However, illness in and of itself was not the only demographic depressant. A second may have been the way the various diseases interacted with one another and, even more importantly, the way they interacted with the recurring cycles of pestilence. Professor Ann Carmichael thinks one reason the influenza and smallpox outbreaks claimed so many lives is that the ill, especially the very old and young, were deprived of basic life-saving measures, like the delivery of food and normal sanitary care because the plague had killed their caregivers.

A "birth dearth" may also have contributed to the demographic decline. In the century after the Black Death, reproductive patterns seemed to have changed; women began to marry younger, and, paradoxically, females who marry before twenty tend to be less fertile over a lifetime than women who marry later. The recurring waves of disease also helped to shrink the pool of potential parents.

As the population declined, the character of medieval society began to change. For one thing, an early death became a near certainty. "To the best of our knowledge," says historian David Herlihy, "in the good years of the thirteenth century, life expectancies were 35 to 40. The ferocious epidemics of the late fourteenth century cut that figure to below 20." As the population began to stabilize again around the year 1400, Professor Herlihy thinks, people may have begun living to thirty or so. However, since high medieval infant and childhood mortality rates had a distorting effect, real-world life expectancies were not quite as dismal as Professor Herlihy's numbers make them sound. Around 1370 or 1380, a healthy twelve-year-old peasant boy in Essex, England, still had an additional forty-two years of life left. In other words, the boy could expect to live to fifty-four. But, thanks to the recurring waves of illness, by the early fifteenth century, that figure had fallen to fifty-one, and by the middle of the century, to forty-eight. In less than a hundred years, the boy had lost about 14 percent of his life span. Despite the advantages of better diet and better housing, the English nobility fared little better. In 1400 the average English peer was living eight years less than his great-grandfather had in 1300.

Post–Black Death society was also an old society. One of the paradoxical artifacts of negative demographic growth is that as population shrinks, median age increases. This is happening in modern Europe today. According to *The Economist* magazine, if current demographic trends hold—that is, if the European birth rate remains at or just below replacement rates for the dead—by the year 2050 the median age on the continent will be an astonishing fifty-two. Figures for the late Middle Ages paint a similar picture. In 1427 Florence had the same percentage of sixty-somethings as a modern low-birth-rate Western nation—15 percent. Even more suggestive are figures from the convent of Longchamp, near Paris. In 1325 the percentage of nuns at the convent who were sixty or older was already high—24 percent. By 1402, after a half century of epidemic disease, the figure had risen to 33 percent. Even more telling, in the almost eighty years between 1325 and 1402, the percentage of nuns at Longchamp between twenty and sixty—that is, the percentage in the most productive, vital years of life—declined from almost 50 percent to 33 percent.

The consequences of the Black Death cannot be properly understood unless set against this backdrop of severe, chronic population decline and lack of a vigorous young workforce. One of the most eye-catching of the consequences was a severe decline in the continent's physical infrastructure. Circa 1400, Europe was beginning to resemble medieval Rome: there were hulking pockets of survivors surrounded by untended fields, unmended fences, unrepaired bridges, abandoned farms, overgrown orchards, half-empty villages, and crumbling buildings, and hovering over everything was the oppressive sound of silence. Indeed, the continent's physical deterioration became so pronounced, it entered the cultural vocabulary of the post–Black Death era. One of the everyday sentences fifteenth-century English schoolboys were required to translate into Latin was: "[T]he roof of an old house had almost fallen on me yesterday."

Another consequence of the chronic worker shortage was that the cost of labor—and the cost of everything labor made—increased dramatically. Matteo Villani, who was a snob as well as a moralist, complained that "serving girls . . . want at least 12 florins per year and the more arrogant among them 18 or 24 florins, and . . . minor artisans working with their hands want three times . . . the usual pay, and laborers on the land all want oxen and . . . seed, and want to work the best lands and abandon all others." A thousand miles to the north, the cackling English monk Henry Knighton snorted that "all essentials were so expensive that something which had previously cost one quid, was now worth four quid or five quid." Across the channel in France, prices were so high, Guillaume de Machaut wrote a poem about inflation.

For many have certainly
Heard it commonly said
How in one thousand three hundred and forty nine
Out of one hundred there remained but nine.
Thus, it happened that for the lack of people
Many a splendid farm was left untilled,
No one plowed the fields
Bound the cereals and took in the grapes,
Some gave triple salary
But not for one denier was twenty [enough]
Since so many were dead . . .

Around 1375, food prices began to stabilize again and then to fall, as the demand for foodstuffs declined along with the population. This produced another sentence for fifteenth-century English schoolboys to translate into Latin: "No man now alive . . . can remember that ever he saw wheat or peas or corn or any other foodstuff . . . cheaper than we see now." However, the price of nearly everything except food either continued to rise or else stabilized at a high level—and this brought about a change in the European social structure so unprecedented, an amazed chronicler described it "as an inversion of the natural order."

In the fifty years after the Black Death, the medieval world's traditional economic winners and losers exchanged places. The new losers, the landed gentry, began to see their wealth shredded by the scissors of low food prices and high labor costs; the new winners, the people at the bottom, saw their one marketable asset—labor—increase dramatically in value, and with it their standard of living rise. Here is Matteo Villani looking down his nose again: "The common people, by reason of the abundance and superfluity that they found, would no longer work at their accustomed trades; they wanted the dearest and most delicate foods . . . while children and common women clad themselves in all the fair and costly garments of the illustrious who had died." Down the road in Siena, Agnolo di Tura, newly remarried and prospering, was also complaining about the greed of the lower orders. Declared the former shoemaker, "the workers of the land and the orchards, because of their great extortions and salaries, totally destroyed the farms of the citizens of Siena."

Peasants were often the biggest winners among the poor. Serfdom, in decline before the mortality, now began to disappear entirely. In the second half of the fourteenth century, a man could simply up and leave a manor, secure in the knowledge that wherever he settled, someone would hire him; alternately, the peasant could use his new leverage to extract rent reductions or obtain relief from hated feudal obligations such as the heriot—or death tax—from a hard-pressed lord. And since there was now a great deal of excess farmland available, the peasant could often pick and choose his land. In the half century after the Black Death, crop yields rose, not because agriculture improved but because now only the best land was being farmed. One measure of the new peasant prosperity was a change in inheritance patterns. Before the Black Death, peasant holdings were so small, there was not enough land for anyone but the eldest son. By 1450 peasants were often prosperous enough to leave a parcel of land to all their children—including, increasingly, their daughters.

The labor shortage even benefited the itinerant laborers who moved from manor to manor, picking up whatever work was available. Then as now, a migrant worker's salary and working conditions put him at the bottom of the economic ladder. But by the summer of 1374 a migrant worker named Richard Tailor sensed that a new day was dawning for men like him. Thus, on July 3, when his employer, William Lene, offered what Tailor felt was an inadequate wage for plowman's work, Tailor in effect told Lene: "Take this job and shove it." Walking out at the start of harvest season, Tailor made more in the next two months, August and September 1374, than the annual wage Lene had offered him—fifteen shillings versus a mere thirteen shillings, four quid.

Women were also significant economic winners in the new social order. The labor shortage opened up traditionally well-paying male occupations like metalwork and stevedoring, though the women who worked in these occupations were not paid as much as their male counterparts, and the work itself could be dangerous. In 1389, on a road near Oxford, a stevedore named Joan Edwaker was killed when her wagon tumbled over and she was crushed to death. A more typical path to empowerment was professional advancement in a traditional female field. Women cloth workers, for example, often rose from the ranks of low-paying wool combers to higher-paying weavers. By 1450 brewing—a female-dominated profession to begin with—had become virtually *all* female. Additionally, many widows took over family shops or businesses—and, not uncommonly, ran them better than their dead husbands. *Y. pestis* turns out to have been something of a feminist.

The poet who wrote

The world is changed and overthrown
That is well-nigh upside down
Compared with days of long ago.

ably articulated the feelings of the economic losers in the post–Black Death era, the landed magnates. Caught between falling land prices—a by-product of falling food prices—and rising labor costs, a great many lords simply abandoned the land. Renting out their estates, they lived off the proceeds. Magnates more committed to the land attempted to make a go of it by switching to less labor-intensive forms of farming. Abandoning grain, they concentrated on sheep and cattle. However, as a group, the ruling classes were far more interested in stuffing the genie of social change back in the bottle than they were in renting out their property or finding ways around high wage costs. After the mortality, "the ruling groups temporarily closed ranks and used the power of the state to defend the interests of the rich in the most blatant manner," says historian Christopher Dyer.

In 1349 and again in 1351, Edward III froze wages at preplague levels; new laws also made it illegal to refuse employment or to break a labor contract. In 1363 a new set of sumptuary laws banned the silk, silver buckles, and fur-lined coats the peasants had grown so fond of—as well as any other item of clothing that smacked of putting on airs or getting above oneself. To sop up some of the lower orders' excess income, in the late 1370s the poll tax was extended to previously exempt groups like unskilled laborers and servants. In 1381 resentment at attempts to reinstate "the days of long ago" helped to ignite the Peasants' Revolt.

Conflicts between the landed magnates and the newly empowered peasantry and laboring class also led to unrest and rebellion on the continent. In France, there were insurrections in 1358, 1381, and 1382, and in Ghent, in 1379.

Agriculture was not the only industry to be burdened by chronic labor shortages. In 1450 industrial Europe was producing fewer goods than it had in 1300. Particularly hard hit was the principal industry of the Middle Ages, the Flemish-dominated cloth industry, which was built around the production of cheap cloth for a mass market. Post–Black Death, there were often not enough consumers to provide a mass market. In addition, tastes changed. As people grew more prosperous, demand for plain, drab Flemish cloth declined in favor of fancier, more vivid, and sophisticated clothing.

Depopulation also had an important effect on technological innovation. The sharp decline in the workforce was an impetus for the development of labor-saving devices in many fields, including book production. During the thirteenth and fourteenth centuries, the demand for books increased steadily, propelled by a growing class of merchants, university-trained professionals, and craftsmen. But making a book in the Middle Ages was a very labor-intensive project; it required several copyists, each of whom would write out a section of the book, called a quire. In the preplague era of low wages, this method could still produce an affordable product, but not in the high-wage postplague era. Enter Johann Gutenberg, an ambitious young engraver from Mainz, Germany. In 1453, at the near-centenary of the mortality, Gutenberg introduced his printing press to the world. Chronic manpower shortages also fostered innovation in mining—new water pumps allowed fewer miners to dig deeper mines—and in the fishing industry, where new methods of salting and storing fish allowed the shrunken fishing fleets of the post–Black Death era to remain at sea longer. In the shipbuilding industry, craftsmen found ways to increase the size of vessels while reducing the size of crews. Labor shortages and high wages also helped to spur changes in the nature of warfare. As the salaries of soldiers increased, war became more expensive. This spurred the development of firearms. Weapons like the musket and cannon meant that the new high-wage soldier would provide more bang for the military buck.

There were also a number of innovations in the medical profession, which like the Church emerged from the plague with its prestige damaged. One was a greater emphasis on practical, clinically oriented medicine, a change reflected in the growing influence of the surgeon and the declining influence of the university-trained physician, who knew a great deal about Aristotle but not much about hernias and hangnails. Anatomy texts also began to become more accurate as autopsies became more common. In the new medical schools, there was a shift in emphasis toward the practical physical sciences.

These changes helped to set the stage for what today is called the scientific method. Increasingly after the Black Death, the physician—rather than deducing a conclusion from pure reason—posited a theory, tested the theory against observable fact, and rigorously analyzed the results to see if they supported the theory.

In the post–Black Death era, the hospital also began to move toward its modern form. The chief purpose of the preplague hospital was to isolate the sick—to remove them from society so that they would neither offend nor infect. "When a sick person entered the hospital he was treated as if he were dead," says Professor Gottfried. "His property was disposed of and, in many regions, a quasi-requiem mass was said for his soul." After the plague, hospitals at least attempted to cure the ill, though any patient fortunate enough to emerge from a hospital of the late Middle Ages in sound health probably owed more to good genes and good luck than to his medical treatment. One noteworthy post–Black Death innovation was the ward system: patients with specific maladies began to be housed together. Patients with broken bones were put in one ward, those with degenerative diseases in another.

The Black Death also played a major role in the birth of public health. One early innovation in the field was the municipal health board, such as those Florence and Venice established in 1348 to oversee sanitation and the burial of the dead. Later the boards would grow more sophisticated. In 1377 Venice established the first public quarantine in its Adriatic colony of Ragusa (Dubrovnik). The lazaretto, or plague house, a Florentine creation, was another early public health landmark. These were part hospital, part nursing home, and, not uncommonly, part prison.

Post–Black Death Europe also began to develop new ideas about how illness spread. It probably was not happenstance that the first systematic theory of contagion was developed by Giovanni Fracastoro, a practicing public health physician in Florence.

The plague also changed medieval higher education. Cambridge established four new colleges: Gonville Hall in 1348, Trinity Hall in 1350, Corpus Christi in 1352, and Clare Hall in 1362, while Oxford created two new schools, Canterbury and New College. Post–Black Death Florence, Prague, Vienna, Cracow, and Heidelberg also established new universities. In many cases, the charters of the schools reflected their tortured beginnings. Many mention the decay of learning and the shortage of priest-educators after the plague as the reason for their founding.

The long century of death that followed the medieval plague also had a profound effect on religious sentiment. People began to long for a more intense, personal relationship with God. One expression of the new mood was what Professor Norman Cantor has called the "privatization of Christianity." Chantries (private chapels), which were always common among the nobility, now became common among well-to-do merchants, professional families, and even artisans, who began to build private chapels through their craft guilds. Another expression of the "privatization" was the growing popularity of mysticism. In an age of "arbitrary, inexplicable tragedy," many people sought to create their own private pipeline to God.

As religious feeling intensified, the wills of the rich began to resemble celestial corporate reports. Few ventured as far into the field of "heavenly accounting" as Sir Walter Manny, whose good works included the purchase of a London cemetery and the construction of a chapel on one of its acres for monks to pray for the plague dead. (In a later iteration, the chapel became one of London's most famous landmarks, the Charterhouse.) However, rare was the wealthy man who died without leaving behind enough money to fund several lifetimes' worth of prayer for the repose of his immortal soul.

The upswing in religious feeling was accompanied by a deepening disillusionment with the Church. In the greatest crisis of the Middle Ages, the Church had proved as ineffective as every other institution in medieval society. In addition, it had lost many of its best priests, and those who survived often behaved in ways that brought shame to religious life. In 1351, as the first wave of plague was lifting, a critic wrote a blistering indictment of the clergy. "About what can you preach to the people?" he asked. "If on humility, you yourselves are the proudest of the world, arrogant and given to pomp. If on poverty, you are the most grasping and covetous, . . . if on chastity—but we will be silent on that." The critic was the sitting pope, Clement VI, a man whose own worldliness made him hard to shock. In the decades after 1351, the ordination of ill-trained boys—the ordination age was dropped from twenty-five to twenty—and ill-suited widowers further damaged the clergy's reputation. As William Langland observed in *Piers Plowman,* the only outstanding characteristic of the new clerical recruits seemed to be cupidity.

Parsons and parish priests complained to the Bishop
That their parishes were poor since the pestilence time
And asked leave and license in London to dwell
And sing requiems for stipends, for silver is sweet . . .

Given the amount of criticism leveled at the Church, it is perhaps unsurprising that several new heretical movements became active in the post–Black Death era, including the anticlerical Lollards, an English sect that attacked the ecclesiastical leadership and even questioned the spiritual benefit of the Mass. However, it would be an oversimplification to claim, as a few scholars have, that the Church's failures during the Black Death led inexorably to the Reformation. The establishment of Protestantism in northern Europe, like other large, complex historical movements, was multicausal in origin. Everything from Henry VIII's libido to the political goals of the German princes who supported Martin Luther contributed to the Reformation. The safest conclusion one can make about the plague's contribution is that, by promoting dissatisfaction with the Church, it created fertile ground for religious change.

There is a surer link between the recurrent plagues and epidemics of 1350 to 1450 and the death-obsessed culture of the late Middle Ages. Some of the motifs of the era, such as the dance of death and the *transi tomb,* predate the plague. But it required an era of mass death to transform these exercises in the macabre into major cultural phenomena. As historian Johan Huizinga observed in his classic work, *The Waning of the Middle Ages,* "no other epoch has laid as much stress . . . on the thought of death." Huizinga might have added that no other epoch has also done so little to soften the image of death. Late medieval man not only expected to die, he expected to die hard and ugly. A case in point is the sculpture that adorns of the *transi tomb* (from the Latin verb *transire,* meaning "to pass away") of Cardinal Jean de Lagrange in Avignon. It depicts the dying cardinal without pity; his mouth is agape, his eyes hollow, his cheeks sunken; his rib cage rises out of his withered lower body like a mountainous coastline. The inscription beneath the sculpture reminds the passerby: "We are a spectacle to the world. Let the great and humble, by our example, see to what state they shall be inexorably reduced, whatever their condition, age, or sex. Why then, miserable person, are you puffed up with pride? Dust you are, unto dust you return, rotten corpse, morsel and meal to worms."

Life's impermanence was also the theme of *The Three Living and the Three Dead,* a story that resonates through much late medieval art and literature. The tale, which came in several versions, centers on an encounter between the living and the dead. After the meeting, each of the three living participants comes away drawing a different moral. One man is reminded that the true purpose of this life is to repent and prepare for the next. Another, shaken by the sight and smell of death, loses himself in thoughts of the here and now; while the third man, like the first, chastened by the encounter, delivers a sermon on the transitory nature of earthly glories.

The Dance of Death, another popular cultural motif of the period, offered a different message. It presents death as a great social leveler—a jolly, ghoulish, jitterbugging democrat who insists on dancing with everyone at the party no matter how rich or poor, how highborn or lowborn. *The Dance of Death*'s dramatic possibilities made it a favorite theme of contemporary painters, poets, and dramatists, who often used the theme to make points about the late medieval social order. Thus, in several renditions, a laborer is depicted as welcoming Death as relief from his toil, while the rich and powerful, wedded to earthly pleasures, recoil in horror from the smiling hooded figure when he extends his hand for a dance. Unlike *transi tombs* and the *Three Living and the Three Dead, The Dance of Death* may have originated during the mortality. One source, the *Grands-Chroniques* of the Abbey of St. Denis, Paris, suggests that the theme arose from a 1348 encounter between two monks from the abbey and a group of dancers. When asked why they danced, one man replied, "We have seen our neighbors die and see them die daily, but since the plague has not entered our town, we hope our merrymaking will keep it away and that is why we dance."

Another story dates the origins of the dance of death to a recurrence of the plague along the Rhine in 1374. According to a German chronicler, during the outbreak, groups of the afflicted, some five hundred strong, would perform dances that concluded with the dancers falling to the ground and begging onlookers to trample on their bodies. The trampling was supposed to effect a cure. However, since plague victims often had difficulty just standing unassisted, the story has the whiff of the apocryphal about it.

For all the terrible suffering the plague inflicted, it may have saved Europe from an indefinite future of subsistence existence.

In the autumn of 1347, when the Black Death arrived in Europe, the continent was caught in a Malthusian deadlock. After two and a half centuries of rapid demographic growth, the balance between people and resources had become very tight. Nearly everywhere, living standards were either falling or stagnating; poverty, hunger, and malnutrition were widespread; social mobility rare; technological innovation stifled; and new ideas and modes of thinking denounced as dangerous heresies. The autumn morning that the Genoese plague fleet sailed into Messina harbor, a thick layer of congealed gel lay over an immobilized Europe.

The high mortalities of the Black Death and the era of recurrent disease helped to end the paralysis and allow the continent to recapture its momentum. Smaller population meant a larger share of resources for survivors—and, often as well, a wiser use of resources. After the plague, low-yielding farmland was used more productively as pasturage, and mills, once used largely to grind grain, were now put to a wider range of purposes, including fulling cloth and cutting wood. Human ingenuity also flowered, as people sought ways to substitute machine power for manpower. "A more diversified economy, a more intensive use of capital, a more powerful technology and standard of living—these seem the salient characteristics of the late medieval economy," says historian David Herlihy. "Plague, in sum, broke the Malthusian deadlock . . . , which threatened to hold Europe in its traditional ways for an indefinite future."

Horrific as a century of unremitting death had been, Europe emerged from the charnel house of pestilence and epidemic cleansed and renewed—like the sun after rain.

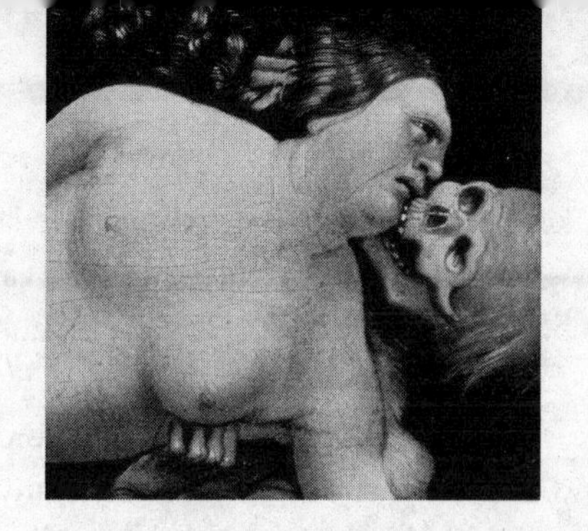

AFTERWORD

The Plague Deniers

FOR THE LAST TWENTY YEARS, A SMALL BUT VOCAL GROUP OF scholars has been challenging the traditional view of the Black Death as a plague pandemic. The "origins" controversy, as it might be called, was ignited in 1984 when Graham Twigg, a respected British zoologist, published *The Black Death: A Biological Reappraisal*. Since then, works such as *The Biology of Plagues,* by Susan Scott, a British sociologist, and her colleague, biologist Christopher J. Duncan; and *The Black Death Transformed,* by Samuel K. Cohn, a professor of medieval history at the University of Glasgow, have kept the controversy roiling. For lack of a better term, these authors—and their supporters—might be called the Plague Deniers, since they believe that the Black Death was caused by a disease other than plague.

The Deniers' case against *Y. pestis* can be reduced to two basic points. The first and weaker part of their argument involves their pet theories about what else could have caused the medieval plague. Anthrax, zoologist Twigg's candidate, has never struck a human community in epidemic form and does not produce buboes (although anthrax victims do develop black boils). Professor Cohn's candidate is a mysterious Disease X, which he does not name, but believes has probably gone extinct. Scott and Duncan go furthest of all, arguing improbably that many of the worst epidemics in Western history, from the fifth-century B.C. Plague of Athens to the Black Death, were caused by an by an Ebola-like illness they call hemorrhagic plague.

The second part of the Deniers' case against *Y. pestis* is far more substantial. Our modern understanding of plague is based on the comprehensive studies that were done during the Third Pandemic. In the hinge decades between the nineteenth and twentieth centuries, Alexandre Yersin identified the plague bacillus; Paul-Louis Simond, a French scientist, the rat-flea mechanism that drives the disease; and the Indian Plague Commission, a creation of the British Raj and one of the great achievements of Victorian medicine, compiled an unprecedentedly detailed profile of *Y. pestis*. Commission officials studied the role climate, sanitation, population density, and—in a few commission reports—nutritional status played in the spread of the plague bacillus and its vectors, the rat and rat flea. The commissioners also looked at which transportation facilities and goods were most often associated with the movement of the disease—grain unsurprisingly proved to be a big magnet for rats.

As the Plague Deniers are quick to point out, the disease that emerged from the commission's findings bears little resemblance to the disease described in the Black Death chronicles. A case in point already mentioned is the widely different dissemination rates of the two pandemics. While the Black Death virtually leaped across Europe, sometimes traveling two to two and a half miles a day, the plague of the Third Pandemic moved at a relatively sluggish ten to twenty miles per year. Another key difference is the astonishing variation in mortality rates. How could a disease that killed at least a third of the population in one appearance (the Black Death) kill under 3 percent of the population in a later outing? Some of the other discrepancies the Plague Deniers cite include:

- *Differences in symptoms.* Cardinal Francis Aidan Gasquet, the great Victorian scholar of the Black Death in England, was among the first experts to call attention to this difference. Writing at the time of the Third Pandemic, the cardinal noted that contemporary accounts of the "ordinary eastern or bubonic plague" rarely mentioned four symptoms that Black Death chroniclers referred to frequently. The four were: a "(1) Gangrenous inflammation of the throat and lungs, (2) violent pains in the region of the chest, (3) vomiting and spitting of blood, and (4) the pestial odor coming from the bodies and breath of the sick."

  Add contagion to the cardinal's list of symptoms, and you have an almost perfect description of the disease Friar Michele da Piazza described tumbling off the Genoese galleys at Messina. "Breath spread the infection . . . , and it seemed as if victims were struck all at once by the affliction . . . and so to speak, shattered by it. . . . [They] violently coughed up blood and after three days of incessant vomiting for which there was no cure, they died."

  Louis Heyligen's description of the Black Death also contains the cardinal's symptom list. "The disease is threefold in its infection," Heyligen wrote," . . . firstly men suffer in their lungs and breathing, and whoever have been corrupted or even slightly attacked cannot by any means escape nor live beyond two days. . . . Many dead bodies have been . . .

dissected and it is found that all that die, thus . . . have had their lungs infected and spat blood. . . ."

Echoing Cardinal Gasquet, the Plague Deniers note that no description of the Third Pandemic, whether written by the Indian Plague Commission or by other Western scientists, contains a list of symptoms comparable to that of Friar Michele and Heyligen, and that includes the symptom of contagion, since until it goes pneumonic, modern plague is spread via rat and flea, not from person to person.

Descriptions of the bubo do appear in accounts of both the Third Pandemic and the Black Death. But as Professor Cohn, a leading Plague Denier, notes, medieval and modern accounts of the plague describe the bubo differently. In modern plague, 55 to 75 percent of the time, the bubo develops in the groin, 10 to 20 percent of the time in the neck. Since the ankle is the most flea-accessible part of the body, this pattern makes sense. However, it is not the pattern described by many Black Death chroniclers. Fourteenth-century accounts usually locate the bubo higher up on the body, behind the ears, for instance, or on the throat, regions difficult for an insect to reach, even one that can jump one hundred times its height.

- *Rat die-offs*. Since the rat flea, *X. cheopis,* does not jump to humans until the local rat population is nearly obliterated, in theory an outbreak of human plague should be preceded by a large rat die-off. And during the Third Pandemic, practice usually followed theory. Preplague rat die-offs were common. However, references to them are exceedingly rare in the literature of the Black Death. Some scholars have tried to explain away the omission by claiming that dead rats were so common on the medieval street, chroniclers thought them unworthy of mention. To put it charitably, that theory seems improbable in the extreme. If an epidemic on the scale of the

Black Death was caused by a rat-borne plague, the streets would have been knee-deep in dead rats and people would have noticed and written about that.

- *Incidence of pneumonic plague*. The disease described by Friar Michele, Louis Heyligen, and other medieval chroniclers sounds like a variant of pneumonic plague, and, judging from the frequent references to blood spitting and hypercontagiousness in early accounts of the plague, the pneumonic disease seems to have been very common in the first six to twelve months of the Black Death, even in regions where the climate should have been hostile to it, like the Mediterranean south. In contrast, in modern outbreaks of the disease, pneumonic plague is uncommon. Some scholars claim that as many as 15 to 25 percent of modern plague cases "go pneumonic." But in Vietnam only 2 percent of the reported cases did.

- *Climate*. The whole issue of climate and plague is perplexing. Plague outbreaks during the Third Pandemic usually reflected the sensitivities of the rat and flea vectors. Outbreaks were rare during the Indian hot season, when the weather was very hot and dry, but common on either side of the hot season, when humidity increased and temperatures moderated—creating conditions favorable to *X. cheopis*. By contrast, the Black Death seemed to be largely immune to climatic effects. While outbreaks were slightly more common in warm weather, as Plague Deniers Scott and Duncan note, in some regions of Europe the mortality reached its peak in December and January. Indeed, *Y. pestis* killed almost as many people in frigid Greenland as it did in temperate Siena.

Before turning to the rejoinders of what might be called the Plague Defenders, a group that includes the majority of historians and almost all microbiologists, mention should be made of two recent discoveries by a team of French scientists. Diagnosing a disease from a list of symptoms in a medieval chronicle or medical tract—a favorite strategy of the Plague Deniers—is fraught with difficulties and imprecisions. One doctor's carbuncle can be another's plague boil. More fundamentally, written evidence ignores the fact that diseases, like people, often change over time. Measles and syphilis look and behave a lot differently today than they did when they first burst into the European population.

DNA is a far more trustworthy diagnostic tool. With that thought in mind, in the late 1990s a group of French paleomicrobiologists removed dental pulp from corpses buried in two plague pits in southern France and tested it. One pit dated from the Black Death, the other from a later recurrence of the plague. In a series of papers published in the *Proceedings of the National Academy of Sciences,* the French investigators reported finding DNA from *Y. pestis* in both samples. The French work has yet to be confirmed by researchers in other laboratories—the final step in scientific acceptance—but Didier Raoult, the lead investigator in the DNA study, is confident of his team's findings. The "medieval Black Death was the plague," he says without equivocation.

But what kind of plague?

Adopting the Russian view and describing the Black Death as an outbreak of marmot plague would help to explain many of the discrepancies that trouble the Plague Deniers. For example, marmot plague's tropism for the lungs would account for the seemingly high incidence of pneumonic disease—even in warm climates, where the weather would not have favored its transmission. Moreover, marmot plague is the only form of rodent plague that is contagious; marmots spread the disease the way humans do, via a marmot version of the cough.

Of course, marmot plague is still marmot plague. But if Dr. Wendy Orent is correct, and at some point the marmot disease evolved into a distinctly human form, given its genetic heritage, such a "humanized" plague might well produce symptoms like chest pain, blood spitting, and, as the lungs and throat became gangrenous and corrupted, a fetid body and breath odor. A humanized version of *Y. pestis* would also explain the absence of rat die-offs. Such an ailment would spread the way Friar Michele and Louis Heyligen describe the Black Death spreading, directly from person to person by way of the breath, though very likely other modes of transmission would have also developed. Thus, Dr. Orent thinks that *P. irritans,* the human flea, would have played an important role in the spread of a "humanized" plague.

Like their Russian counterparts, many American microbiologists reject the arguments of the Plague Deniers, but for different reasons. Most American scientists do not accept the Russian theory of "host" plague strains—that is, that the lethality of a particular strain of *Y. pestis* is shaped by its evolutionary history with particular rodent species. The plague bacillus is only fifteen thousand to twenty thousand years old, says Professor Robert Brubaker, dean of American plague researchers. In evolutionary terms, Dr. Brubaker thinks that is not enough time for the bacillus to have evolved very far away from its original form.

In Dr. Brubaker's view and that of most of his colleagues, the Black Death and the Third Pandemic were classic examples of rat-borne plague. In the American view, differences between the two outbreaks can be explained largely in terms of external factors. One of the most important of these externals is the very different levels of knowledge available to physicians of the fourteenth and late nineteenth centuries. From firsthand observation, medieval Europeans sensed that plague was affected by factors like sanitation, nutrition, and the movement of goods and people, but this practical knowledge was attached to beliefs about the importance of astrology, miasmas, and bodily humors.

"By the late nineteenth century," says Dr. Brubaker, "physicians and scientists understood the principles of contagion . . . [and] had a good working knowledge of how infectious disease spreads and the measures needed to be taken to safeguard public health." One by-product of this new understanding was that seven-hundred-year-old insights could now be transformed into effective public health strategies. The municipal health board of Black Death Florence may have had little success with its sanitary measures, but "the Indian Plague Commissioners believed that they were able to prevent catastrophe [by] imposing aggressive controls on public and hospital sanitation," says historian and physician Ann Carmichael of the University of Indiana.

Effective sanitary measures also played an important role in controlling plague outbreaks in Hong Kong and Canton in the mid-1890s. However, in those localities, physicians cited two other measures with Black Death echoes as also important—adequate nutrition and decent nursing care.

It is noteworthy that when measures like stringent sanitation collapsed, as happened in Bombay during a disease outbreak in 1897, the plague of the Third Pandemic quickly began to behave like the plague of the Black Death. In a Bombay hospital a plague commissioner characterized as rife with "fatigue, destitution, filth, poverty and overcrowding," the death rate reached 64.5 percent in the spring of 1897. Professor Cohn may be correct in saying that no outbreak of the Third Pandemic produced mortalities on the scale of the Black Death, but in the terrible months between August 1896 and February 1897, Bombay lost nineteen thousand people to the disease.

A modern understanding of another aspect of sanitation—personal hygiene—may also help explain why, if the Black Death was an outbreak of rat plague, there were so few rat die-offs. Given the unhygienic state of the medieval body, it is highly likely that *P. irritans,* which preys on people, not rodents, played a major role in spreading the plague from person to person.

The Plague Deniers have long argued that *P. irritans*'s bite, unlike *X. cheopis*'s, transmits too few plague bacilli to make it a very efficient disease vector. However, that weakness may not have mattered during the Black Death. Whatever its form, the medieval plague was extraordinarily virulent; the very high concentrations of bacilli in human blood may have turned even normally weak insect vectors into efficient plague carriers. Furthermore, it is not at all clear that the human flea is that weak a disease vector. Observers as diverse as General Ishii (inventor of the Japanese plague bomb), United States Army Intelligence, and Giovanni Boccaccio have all offered testimonials to *P. irritans*'s efficiency as a disease carrier. The two pigs Boccaccio describes as dropping dead after mauling a blanket almost certainly were killed by the bite of *P. irritans,* which is a pig as well as human flea.

Perhaps the most compelling testimony about the effectiveness of the human flea as a plague vector comes from Dr. Kenneth Gage, chief of the Plague Division at the Centers for Disease Control. From his personal experience fighting the disease in modern Africa, Asia, and South America, Dr. Gage has become convinced that the human flea plays an important but underappreciated role in the spread of plague.

The weakest part of the Black-Death-as-a-purely-rat-borne-plague theory involves the apparently very high incidence of the pneumonic form of the disease. Secondary pneumonic infections occur in bubonic plague, but, at least in modern experience, only infrequently. French scholar Jean-Noël Biraben hypothesizes that the unusually cold weather of the fourteenth century may have been especially "pneumonic-friendly." The problem with the Biraben theory is that, south of the Alps, the weather was still warm when the Black Death arrived.

One possible explanation for the high rate of "pneumonic" Black Death is that two disease strains were at work in 1348 and 1349. The Manchurian outbreak of pneumonic plague in 1911 arose from marmots. Yet the epidemic occurred in the midst of the rat-based bubonic plague of the Third Pandemic. Perhaps something similar occurred during the Black Death? Another possibility is that over the course of the middle years of the fourteenth century, *Y. pestis* underwent a fundamental evolutionary change, as Dr. Orent suggests.

Medievalist and physician Ann Carmichael also has a theory about the medieval plague, one that addresses not only the high incidence of pneumonic disease but, more sweepingly, why, on many clinical and epidemiological indexes, it looks so different from the Third Pandemic. "There may have been something fundamentally different about the nature of the premodern world," says Dr. Carmichael, "something we don't understand and which cannot be duplicated today even in Third World regions."

However, about one thing we can be certain.

Microbiologist Didier Raoult is right; the Black Death was an outbreak of plague.

NOTES

Chapter One: Oimmeddam

2 a handsome town of "beautiful markets": Ibn Battuta in W. Heyd, *Histoire du commerce du Levant au moyen age,* vol. 2 (Leipzig: O. Harrassowitz, 1936), pp. 172–74.

2 Vivaldi brothers: J. R. S. Phillips, *The Medieval Expansion of Europe,* 2nd ed. (Oxford: Clarendon Press, 1998), p. 238.

2 "a city of sea without fish": Eileen Power, *Medieval People* (New York: Harper & Row, 1963), p. 42.

2 "And so many are the Genoese": Steven A. Epstein, *Genoa and the Genoese, 958–1528* (Chapel Hill: University of North Carolina Press, 1996), p. 166.

3 travel from the Crimea to China: Phillips, *The Medieval Expansion of Europe,* p. 100.

3 Description of Caffa: Heyd, *Histoire du commerce du Levant,* pp. 170–74. See also: G. Balbi and S. Raiteri, *Notai genovesi in Oltremare Atti rogati a Caffa e a Licostomo* (Genoa: 1973), sec. 14; R. S. Lopez, *Storie delle Colonie Genovese nel Mediterraneo* (Bologna: 1930).

4 tremendous environmental upheaval: J. F. C. Hecker, *The Epidemics of the Middle Ages,* trans. B. G. Babington (London: Trübner, 1859), pp. 12–15.

4 Mandate of Heaven: Sir Henry H. Howorth, *History of the Mongols, from the 9th to the 19th Century,* vol. 2 (London: 1888), p. 87.

4 "in the Orient": Gabriele de' Mussis, quoted in Stephen D'Irsay, "Defense Reactions During the Black Death," *Annals of Medical History* 9 (1927), p. 169.

4 "Hard by greater India": Louis Heyligen, "Breve Chronicon Cleric Anonymi," excerpted in *The Black Death: Manchester Medieval Sources,* trans. and ed. by Rosemary Horrox (Manchester: Manchester University Press, 1994), pp. 41–42.

5 Dorias of Genoa: Phillips, *The Medieval Expansion of Europe,* p. 102.

5 "eaters of sweet greasy food": René Grousset, *Empire of the Steppes: A History of Central Asia* (New Brunswick, NJ: Rutgers University Press, 1970), p. 249.

5 "The road you take": Francesco Balducci di Pegolotti, in R. S. Lopez and Irving W. Raymond, *Medieval Trade in the Mediterranean World: Illustrative Documents Translated with Introductions and Notes* (New York: Columbia University Press, 1955), pp. 355–58.

5 An ultimatum was sent: A. A. Vasiliev, *The Goths in the Crimea* (Cambridge, MA: Mediaeval Academy of America, 1936), p. 48.

6 "Oh God": Gabriele de' Mussis, "Historia de Morbo," in Horrox, *The Black Death,* p. 17.

6 "In 1346": Ibid., p. 16.

6 "terrible events": Heyligen, "Breve Chronicon Cleric Anonymi," in Horrox, *The Black Death,* p. 42.

6 pestilence raged in the East: Michael W. Dols, *The Black Death in the Middle East* (Princeton: Princeton University Press, 1977), p. 40.

6 Jijaghatu Toq-Temur and his sons: Ibid., p. 41.

6 Hopei province: William H. McNeill, *Plagues and Peoples* (New York: Anchor Books, 1976), p. 173.

7 "a six month ride": Dols, *The Black Death in the Middle East,* p. 41.

7 Lake Issyk Kul: Jean-Noël Biraben, *Les hommes et la peste en France et dans les pays européens et méditerranéens,* vol. 1 (Paris: Mouton & Co., 1975), pp. 49–55.

7 "In the year": J. Stewart, *Nestorian Missionary Enterprise: The Story of a Church on Fire* (Edinburgh: T. & T. Clark, 1928), p. 198.

8 after Issyk Kul: Robert S. Gottfried, *The Black Death: Natural and Human Disaster in Medieval Europe* (New York: Free Press, 1983), p. 36.

8 Russian Chronicle: Ole J. Benedictow, *The Black Death, 1346–1353: The Complete History* (Woodbridge, Suffolk: Boydell Press, 2004), p. 50.

8 "Stunned and stupefied": de' Mussis, "Historia de Morbo," in Horrox, *The Black Death,* p. 17.

9 fabricated the catapults: Benedictow, p. 52.

10 chronicler of Este: Quoted in Philip Ziegler, *The Black Death* (New York: Harper & Row, 1969), p. 16.

10 ancient Indian legend: David E. Stannard, "Disease, Human Migration and History" in *The Cambridge World History of Human Disease* (Cambridge: Cambridge University Press, 1993), p. 35.

11 200 million people: Robert R. Brubaker, "The Genus Yersinia," *Current Topics in Microbiology* 57 (1972): p. 111.

11 Foster scale: Harold Foster, "Assessing the Magnitude of Disaster," *Professional Geographer* 28 (1976): pp. 241–47.

11 David Herbert Donald: David Herbert Donald, "The Ten Most Significant

Events of the Second Millennium," in *The World Almanac and Book of Facts* (Mahwah, NJ: Primedia, 1999), p. 35.

11 Cold War—era study: Jack Hirshleifer, *Disaster and Recovery: The Black Death in Western Europe,* prepared for Technical Analysis Branch United States Atomic Energy Commission (Los Angeles: RAND Corporation, 1966), pp. 1–2.

11 mortality figure is 33 percent: Ziegler, *The Black Death,* p. 230. See also: Maria Kelly, *A History of the Black Death in Ireland* (Stroud, Gloucestershire: Tempus, 2001), p. 41; "The Black Death in the Middle Ages," *Dictionary of the Middle Ages,* ed. by Joseph R. Strayer (New York: Charles Scribner, 1982), p. 244; Horrox, *The Black Death,* p. 3; William Naphy and Andrew Spicer, *The Black Death and History of Plagues, 1345–1730* (Stroud, Gloucestershire: Tempus, 2000), pp. 34–35.

12 "Where are our dear friends?": Francesco Petrarch, "Letter from Parma," in Horrox, *The Black Death*, pp. 248–49.

12 In the Islamic Middle East: Gottfried, *The Black Death,* p. 35.

12 "the voice of existence": Ibn Khaldun, in Dols, *The Black Death in the Middle East,* p. 67.

12 In China: McNeill, *Plagues and Peoples,* p. 174.

12 disaster on the scale of the Black Death: Naphy and Spicer, *The Black Death and History of Plagues,* p. 35.

13 environmental stress in the fourteenth century: M. G. L. Baillie, "Putting Abrupt Environmental Change Back into Human History," in *Environments and Historical Change,* ed. Paul Slack (Oxford: Oxford University Press, 1998), pp. 52–72. See Also Bruce M.S. Campbell, "Britain 1300," *History Today,* June 2000.

13 relationship between plague and earthquakes: Personal communication, Dr. Ken Gage, chief of Plague Control Division, Centers for Disease Control and Prevention, U.S. Centers for Disease Control.

13 "We marvel to see": M. G. L. Baillie, "Marking in Marker Dates: Towards an Archeology with Historical Precision," *World Archeology* 23, no. 2 (Oct. 1991): 23.

13 environmental instability: Gottfried, *The Black Death,* p. 34.

14 Seismic activity in the world's oceans: Personal communication. M. G. L. Baillie, professor, School of Archaeology and Paleoecology, Queen's University, Belfast.

14 risk factors in plague: Robert Pollitzer, *Plague* (Geneva: World Health Organization, 1954); L. Fabian Hirst, *The Conquest of Plague: A Study of the Evolution of Epidemiology* (Oxford: Clarendon Press, 1953); Wu Lien-Teh et al., *Plague: A Manual for Medical and Public Health Workers* (Shanghai: Weishengshu National Quarantine Service, Shanghai Station, 1936); Wu Lien-Teh, *A Treatise of Pneumonic Plague* (Geneva: 1926); Plague Research Commission, "On the Seasonal Prevalence of Plagues in India," *Journal of Hygiene* 8 (1900): 266–301; Plague Research Commission, "Statistical Investigation of Plague in the Punjab, Third Report on Some Factors Which Influence the Prevalence of Plague," *Journal of Hygiene* 11 (1911): 62–156.

14 role of malnutrition: David Herlihy, in *The Black Death and the Transformation*

of the West, ed. by Samuel K. Cohn, Jr. (Cambridge, Mass.: Harvard University Press, 1997), p. 34.

16 "Malthusian deadlock": Herlihy, in *The Black Death and the Transformation of the West,* p. 34.

16 died of starvation: William Chester Jordan, *The Great Famine: Northern Europe in the Fourteenth Century* (Princeton: Princeton University Press, 1996), pp. 118, 147.

16 medieval city a human cesspool: Jean-Pierre Leguay, *La rue au Moyen age* (Paris: Éditions Ouest-France, 1984).

17 "I shall undress myself": Giovanni Boccaccio, *The Decameron,* trans. G. H. McWilliam (London: Penguin Books, 1972), p. 308.

17 "boiled over": Terence McLaughlin, *Coprophilia, Or a Peck of Dirt* (London: Cassell, 1971), p. 19.

17 the medieval countryside: Benedictow, *The Black Death, 1346–1353,* pp. 33, 34.

18 "Corrupted air": "The Report of the Paris Medical Faculty, October 1348," in Horrox, *The Black Death,* p. 158.

18 Gentile da Foligno: "Tractatus de pestilenta," in *Archive für Geschichte der Medizin,* ed. by Karl Sudhoff (Berlin: 1912), p. 84.

18 "a troupe of ladies": Henry Knighton, "Chronicon Henrici Knighton," in Horrox, *The Black Death,* p. 130.

19 grain or cloth shipments: Graham Twigg, *The Black Death: A Biological Reappraisal* (London: Batsford Academic and Educational, 1984), p. 21.

19 in an infected flea: Robert R. Brubaker, "Yersinia Pestis," in *Molecular Medical Microbiology,* ed. by M. Sussman (London: Academic Press, 2001), pp. 2033–2058.

20 incubation period: Robert Perry and Jacqueline D. Fetherstone, "Yersinia Pestis—Etiologic Agent of Plague," *Clinical Microbiology Review* 10, no. 1 (Jan. 1997): p. 58.

20 God's tokens: Benedictow, *The Black Death, 1346–1353: The Complete Story,* p. 26.

21 Infrequency of plague symptoms: personal communication with Dr. Kenneth Gage, chief of the Plague Control Division, CDC.

22 Difficulty in transmitting pneumonic plague: Benedictow, p. 28.

22 Survival time in septicemic plague: Ibid., p. 26.

23 term Black Death: Jon Arrizabalaga, "Facing the Black Death: Perceptions and Reactions of University Medical Practitioners," in *Practical Medicine from Salerno to the Black Death,* ed. Luis Garcia-Ballester, Robert French, Jon Arrizabalaga, and Andrew Cunningham (Cambridge: Cambridge University Press, 1994), pp. 242–243.

23 the "Big Death": Samuel K. Cohn, Jr., *The Black Death Transformed: Disease and Culture in Early Renaissance Europe* (London: Arnold, 2002), p. 104.

23 recent research suggests: J. Clairborne Stephens et al., "Dating the Origin of the CCR5-Δ32 AIDS-Resistance Allele by the Coalescence of Haplotypes," *American Journal of Human Genetics* 62 (1998): 1507–15. See also: Cohn, *The Black Death Transformed,* p. 250.

25 "Unusual conjuction": "The Report of the Paris Medical Faculty, October 1348," in Horrox, *The Black Death*, p. 158.

25 "No fellow human being": "Letter of Edward III to Alfonso, King of Castile," in Horrox, *The Black Death,* p. 250.

25 "black smoke": W. Rees, "The Black Death in England and Wales," *Proceedings of the Royal Society of Medicine* Vol. 16 (part 2 [1920]): 134.

25 "waiting among the dead": In *Annalium Hibernae Chronicon,* ed. by R. Butler (Dublin: Irish Archaeological Society, 1849), p. 37.

26 "the sick hated": Marchionne di Coppo Stefani, *Cronica friorentino,* ed. by Niccolo Rodolico, RIS, XXX/1 (Città di Castello: 1903).

26 "end of the world": Agnolo di Tura del Grasso, "Cronaca sense attribuita ad Agnolo di Tura del Grasso," in *Cronache senesi,* ed. by A. Lisini and F. Iacometti, RIS, XV/6 (Bologna: 1931–37), p. 555.

26 "each grave": Ordinances of Pistoia 1348, quoted in Horrox, *The Black Death,* pp. 195–203.

26 "turned and fled": Neuburg Chronicle, in Ziegler, *The Black Death,* p. 84.

27 "lashed themselves": Thomas Walsingham, *Historia Anglicana 1272–1422,* H.T. Riley (ed.), 2 vols, Rolls Series, 1863–64, Vol 1, p. 275.

27 "no mortal": Hecker, p. 13.

Chapter Two: "They Are Monsters, Not Men"

30 *La Practica della Mercata*: R. S. Lopez and Irving W. Raymond, *Medieval Trade in the Mediterranean World: Illustrative Documents Translated with Introductions and Notes* (New York: Columbia University Press, 1955), pp. 355–58.

30 travel time to Mongolia and China: J. R. S. Phillips, *The Medieval Expansion of Europe* (Oxford: Clarendon Press, 1998) p. 100.

30 "They [are] like beasts": René Grousset, *Empire of the Steppes: A History of Central Asia* (New Brunswick, N.J.: Rutgers University Press, 1970), p. 249.

31 Description of Mongols: William of Rubruck, *The Mission of Friar William of Rubruck: His Journey to the Court of the Great Khan Möngke, 1253–1255,* trans. by Peter Jackson (London: Hakluyt Society, 1990), pp. 89–90.

31 "discovery of Asia": René Grousset, *Histoire de l'Asie* (Paris: 1922), p. 130.

31 Dog Men: Ibn Battuta, in *Cathay and the Way Thither: Being a Collection of Medieval Notices of China,* trans. and ed. by Colonel Sir Henry Yule (London: Hakluyt Society, 1913–16), p. 94.

31 Prester John: Robert Marshall, *Storm from the East: From Genghis Khan to Khubilai Khan* (Berkeley: University of California Press, 1994), p. 121.

31 "delighted, yea": J. Stewart, *Nestorian Missionary Enterprise: The Story of a Church on Fire* (Edinburgh: T. & T. Clark, 1928), p. 7.

32 "I myself am": Eileen Power, "The Opening of the Land Routes to Cathay," in *Travel and Travellers of the Middle Ages,* ed. by Arthur Percival Newton (London: K. Paul, Trench, Trubner & Co., 1926), p. 147.

32 William's discoveries: Rubruck, *The Mission of Friar William of Rubruck,* p. 50.
32 defended the Western concept: Ibid., p. 229.
32 second wave of European visitors: Power, "The Opening of the Land Routes to Cathay," p. 128.
32 Italian colonies: Phillips, *The Medieval Expansion of Europe,* pp. 104–5.
33 "worth more": Power, "The Opening of the Land Routes to Cathay," p. 137.
33 route led down: Ibid., pp. 140–41.
33 Hangchow, Venice of the East: Stewart, *Nestorian Missionary Enterprise,* p. 193. See also: Power, "The Opening of the Land Routes to Cathay," pp. 134–35.
33 singing virgins: Howorth, *History of the Mongols,* p. 310.
33 route across the northern steppe: Phillips, *Medieval Expansion of Europe,* p. 99, and Power, "The Opening of the Land Routes to Cathay," p. 142. See also McNeill, *Plagues and Peoples,* p. 163.
33 tarabagan colonies: Hirst, *Conquest of Plague,* p. 189.
33 *Memories of a Hunter in Siberia:* A. K. Tasherkasoff, *Memories of a Hunter in Siberia*. Wu Lien-Teh et al., *Plague: A Manual for Medical and Public Health Workers* (Shanghai: Weishengshu National Quarantine Service, Shanghai Station, 1936), p. 198.
34 "tarabagan gardens": Ibid., p. 7.
34 marmot plague: Wendy Orent, *Plague: The Mysterious Past and Terrifying Future of the World's Most Dangerous Disease* (New York: Free Press, 2004), pp. 56–60, 158.
34 "I only want one strain": Ibid., p. 58.
34 seems to have originated: Wu Lien-Teh, "The Original Home of the Plague," *Japan Medical World* 4, no. 1 (January 15, 1924): 7. See also: Orent, *Plague,* pp. 55–60.
34 big bang: Dr. Robert R. Brubaker, Professor of Microbiology at Michigan State University, personal communication.
34 *Y. pestis* is only: Mark Achtman et al., "Yersinia pestis: The Cause of the Plague Is a Recently Emerged Clone of Yersinia Pseudotuberculosis," *Proceedings of the National Academy of Sciences* 46, no. 24 (November 23, 1999): 14043–48.
35 *Y. pestis* has all the properties: Brubaker, "Yersinia Pestis," pp. 2033–2058.
35 lions: Samuel K. Cohn, Jr., *The Black Death Transformed: Disease and Culture in Early Renaissance Europe* (London: Arnold, 2002), p. 132.
35 different flea species: Robert Perry and Jacqueline D. Fetherstone, "Yersinia Pestis—Etiologic Agent of Plagues," *Clinical Microbiology Review* 10, no. 1 (Jan. 1997): p. 52.
36 General Shiro Ishii: Thomas W. McGovern, M.D., and Arthur M. Friedlander, M.D., "Plague," in *Military Aspects of Chemical and Biological Warfare* (Washington, DC: Office of the Surgeon General at TNN, 1997), pp. 483–85.
36 "one of Ishii's greatest achievements": Ibid., p. 485.
36 number of animals are also resistant: Perry and Fetherstone, "Yersinia Pestis," p. 53.

36 allele that protects: Cohn, *The Black Death Transformed,* p. 252.
36 partial immunity: Perry and Fetherstone, "Yersinia Pestis," p. 52.
37 surge years: Hirst, *The Conquest of Plague,* p. 214.
37 sun spot cycles: Ibid.
38 local hunters: Ibid., p. 217.
38 pneumonic plague broke out: Graham Twigg, *The Black Death: A Biological Reappraisal* (London: Batsford Academic and Educational, 1984), pp. 164–65.
38 looking for new pastureland: Robert S. Gottfried, *The Black Death: Natural and Human Disaster in Medieval Europe* (New York: Free Press, 1983), p. 34.
38 wind patterns: Ibid., p. 14.
39 "Assuredly the far-flung": McNeill, *Plagues and Peoples,* p. 175.
39 water temperatures: Stewart, *Nestorian Missionary Enterprise,* p. 199.
39 "We have found": Ibid., p. 193.
39 "This is the grave": Ibid., pp. 212–13.
42 "The Japanese . . . have": Thomas Butler, *Plague and Other Yersinia Infections* (New York: Plenum Medical Book Co., 1983), p. 17.
42 "The pulp of the buboes": Ibid., p. 18.
42 "disease of rats": William Ernest Jennings, *A Manual of Plague* (London: Rebman, 1903), pp. 39–40.
42 port of Pelusium: Jean-Noël Biraben and Jacques Le Goff, "The Plague in the Early Middle Ages," in *Biology of Man in History,* ed. by Robert Foster and Orest Ranum (Baltimore: Johns Hopkins University Press, 1975), p. 58.
42 trade route from Egypt: Ibid.
42 tree ring dates: M. G. L. Baillie, "Putting Abrupt Environmental Change Back into Human History," in *Environments and Historical Change,* ed. by Paul Slack (Oxford: Oxford University Press, 1998), pp. 55–56.
43 "The streams are": Georges Duby, *The Early Growth of the European Economy,* trans. by Howard B. Clarke (Ithaca, NY: Cornell University Press, 1974),p. 10.
43 "History of Evagrius Scholastica Ecclesiastica," trans. M. Whitley (Liverpool: University of Liverpool, 2000), pp. 229–33.
43 name badges: P. Allen, "The 'Justinianic' Plague in Byzantion," *Revue Internationale des Études Byzantines* 49 (1979):5–20.
43 "In every field": Ibid., p. 12.
43 "soon no coffins": Biraben and Le Goff, "The Plague in the Early Middle Ages," p. 57.
43 smallpox and measles outbreaks: McNeill, *Plagues and Peoples,* pp. 131–32.
44 if not disease-free: Gottfried, *The Black Death,* p. 12.
44 population plunged precipitously: "Demography," in Joseph Strayer, ed., *The Dictionary of the Middle Ages* (New York: Scribner, 1982), p. 140.
44 twenty thousand residents: David Herlihy, "Ecological Conditions and Demographic Changes," in *Western Europe in the Middle Ages* (Boston: Houghton Mifflin, 1977), p. 4.
44 dense woodland: "Demography," in *The Dictionary of the Middle Ages,* p. 140.

44 Little Optimum: "Climatology," Ibid., p. 456.
45 European farms began to produce: Gottfried, *The Black Death,* p. 25.
45 horse collar: Herlihy, "Ecological Conditions and Demographic Changes," p. 18.
45 *carruca* plow: Ibid., p. 17.
45 "river throws itself": David Levine, *At the Dawn of Modernity: Biology, Culture, and Material Life in Europe After the Year 1000* (Berkeley: University of California Press, 1996), p. 169.
46 Population estimates: "Demography," in *The Dictionary of the Middle Ages,* p. 141.
46 urban life reawakened: "Demography," Ibid., p. 141.
46 medieval countryside: "Demography," Ibid.
47 village of Broughton: Edward Britton, *The Community of the Vill: A Study in the History of the Family and Village Life in Fourteenth-Century England* (Toronto: Macmillan of Canada, 1977), p. 138.
47 Europeans burst out of: "Demography," in *The Dictionary of the Middle Ages,* p. 140.
48 prostitutes-for-a-day: Joseph and Frances Gies, *Life in a Medieval City* (New York: Harper & Row, 1969), p. 86.
48 local tolls: Power, "The Opening of the Land Routes to Cathay," p. 137.
48 "rulers of half": Eileen Power, *Medieval People* (New York: Harper & Row, 1963), p. 42.
49 Sorceress of Ryazan: Marshall, *Storm from the East,* p. 97.
49 "For our sins": Phillips, *The Medieval Expansion of Europe,* p. 62.
50 held a *kuriltai*: Marshall, *Storm from the East,* p. 88.
50 fisheries of Yarmouth: Power, "The Opening of the Land Routes to Cathay," p. 127.
50 "Old Man of the Mountain": Phillips, *The Medieval Expansion of Europe,* p. 63.
50 "monsters rather than men": Ibid.
50 "You personally, as the head": Ibid., p. 60.
51 "great numbers of Pharaoh's rats": Marco Polo in Wu Lien-Teh, *Plague: A Manual for Medical and Public Health Workers,* p. 199.

Chapter Three: The Day Before the Day of the Dead

54 Broughton had some 268 residents: Edward Britton, *The Community of the Vill: A Study in the History of the Family and Village Life in Fourteenth-Century England* (Toronto: Macmillan of Canada, 1977), p. 138.
54 Broughton was anglicizing itself: Ibid., pp. 11–12.
54 John's great aunt Alota: Ibid., p. 29.
55 John was fined for drinking: Ibid., pp. 42–43.
55 receive an *alebedrep*: Barbara A. Hanawalt, *The Ties That Bound: Peasant Families in Medieval England* (Oxford: Oxford University Press, 1986), p. 58.
55 spinal deformations: Brian M. Fagan, *The Little Ice Age: How Climate Made History, 1300–1850* (New York: Basic Books, 2000), p. 33.

55 die young: David Herlihy, "The Generation in European History," in *The Social History of Italy and Western Europe, 700–1500,* vol. 12 (London: Variorum Reprints, 1978), p. 351.
55 heriot: Hanawalt, *The Ties That Bound,* p. 110.
56 acreage under plow: Christopher Dyer, *Making a Living in the Middle Ages: The People of Britain 850–1520* (New Haven: Yale University Press, 2000), p. 239.
56 decline in productivity: Ibid.
56 officials in west Derbyshire: Ibid., p. 236.
56 rents in central London: Ibid., p. 243.
57 "Many . . . went hungry": David Herlihy, in *The Black Death and the Transformation of the West,* ed. Samuel K. Cohn, Jr. (Cambridge, Mass.: Harvard University Press, 1997), p. 38.
57 weather was changing: "Climatology," in *The Dictionary of the Middle Ages,* pp. 454–55. See also Philip Ziegler, *The Black Death* (Harper & Row, 1969), p. 32.
58 "The ice now comes: "Climatology," in *The Dictionary of Middle Ages,* p. 455.
58 Little Ice Age: Fagan, *The Little Ice Age,* pp. 48–49.
58 poor and mediocre harvests: Ian Kershaw, "The Great Famine and Agrarian Crisis in England, 1315–1322," *Past and Present* 59 (May 1975), p. 7.
58 "inundation of waters": William Chester Jordan, *The Great Famine: Northern Europe in the Fourteenth Century* (Princeton: Princeton University Press, 1996), p. 24.
58 Flanders experienced some of the worst downpours: Henry Lucas, "The Great European Famine of 1315, 1316, and 1317," *Speculum* 5 (1930): 348.
59 near the English village of Milton: Ibid., p. 346.
59 poor huddled under trees: Jordan, *The Great Famine,* p. 143.
59 "cries that were heard": Ibid., p. 141.
59 "dearness of wheat": Ibid., p. 135.
60 cost of wheat: Lucas, "The Great European Famine," p. 352.
60 year's worth of barley: Dyer, *Making a Living in the Middle Ages,* p. 230.
60 "extracted the bodies": John de Trokelowe in John Aberth, *From the Brink of the Apocalypse: Confronting Famine, War, Plague, and Death in the Later Middle Ages* (New York: Routledge, 2000), p. 13.
60 "Incarcerated thieves": Ibid., p. 14.
60 "parents, after slaying their children": Jordan, *The Great Famine,* p. 148.
60 accounts of cannibalism: Ibid., pp. 149–50.
60 exiled for stealing food: Ibid., p. 271.
60 Adam Bray: Dyer, *Making a Living in the Middle Ages,* p. 231.
61 "certain malefactors": Lucas, "The Great European Famine," p. 360.
61 "serenity of the air": Aberth, *From the Brink of the Apocalypse,* p. 35.
61 Bolton Abbey: Dyer, *Making a Living in the Middle Ages,* p. 229.
61 "most savage, atrocious death": Jordan, *The Great Famine,* p. 143.
61 In Antwerp: Lucas, "The Great European Famine," p. 367.
61 In Erfurt: Jordan, *The Great Famine,* p. 144.
61 In Louvain: Ibid., p. 144.

61 In Tournai: Aberth, *From the Brink of the Apocalypse,* p. 54.
61 animals began to die: Kershaw, *The Great Famine and Agrarian Crisis,* pp. 20–21. See also: Jordan, *The Great Famine,* p. 36.
62 "It is a dysentery type illness: Ibid., p. 14.
62 vitamin deficiencies: Aberth, *From the Brink of the Apocalypse,* pp. 14–15.
62 half-million people died: Dyer, *Making a Living in the Middle Ages,* p. 235.
62 Flanders and Germany: Jordan, *The Great Famine,* p. 148.
63 "Think how their bodies": Giovanni Morelli in Herlihy, in *The Black Death and the Transformation of the West,* p. 33.
63 "poorly nourished": Simon Couvin in Herlihy, in *The Black Death and the Transformation of the West,* p. 33.
63 question the link: Biraben, *Les hommes et la peste en France et dans les pays européens et méditerranéens,* vol. 1 (Paris: Mouton, 1975), pp. 131–32.
63 periods of dearth: Cohn, *The Black Death Transformed: Disease and Culture in Early Renaissance Europe* (London: Arnold, 2002), p. 32.
63 "A famine": Jordan, *The Great Famine,* p. 186.
64 fetal malnutrition is also a factor: S. E. Moore, A. C. Cole, et al., "Prenatal or Early Postnatal Events Predict Infectious Deaths in Young Adulthood in Rural Africa," *International Journal of Epidemiology* 28, no. 6 (December 1999): 1088–1095.
64 mortality pattern: Jordan, *The Great Famine,* pp. 186–87.
64 confronted a peddler: Ernest L. Sabine, "City Cleaning in Medieval London," *Speculum* 12, no. 1 (1937): 29.
65 "I make bold": *Stefan's Florilegium,* ed. by Mark Harris, May 19, 1997, stefan@florilegium.org. See also: www.florilegium.org.
65 antirodent remedies: Ibid.
65 Princess Asaf-Khan: Hirst, *The Conquest of Plague,* p. 124.
65 "Dead rats in the east": Shi Tao-nan in Wu Lien-Teh et al., *Plague,* p. 12.
66 incredible powers of reproduction: J. Laurens Nicholes, *Vandals of the Night* (Los Angeles, 1948), pp. 18–19.
66 remarkable qualities: Robert Pollitzer, *Plague* (Geneva: World Health Organization, 1954), p. 286.
66 reconnaissance lesson: Nicholes, *Vandals of the Night,* p. 22.
67 rats have been observed laughing: Jaak Panksepp and Jeffrey Burgdorf, "'Laughing' Rats and the Evolutionary Antecedents of Human Joy?" *Physiology and Behavior* 79 (2003): 533–47.
67 rat migrations: Pollitzer, *Plague,* p. 294.
67 rat first appeared in the West: F. Audoin-Rouzeau, "Le rat noir *(Rattus rattus)* et la peste dans l'occident antique et médiéval," *Bulletin de la Société de Pathologie Exotique* 92, no. 5 (1999): 125–35.
68 "horse dung": Sabine, "City Cleaning in Medieval London," p. 26.
69 "Every seat": Lucinda Lambton, *Temples of Convenience and Chambers of Delight* (New York: St. Martin's Press, 1995), p. 9.
69 public sanitation systems: Sabine, "City Cleaning in Medieval London," p. 21.

70 "dung, lay-stalls": *Memorials of London and London Life in the XIIIth, XIVth, and XVth Centuries,* ed. by H. T. Riley (London: London, Longmans, Green and Co., 1868), p. 295.
70 municipal sanitation workers: Sabine, "City Cleaning in Medieval London," p. 23.
70 attacked by an assailant: Ibid., p. 30.
71 two women in Billingsgate: Ibid.
71 "Filth [is] being": Philip Ziegler, *The Black Death* (New York: Harper & Row, 1969), p. 156.
71 unfortunate English peasant: Graham Twigg, *The Black Death: A Biological Reappraisal* (London: Batsford Academic and Educational, 1984), p. 102.
71 rat count: Ibid., p. 105.
71 "meanest Roman": Edward Gibbon in McLaughlin, *Coprophilia,* p. 7.
72 "To those who are well": Ibid., p. 11.
72 St. Agnes: Ibid.
72 St. Francis: Ibid., p. 7.
72 "civil and mannerly": Ibid., p. 86.
72 "Hi, the fleas": *Stefan's Florilegium,* ed. by Mark Harris, May 19, 1997, stefan@florilegium.org.
72 battle changed: Aberth, *From the Brink of the Apocalypse,* p. 63.
73 village of Coutrai: Clifford J. Rogers, "The Age of the Hundred Years War," in *Medieval Warfare: A History,* ed. by Maurice Keen (Oxford: Oxford University Press, 1999), p. 137.
74 larger armies produced larger concentrations: Aberth, *From the Brink of the Apocalypse,* p. 63.
74 "A castle can hardly be taken": Rogers, "The Age of the Hundred Years War," p. 136.
74 "humble and innocent": Aberth, *From the Brink of the Apocalypse,* p. 84.
75 "dismal devastation": Ibid., p. 86.
75 "Many people [have been] slaughtered": Rogers, "The Age of the Hundred Years War," p. 152.
75 old Soviet army, which fought in Afghanistan: Lieutenant Colonel Lester W. Grant and Major William A. Jorgensen, "Medical Support in a Counter-Guerrilla War: Epidemiologic Lessons Learned in the Soviet-Afghan War," *U.S. Army Medical Department Journal,* May–June 1995, pp. 1–11.
76 plague between 1966 and 1974: Dr. Evgeni Tikhomirov, in *Plague Manual: Epidemiology, Distribution, Surveillance and Control,* ed. by David. T. Dennis and Kenneth L. Gage (Geneva: World Health Organization, 1999), pp. 23, 24.
76 lived in dirt bunkers: L. J. Legters, A. J. Cottingham, and D. H. Hunter, "Clinical and Epidemiologic Notes on a Defined Outbreak of Plague in Vietnam," *American Journal of Tropical Medicine and Hygiene* 19, no. 4 (1970): 639–52.
76 village of Dong Ha: Lieutenant Commander Frederick M. Burkle, Jr., "Plague as Seen in South Vietnamese Children," *Clinical Pediatrics* 12, no. 5 (May 1973): 291–98.

Chapter Four: Sicilian Autumn

80 "sickness clinging to": Philip S. Ziegler, *The Black Death* (New York: Harper & Row, 1969), p. 40.

80 "Speak, Genoa:" de' Mussis, "Historia de Morbo," in *The Black Death: Manchester Medieval Sources,* trans. and ed. by Rosemary Horrox (Manchester: Manchester University Press, 1994), p. 19.

80 "three galleys": Louis Heyligen, "Breve Chronicon Clerici Anonymi," in Horrox, *The Black Death*, p. 42.

80 "full of infected sailors": Giovanni Villani, quoted in Robert S. Gottfried, *The Black Death: Natural and Human Disaster in Medieval Europe* (New York: Free Press, 1983), p. 53.

80 Genoese fleet: Albano Sorbelli, ed., *Corpus Chronicorum Bononiensium,* RIS, XVIII/I, 2 vol. (Città di Castello: 1910–38), Chronica B., p. 584.

81 scribe who estimated: Gottfried, *The Black Death,* p. 38.

81 "Men inhumanely": C. S. Bartsocas, "Two Fourteenth-Century Greek Descriptions of the Black Death," *Journal of the History of Medicine and Allied Sciences* 21, no. 4 (Oct. 1966): 394–95.

81 "Upon arrival": Ibid., p. 395.

81 *Y. pestis* followed the trade routes: Michael W. Dols, *The Black Death in the Middle East* (Princeton: Princeton University Press, 1977), pp. 36–39.

82 island immediately rose up: Hecker, *Epidemics of the Middle Ages,* p. 13.

82 "Ships were dashed": Ibid.

82 "pestiferous wind": Ibid.

83 "future tense of verbs": Leonardo Sciascia, *La Sicile comme métaphore* (Paris: Editions Stock, 1979), p. 53.

83 "In October 1347": Michele da Piazza, "Bibliotheca Scriptorum qui res in Sicilica getas sub aragonum imperio retulere," excerpted in Horrox, *The Black Death,* p. 36.

84 "disease in their bodies": Jean-Noël Biraben, *Les hommes et la peste en France et dans les pays européens et méditerranéens,* vol. 1 (Paris: Mouton, 1975), pp. 49–55.

84 "a sort of boil": Ibid.

84 voyage to Italy: Mark Wheelis, "Biological Warfare at the 1346 Siege of Caffa," *Emerging Infectious Diseases* 8, no. 9 (2002): 974–75.

85 "The disease bred": da Piazza, "Bibliotheca Scriptorum," in Horrox, *The Black Death,* p. 36.

85 "Cats and . . . livestock": Ibid.

85 "a black dog": Ibid., p. 38.

85 "earth gaped wide": Ibid., pp. 38–39.

86 "With his friends": Ziegler, *The Black Death,* p. 133.

86 "a man, wanting to make his will": de' Mussis, "Historia de Morbo," in Horrox, *The Black Death,* p. 21.

86 "Don't talk to me": da Piazza, "Bibliotheca Scriptorum," in Horrox, *The Black Death,* p. 39.

86 "stupid idea": Ibid., pp. 38–39.
87 "see him dead": Ibid., p. 37.
87 Duke Giovanni: Ibid., p. 41.
88 third of Sicily: Ziegler, *The Black Death,* p. 62.
88 "nature of a donkey": Quoted in Benjamin Z. Kedar, *Merchants in Crisis: Genoese and Venetian Men of Affairs and the Fourteenth-Century Depression* (New Haven: Yale University Press, 1976), p. 9.
88 "heading toward the Atlantic": Heyligen, "Breve Chronicon Clerici Anonymi," in Horrox, p. 42.
89 ecological upheaval in Italy: J. C. L. Sismondi, *Histoire des Républiques Italiennes du Moyen Age,* vol. 4 (Paris: 1826), p. 11.
89 "severe shortage": in Ziegler, *The Black Death,* p. 44.
89 poisonous gas: Hecker, *Epidemics of the Middle Ages,* p. 14.
89 "true of Italy": Ziegler, *The Black Death,* p. 45.
90 "fine circuit of walls": Anonimo Genovese, in *Poete del Duecento,* vol. 1, ed. by G. Contini (Milan and Naples: 1961), p. 751.
90 reconstruction of the timeline: Biraben, *Les hommes et la peste en France,* pp. 53–55.
90 during this second visit: Ibid.
91 monetary bequest: Steven A. Epstein, *Genoa and the Genoese, 958–1528* (Chapel Hill: University of North Carolina Press, 1996), p. 211.
91 De Benitio and his colleagues: Ibid., pp. 211–12.
92 "Venetians are like pigs": Quoted in Kedar, *Merchants in Crisis: Genoese and Venetian Men of Affairs and the Fourteenth-Century Depression,* p. 9.
92 all-day parade: Martino da Canale, in Eileen Power, *Medieval People* (New York: Harper & Row, 1963), pp. 43–45.
93 instruction from municipal authorities: Mario Brunetti, "Venezia durante la Peste del 1348," *Ateneo Veneto* 32 (1909): 295–96.
94 "*Corpi morti!*": D'Irsay, "Defense Reactions During the Black Death," *Annals of Medical History* 9 (1927) p. 171.
94 five feet deep: Ibid., p. 297.
94 banned *gramaglia*: Ibid.
94 often mentioned as the source: Robert S. Gottfried, *The Black Death: Natural and Human Disaster in Medieval Europe* (New York: Free Press, 1983), p. 48.
95 Ragusa: Francis Aidan Gasquet, *The Black Death of 1348 and 1349* (London: George Bell and Sons, 1908), p. 65.
95 killed about 60 percent: Frederick C. Lane, *Venice, a Maritime Republic* (Baltimore: Johns Hopkins University Press, 1973), p. 169.
95 "rather die here": D'Irsay, "Defense Reactions During the Black Death," p. 174.
95 "At the beginning": *Cronica di Pisa di Ranieri Sardo,* ed. by Ottavio Banti, *Fonti per la Storia d'Italia* 99 (1963).
96 preparations for the coming onslaught: Ann G. Carmichael, *Plague and the Poor in Renaissance Florence* (Cambridge: Cambridge University Press, 1986), p. 99.
96 "bodies . . . shall not be removed": "Gli Ordinamenti Sanitari del Commune di

Pistoia contro la Pestilenza del 1348," in *Archivio Storico Italiano,* ed. by A. Chiappelli, series 4, 20 (1887), pp. 8–12.

96 "it shall be understood": Ibid., pp. 11–12.

96 Gentile da Foligno: Lynn Thorndike, *A History of Magic and Experimental Science,* vol. 3 (New York: Columbia University Press, 1931), pp. 237–39.

97 plague tract: Ibid., p. 243.

97 "unprecedented." Ibid.

97 *studium generale*: William M. Bowsky, "The Impact of the Black Death upon Sienese Government and Society," *Speculum* 39, no. 1 (Jan. 1964): p. 13.

98 In Orvieto: Elizabeth Carpentier, *Une ville devant la peste: Orvieto et la peste noire de 1348* (Paris: 1962), pp. 79–81.

98 thinks 50 percent: Ibid., p. 135.

99 only 29 percent: David Herlihy, "Plague, Population and Social Change in Rural Pistoia, 1201–1430," *Economic History Review* 18, no. 2 (1965): p. 231.

99 In neighboring Bologna: Shona Kelly Wray, "Last Wills in Bologna During the Black Plague," unpublished Ph.D. thesis (Boulder: University of Colorado, 1998), p. 165.

Chapter Five: Villani's Last Sentence

102 "grown to vigor": Giovanni Villani, in Ferdinand Schevill, *History of Florence, from the Founding of the City Through the Renaissance* (New York: Frederick Ungar, 1961), p. 239.

102 "full of infected": Ibid.

102 Villani biography: Louis Green, *Chronicle into History: An Essay on the Interpretation of History in Florentine Fourteenth-Century Chronicles* (Cambridge: Cambridge University Press, 1992), pp. 1–20.

103 "appetite of women": Ibid., p. 13.

103 appetite for disastrous and apocalyptic events: Ibid.

103 "presence of the father": Giovanni Villani, in Schevill, *History of Florence,* p. 222.

103 defaulted on his loans: Giovanni Villani, in Gene A. Brucker, *Florence: The Golden Age, 1138–1737* (Berkeley: University of California Press, 1998), p. 251.

103 "plague was . . . foretold": Giovanni Villani, in Green, *Chronicle into History,* p. 37.

104 "sign of future and great events": Ibid., p. 38.

104 "dry or oily substance": Boccaccio, *Decameron,* trans. G. H. McWilliam (London: Penguin, 1972), p. 6.

104 Clerical deaths: Aliberto B. Falsini, "Firenze dopo il 1349; le Consequenze della Pestra Nera," *Archivo Storico Italiano* 130 (1971): p. 437.

104 "plague lasted": Giovanni Villani in Schevill, *History of Florence,* p. 240.

105 "Dear ladies": Boccaccio, *Decameron,* pp. 14–16.

105 "two pigs": Ibid., p. 6.

106 "dropped dead": Ibid., p. 11.
106 "one citizen avoiding another": Ibid., pp. 8–9.
106 "Countless numbers": Ibid., p. 9.
106 "a practice": Ibid., p. 9.
106 "A great many people": Ibid.
107 "once been the custom": Ibid., pp. 9–10.
107 "rare for bodies": Ibid., p. 10.
108 "Such was the multitude of corpses": Ibid., p. 12.
108 "nothing was more senseless": Giulia Calvi, *Storie di anno di peste . . .* (Milan: Bompiani, 1984), pp. 108–9.
108 "public and heroic event": Caroline Walker Bynum, "Disease and Death in the Middle Ages," *Culture, Medicine and Psychiatry* 9 (1985): 97–102.
109 "Some people": Boccaccio, *Decameron,* p. 7.
109 "a middle course": Ibid., p. 8.
109 "Some again": Ibid.
109 chronicle of another Florentine: Marchione di Coppo Stefani, *Cronica fiorentino,* ed. Niccolo Rodolico, RIS, XXX/1 (Città di Castello, 1903), pp. 229–32.
110 dinner parties: Ibid., p. 230.
110 "They could not": Stefani, pp. 229–32.
111 "cost of things grew": Ibid., p. 231.
111 death rate: Anne G. Carmichael, *Plague and the Poor in Renaissance Florence* (Cambridge: Cambridge University Press, 1986), p. 60.
111 "public order held": Falsini, "Firenze dopo il 1349," p. 439.
111 two buboes: Giovanni Villani, in Schevill, *History of Florence,* p. 240.
111 even chickens being stricken: Stefani, *Cronica florentino,* p. 230.
111 "must not deceive": Quote in Cohn, *The Black Death Transformed,* p. 14.
112 Travel rates: Graham Twigg, *The Black Death: A Biological Reappraisal* (London: Batsford Academic and Educational, 1984), p. 139.
112 plague ward: Cohn, *The Black Death Transformed,* pp. 27–28.
112 "aerial spirit": J. Michon, *Documents inédits sur la grande peste de 1348* (Paris: J.-B. Baillère et Fils, 1860), p. 46.
112 produced symptoms uncommon: Francis Aidan Gasquet, *The Black Death of 1348 and 1349* (London: George Bell and Sons, 1908), pp. 8–9.
112 dead within three days: Giovanni Villani, in Schevill, *History of Florence,* p. 240.
113 never exceeded *3 percent*: Cohn, *The Black Death Transformed,* p. 2.
113 possibly anthrax: Twigg, *Black Death,* pp. 220–21. See also: Susan Scott and Christopher Duncan, *Biology of Plagues: Evidence from Historical Populations* (Cambridge: Cambridge University Press, 2001) pp. 7, 14, 362–63.
113 DNA from *Y. pestis*: Didier Raoult et al., "Molecular Identification by 'Suicide' PCR of Yersinia pestis as the Agent of Medieval Black Death," *Proceedings of the National Academy of Sciences* 97, no. 7 (Nov. 2000): 12800–03.
114 "*La mortalita*": Agnolo di Tura, *Cronaca senese,* ed. Alessandro Lisini and F. Iacometti (Bologna, 1931–1937), p. 555.

114 rough countrymen: "The Impact of the Black Death upon Sienese Government and Society," *Speculum* 29 (1) (January 1964), p. 14.
115 coastal village called Talamone: William Bowsky, *A Medieval Italian Commune: Siena Under the Nine, 1287–1355* (Berkeley: University of California Press, 1981), p. 6.
115 "grown in population": Agnolo di Tura, *Cronaca senese,* p. 413.
115 "enlargement of the city cathedral": Ibid., p. 490.
116 "more beautiful": Ibid., p. 525.
116 gifts Agnolo bought: Bowsky, "Impact of the Black Death," p. 4.
117 so many houses are listed: Ibid., p. 4n.
117 official Sienese reaction: Ibid., pp. 14–15.
117 "Rooms were constructed": Agnolo di Tura, *Cronaca senese,* p. 488.
118 "parts of Siena": Ibid., p. 555.
118 "they put in the same trench": Ibid.
118 "members of a household": Ibid.
118 "some of the dead": Ibid.
119 "*And I, Agnolo*": Ibid.
119 "52,000 persone": Bowsky, "Impact of the Black Death," p. 17.
119 preplague population: Ibid., p. 10.
119 cathedral renovation: Philip Ziegler, *The Black Death* (New York: Harper & Row, 1969), p. 58.
119 Cola di Rienzo: Ferdinand Gregorovius, *History of the City of Rome in the Middle Ages,* trans. by A. Hamilton (Chicago: University of Chicago Press, 1971), pp. 350–55.
120 "Great God": Ibid., p. 306.
120 "It is better to die": Morris Bishop, *Petrarch and His World* (Bloomington: Indiana University Press, 1963), p. 264.
120 X-rated libido: Diana Wood, *Clement VI: The Pontificate and Ideas of an Avignon Pope* (Cambridge: Cambridge University Press, 1989), p. 7.
120 "met a God": Bishop, *Petrarch and His World,* p. 257.
120 population had fallen: Christopher Hibbert, *Rome: The Biography of a City* (New York: Viking Press, 1985), p. 92.
120 pilfered marble: Bishop, *Petrarch and His World,* p. 119.
121 cow pastures: Ibid., p. 122.
121 "no one to govern": Gregorovius, *History of the City of Rome,* p. 245.
122 fantasy version of himself: Ibid., p. 270.
122 "Love and I": Francesco Petrarch, quoted in Bishop, *Petrarch and His World,* p. 152.
122 "You say": Ibid., p. 68.
123 Laura de Sade: Ibid., p. 64.
123 "sanctified conversation": Ibid., p. 257.
123 dressed in full knight's armor: Ibid., p. 259.
123 remarkable oration: Gregorovius, *History of the City of Rome,* p. 274.

124 "O Tribune": Bishop, *Petrarch and His World,* p. 260.
124 dressed in scarlet: Ibid., p. 261.
125 St. Peter's Basilica: Gregorovius, *History of the City of Rome,* p. 250.
125 "Forgive us our trespasses": Ibid., p. 289.
126 dipped it in Colonna's blood: Ibid., p. 308.
126 "A long farewell": Bishop, *Petrarch and His World,* p. 265.
126 "can't go on": de' Mussis, "Historia de Morbo," in *The Black Death: Manchester Medieval Sources,* trans. and ed. by Rosemary Horrox (Manchester: University of Manchester Press, 1994), p. 23.

Chapter Six: The Curse of the Grand Master

128 "Crimes that defile": Malcolm Barber, *The Trial of the Templars* (Cambridge: Cambridge University Press, 1978), p. 45.
128 Geoffroi de Charney: Ibid., p. 3.
128 hauled off to royal prisons: Ibid, p. 45.
128 "a bitter thing": Ibid.
129 the king's peace: Jonathan Sumption, *The Hundred Years War: Trial by Battle,* vol. 1 (Philadelphia: University of Pennsylvania Press, 1991), p. 23.
129 largest treasury: Barbara W. Tuchman, *A Distant Mirror: The Calamitous 14th Century* (New York: Ballantine Books, 1978), p. 42.
129 Gerard de Pasagio: Barber, *Trial of the Templars,* p. 56.
129 Bernard de Vaho: Ibid.
130 Templars were in tatters: Tuchman, *A Distant Mirror,* p. 43.
130 "burned to death": Guillaume de Nangis, "Chronique latine de Guillaume de Nangis de 1113 à 1300, avec les continuations de cette chronique de 1300 à 1368," in *Société de l'histoire de France,* ed. by H. Géraud, vol. 1 (Paris: J. Renouard et Cie, 1843), pp. 402–03.
130 called down a curse: Tuchman, *A Distant Mirror,* p. 44.
130 Villani mentions it: Barber, *Trial of the Templars,* p. 242.
130 "gorged, contented and strong": Jean Froissart, *The Chronicle of J. Froissart,* ed. by S. Luce, trans. by Sir John Bourchier and Lord Berners (London: D. Nutt, 1901–1903), p. 117.
130 "prating Frenchmen": "Summa curiae regis," *Archiv für kunde österreichiche Geschichtsquellen,* vol. 14, ed. by H. Stebbe (Vienna: K. K. Hofund Statsdruckerei, 1855), p. 362.
130 "government of the earth": Jean de Jandun, "Traite des louanges de Paris," in *Paris et ses historiens aux XIVe et XVe siècles; documents et écrits originaux recueillis et commentés par Le Roux de Lincy,* ed. by Le Roux de Lincy (Paris: Imprimerie impériale, 1867), p. 60.
130–31 Description of France: Sumption, *The Hundred Years War,* pp. 12–26.
131 French culture: Ibid., p. 14.
131 Death of Philip the Fair's sons: Tuchman, *A Distant Mirror,* pp. 44–46.

132 population: Daniel Lord Smail, "Mapping Networks and Knowledge in Medieval Marseille, 1337–1362," unpublished Ph.D. thesis (Ann Arbor: University of Michigan, 1994), p. 6.
133 Place des Accoules: Ibid.
133 Madame Gandulfa's wandering drainpipe: Ibid., p. 5.
134 notary named Jacme Aycart: Ibid., p. 53.
134 November 1: Jean-Noël Biraben, *Les hommes et la peste en France et dans les pays européens et méditerranéens,* vol. 1 (Paris: Mouton & Co., 1975), pp. 49–55.
134 "were infected": Heyligen, "Breve Chronicon Clerici Anonymi," in *The Black Death: Manchester Medieval Sources,* trans. and ed. by Rosemary Horrox (Manchester: University of Manchester Press, 1994), p. 42.
135 three galleys: Ibid.
135 "The infection that these galleys": Ibid., p. 15.
135 "four of five": Ibid., p. 43.
135 "unbelievable": Gilles li Muisis, "Receuil des chroniques de Flandre," in Horrox, *The Black Death,* p. 46.
136 Jacme de Podio: Smail, "Mapping Networks and Knowledge in Medieval Marseille," p. 52.
136 man who had spent hours: Ibid., p. 55.
137 "residents accommodated the effects": Daniel Lord Smail, "Accommodating the Plague in Medieval Marseille," *Continuity and Change* 11, no. 1 (1996): p. 12.
137 Jewish moneylender: Ibid., p. 30.
137 "If the plague had": Ibid, p. 13.
138 residents of Toulon: J. Shatzmiller, "Les Juifs de Provence pendant la Peste noire," *Revue des Études Juives* 133 (1974): 457–80.
138 "throw their children": Jean de Venette, "Chronique Latin de Guillaume de Nangis avec les continuations de cette chronique," in Horrox, *The Black Death,* p. 56.
138 "There is no one left": Shatzmiller, "Les Juifs de Provence," p. 471.
139 described in meticulous detail: "Strassburg Urkundenbuch," in Horrox, *The Black Death,* pp. 211–19.
140 interrogation oath: Jacob R. Marcus, *The Jew in the Medieval World: A Source Book, 315–1791* (New York: JPS, 1938), pp. 49–50.
140 "Jewish" tortures: Ibid.
141 every Jew between Bordeaux and Albi: Tuchman, *A Distant Mirror,* p. 41.
141 Mediterranean heritage of tolerance: Shatzmiller, "Les Juifs de Provence," pp. 475–80.
141 seven churches: T. Moore, *Historical Life of Joanna of Sicily, Queen of Naples and Countess of Provence,* vol. 1 (London: Baldwin, Cradock, and Joy, 1824), p. 304.
141 eleven houses of ill repute: Morris Bishop, *Petrarch and His World* (Bloomington: Indiana University Press, 1963), p. 48.
142 powdered remains: Tuchman, *A Distant Mirror,* p. 43.
142 French mistress: Ibid., p. 26.

142 Dispensations: Ibid., p. 27.
143 "The simple fishermen": Iris Origo, *The Merchant of Prato* (New York: Alfred A. Knopf, 1957), p. 8.
143 dinner party Clement V gave: Eugene Müntz, "L'argent et le luxe à la cour pontificate d'Avignon," *Revue des Questions Historiques* 66 (1899): 403.
143 ninety-six tons: Bishop, *Petrarch and His World,* p. 42.
143 country walkabout: Ibid, p. 45.
144 "No sovereign exceeded": F. Moore, *Historical Life of Joanna of Sicily*, p. 365.
144 magnificent papal palace: Diana Wood, *Clement VI: The Pontificate and Ideas of an Avignon Pope* (New York: Cambridge University Press, 1989), pp. 54, 55.
144–45 Operation of palace: Bishop, *Petrarch and His World,* p. 45. See also: Tuchman, *A Distant Mirror,* pp. 27–29.
145 "my predecessors": G. Mollat, *The Popes at Avignon, 1305–1378* (New York: Harper & Row, 1963), p. 38.
145 New England town: Bishop, *Petrarch and His World,* p. 48.
145 "the most dismal": Francesco Petrarch, *Prosa,* ed. by G. Martelloti, P. G. Ricci, and E. Carrara (Milan and Naples: Riccardi, 1955), p. 120.
145 "*Avenio, cum vento*": Bishop, *Petrarch and His World,* p. 48.
145 lack of adequate infrastructure: Ibid., p. 47.
146 "A field full of": St. Birgitta in Tuchman, *A Distant Mirror,* p. 29.
146 "Babylon of West": Ibid.
146 "good looks": Francesco Petrarch, "Letter to Posterity," in *Petrarch: The First Modern Scholar and Man of Letters,* trans. James Harvey Robinson (New York: G. P. Putnam's Sons, 1898), p. 15.
146 "There is no peace": Petrarch in Bishop, *Petrarch and His World,* p. 155.
147 "Watching a woman undress": St. Clair Baddeley, *Queen Joanna I of Naples, Sicily and Jerusalem, Countess of Provence, Forcalquier and Piedmont: An Essay on Her Times* (London: W. Heinemann, 1893), p. 85.
147 de Sades are a prominent Avignon family: Bishop, *Petrarch and His World,* p. 64.
147 "cannot have taken": Ibid., p. 83.
147 Louis Heyligen: *Un ami de Petrarque: Louis Sanctus de Beringen* (Paris/Rome, 1905). See also: Andries Welkenhuysen, "La Peste en Avignon (1348), décrite par un témoin oculaire, Louis Sanctus de Beringen" in *Pascua mediaevalia: Studies voor Prof. Dr. J. M. de Smet,* ed. by R. Lievens et al. (Louvain: 1983), pp. 452–92.
148 Guy de Chauliac: E. Nicaise, *La Grande Chirurgie de Gui de Chauliac* (Paris: Ancienne Librairie Germer Baillière, 1890), introduction. See also: Jordan D. Haller, "Guy de Chauliac and His Chirurgia Magna," *Surgical History* 55 (1964): 337–43.
149 "Plague!": Albert Camus, *The Plague,* trans. by Stuart Gilbert (New York: Vintage, 1991), p. 40.
150 "outlying districts": Ibid., p. 58.

150 "They say": Heyligen, "Breve Chronicon Clerici Anonymi," in Horrox, *The Black Death,* pp. 41, 42.
151 "powders or unguents": Guy de Chauliac, in Anna M. Campbell, *The Black Death and Men of Learning* (New York: Columbia University Press, 1931), p. 3.
151 Avignon's pigs: J. Enselme, "Glosse sur le passage dans la ville Avignon," *Revue Lyonnaise de Médecine* 18, no. 18 (Nov. 1969): p. 702.
151 "small still flame": Camus, *The Plague,* p. 90.
151 "infected lungs": Heyligen, "Breve Chronicon Cleric Anonymi," in Horrox, *The Black Death,* p. 42.
152 "Priests do not": Ibid., pp. 42–44.
152 "Laura": Francesco Petrarch, quoted in R. Crawfurd, *Plague and Pestilence in Literature and Art* (Oxford: Clarendon Press, 1914), pp. 115, 116.
153 "She closed her eyes": Quoted in Bishop, *Petrarch and His World,* p. 275.
153 "attended by 2,000": Heyligen, "Breve Chronicon Clerici Anonymi," in Horrox, *The Black Death,* p. 44.
153 "None of us": Camus, *The Plague,* p. 181.
154 "fair and noble": Moore, *Historical Life of Joanna of Sicily,* p. 302.
154 "Her figure": Ibid., p. 312.
154 wonders of the medieval world: Baddeley, *Queen Joanna I of Naples,* p. 110.
155 "great harlot": Louis of Hungary in Thomas Caldecot Chubb, *The Life of Giovanni Boccaccio* (Port Washington, N.Y.: Kennikat Press, 1969), p. 130.
155 Andreas's murder: Baddeley, *Queen Joanna I of Naples,* pp. 50–52.
156 "in name only": Ibid., p. 43.
156 "she-wolf": Chubb, *Life of Giovanni Boccaccio,* p. 131.
156 "Your former ill faith": Louis of Hungary in Baddeley, *Queen Joanna I of Naples,* p. 61.
156 "quietly triumph": Ibid., p. 85.
157 court assembled: Moore, *Historical Life of Joanna of Sicily,* p. 310.
157 "Queen of Naples": Ibid., p. 315.
157 "came pale and slowly": Baddeley, *Queen Joanna I of Naples,* pp. 88–89. See also: Moore, *Historical Life of Joanna of Sicily,* pp. 309–11.
158 "above suspicion": Moore, *Historical Life of Joanna of Sicily,* p. 313.
158 "beloved daughter": Clement VI in Baddeley, *Queen Joanna I of Naples,* p. 91.
158 purchased Avignon: Ibid., pp. 92–93.
159 "They say that my lord": Heyligen, "Breve Chronicon Clerici Anonymi," in Horrox, *The Black Death: Natural and Human Disaster in Medieval Europe* (New York: Free Press, 1982), p. 45.
159 twenty-four million: Robert S. Gottfried, *The Black Death: Natural and Human Disaster in Medieval Europe* (New York: Free Press, 1983), p. 77.
159 two roaring fires: Philip S. Ziegler, *The Black Death* (New York: Harper & Row, 1969), p. 67.
160 "avoid infamy": Guy de Chauliac, in Campbell, *The Black Death and Men of Learning,* p. 3.

160 "The mortality": Ibid., p. 2.
161 in the 50 percent range: Ziegler, *The Black Death*, p. 66.
161 unacceptably high: Herman Kahn, *On Thermonuclear War* (Princeton: Princeton University Press, 1961), p. 30.

Chapter Seven: The New Galenism

163 In Paris: Cornelius O'Boyle, "Surgical Texts and Social Concepts: Physicians and Surgeons in Paris, c. 1270 to 1430," in *Practical Medicine from Salerno to the Black Death,* ed. by L. Garcia-Ballester, Roger French, Jon Arrizabalaga, and Andrew Cunningham (Cambridge: Cambridge University Press, 1994), p. 158.
164 formal medical schools: Michael McVaugh, "Bedside Manners in the Middle Ages," *Bulletin of the History of Medicine* 71, no. 2 (1997): 203.
164 "within a week": Ibid.
164 "Nowhere a better": Geoffrey Chaucer, "The Physician," in the prologue to *The Canterbury Tales* (New York: Penguin Books, 2003).
165 reinterpretation and expansion: Luis Garcia-Ballester, Introduction, in Garcia-Ballester, et. al., *Practical Medicine from Salerno to the Black Death,* p. 10.
165 William the Englishman: McVaugh, "Bedside Manners in the Middle Ages," p. 204.
165 "Avicenna, Averroes": Chaucer, "The Physician."
166 based on the New Galenism: McVaugh, "Bedside Manners in the Middle Ages," p. 204.
166 licensure: O'Boyle, "Surgical Texts and Social Concepts," pp. 163–64.
166 patients came forward to testify: Pearl Kibre, "The Faculty of Medicine at Paris, Charlatanism, and Unlicensed Medical Practices in the Later Middle Ages," *Bulletin of the History of Medicine* 27, no. 1 (Jan.–Feb. 1953): 9.
167 John of Padua: Ibid., p. 8.
167 medical pecking order: O'Boyle, "Surgical Texts and Social Concepts," p. 163.
168 medical etiquette: McVaugh, "Bedside Manners in the Middle Ages," p. 208.
168 "Your visit means": Ibid., p. 210.
168 "that may do some good": Ibid., p. 214.
168 theory of the four humors: Edward J. Kealey, *Medieval Medicus: A Social History of Anglo-Norman Medicine* (Baltimore: Johns Hopkins University Press, 1981), p. 16.
169 "health is primarily": Mark D. F. Shirley, "The Mediaeval Concept of Medicine," www.durenmar.de/articles/medicine.html, accessed June 26, 2004. See also: *Hippocratic Writings,* ed. by G. E. R. Lloyd (London: Harmondsworth, Penguin, 1978), p. 262.
169 Contagion: Paul Slack, "Responses to Plague," from *In Time of Plague,* Arlen Mack, ed. (New York: New York University Press, 1991), p. 115.
169 "the first cause": "The Report of the Paris Medical Faculty, October 1348," in *The Black Death: Manchester Medieval Sources,* trans. and ed. by Rosemary Horrox (Manchester: Manchester University Press, 1994), pp. 159–60.

170 "many of the vapors": Ibid., p. 161.
170 "seasons have not succeeded": Ibid., pp. 161–62.
170 twenty-four plague tracts: Dominick Palazotto, "The Black Death and Medicine: A Report and Analysis of the Tractaes," unpublished Ph.D. thesis (Lawrence: University of Kansas, 1973), p. 28.
171 "an acute disease": Anna M. Campbell, *The Black Death and Men of Learning* (New York: Columbia University Press, 1931), p. 78.
171 defense against plague: Ibid., pp. 65–66.
172 Ibn Khatimah and his fellow Spanish Arab: Ibid., p. 27.
172 aromatic substances: Ibid., pp. 67, 68.
172 "where bodies have open pores": Bengt Knutsson, "A Little Book for the Pestilence," in Horrox, *The Black Death,* p. 175.
173 antidotes: Campbell, *The Black Death and Men of Learning,* p. 71.
173 good diet: Ibid., pp. 72, 74.
174 "accidents of the soul": Ibid., p. 77.
175 "deed of arms": Barbara Tuchman, *A Distant Mirror: The Calamitous 14th Century* (New York, Ballantine Books, 1978), p. 82.
175 D-day beaches: Jonathan Sumption, *The Hundred Years War: Trial by Battle*, vol. 1 (Philadelphia: University of Pennsylvania Press, 1991), p. 500.
175 "could be seen by anyone": Jean de Venette, *The Chronicle of Jean de Venette,* trans. by Jean Birdsall, ed. by Richard A. Newhall (New York: Columbia University Press, 1953), p. 41.
176 personal combat: Jonathan Sumption, *The Hundred Years War: Trial by Battle,* Vol. 1 (Philadelphia: University of Pennsylvania Press, 1991), p. 519.
176 "My good people": Jean Froissart, *Stories from Froissart,* ed. by Henry Newbolt (New York: Macmillan, 1899), p. 16.
176 "in our neighborhood": Peter Damouzy, excepted in *Histoire Littéraire de la France,* ed. by A. Coville, vol. 37 (Paris: 1938), pp. 325–27.
176 "led to their own cemetery": Norman F. Cantor, *In the Wake of the Plague: The Black Death and the World It Made* (New York: Free Press, 2001), p. 157.
177 "In August": de Venette, *Chronicle of Jean de Venette,* p. 51.
178 Hôtel-Dieu: Michel Félibien, *Histoire de la Ville de Paris,* vol. 1 (Paris: Chez G. Desprez et J. Desessartz, 1725), pp. 380–95.
178 Notre Dame: Ian Robertson, *Paris and Versailles* (New York: Blue Guides, W. W. Norton, 1989), pp. 60–70.
178 university faculty: Campbell, *The Black Death and Men of Learning,* pp. 156–57.
178 Philip VI: Raymond Cazelles, *La Société politique et la crise de la Royauté sous Philippe de Valois* (Paris: Librairie de Agencies, 1958).
179 building fund: M. Mollat, "La Mortalité à Paris," *Moyen Age* 69 (1963): 502–27.
180 "For a considerable period": de Venette, "Chronique Latin de Guillaume de Nangis," in Horrox, *The Black Death,* pp. 55–56.

180 mortality figures into question: Philip Ziegler, *The Black Death* (New York: Harper & Row, 1969) p. 79.
180–181 estimates of other contemporaries: Samuel K. Cohn, Jr., *The Black Death Transformed: Disease and Culture in Early Renaissance Europe* (London: Arnold, 2002), pp. 89, 90.
181 Richard the Scot: Ibid., p. 89.
181 "Nothing like it": de Venette, "Chronique Latin de Guillaume de Nangis," in Horrox, *The Black Death,* p. 55.
181 "they no longer know or care": George Deaux, *The Black Death, 1347* (New York: Weybright & Talley, 1969), p. 71.
182 plague flag: L. Porquet, *La Peste en Normandie* (Vive: 1898), p. 77.
182 "marvelously great": Augustin Thierry, *Recueil des Monuments inédits de l'Histoire du Tiers Etat,* vol. 1, p. 544.
182 "Master Jean Haerlebech": Gilles li Muisis, "Receuil des chroniques de Flandre," in Horrox, *The Black Death,* p. 48.

Chapter Eight: "Days of Death Without Sorrow"

183 "a yeoman": Ranulf Higden, in Maurice Collis, *The Hurling Time* (London: Faber & Faber, 1958), p. 42.
183 "dress in clothes": John of Reading, "Chronica Johannis de Reading et Anonymi Cantuariensis 1346–1367," in *The Black Death: Manchester Medieval Sources,* trans. and ed. by Rosemary Horrox (Manchester: Manchester University Press, 1994), p. 131.
184 Edward II: Michael Prestwich, *The Three Edwards: War and State in England, 1272–1377* (London: Weidenfeld and Nicolson, 1980), pp. 53–99.
184 "Good son": Ibid., p. 113.
184 "new sun": Thomas Walsingham, p. 20; "Rank to Rank," J. Froissart, in Collis, *The Hurling Time,* p. 29.
185 eight million sheep: R. A. Pelham, "The Fourteenth Century," in *An Historical Geography of England Before 1800,* ed. by H. C. Darby (Cambridge: Cambridge University Press, 1951), p. 240.
185 industrial economy: Christopher Dyer, *Making a Living in the Middle Ages* (New Haven, Conn.: Yale University Press, 2002), p. 215. See also R. A. Pelham, "The Fourteenth Century," pp. 249, 258.
185 "no woman": Walsingham, in Collis, *Hurling Times,* p. 40.
185 Frenchman was effeminate: Prestwich, *The Three Edwards,* p. 211.
185 "scarcely a day": Higden, "Polychronicon," in Horrox, *The Black Death,* p. 62.
186 "The life of men": William Zouche, Archbishop of York, *Historical Letters and Papers from the Northern Registers,* in Horrox, *The Black Death,* pp. 111–12.
186 "neighboring kingdom": Ralph of Shrewsbury, "Register of Bishop Ralph of Shrewsbury," in Horrox, *The Black Death,* p. 112.
186 "One news": John Ford, *The Broken Heart* (London, 1633), Act V, scene III.

186 perhaps 50 percent: Dyer, *Making a Living in the Middle Ages,* p. 272.

186 "Waiting among the dead": John Clynn, "Annalieum Hibernae Chronicon," in Horrox, *The Black Death,* p. 84.

187 historical evidence points to Melcombe: Grey Friar's Chronicle, "A Fourteenth Century Chronicle from the Grey Friars at Lynn," *English Historical Review* 72 (1957): 274; Malmesbury Abbey, "Polychronicon," in Horrox, *The Black Death,* p. 63.

187 Description of Melcombe: *Victoria County History, Dorset,* vol. 2 (London: Constable; 1908), p. 123.

188 requisitioning twenty ships: *Victoria County History, Dorset,* vol. 2, p. 186.

188 Six Burghers: Jean le Bel, in Collis, *The Hurling Times,* p. 37.

189 "English ladies": Walsingham, in Collis, *The Hurling Times,* p. 41.

189 "so depopulated": Francis Aidan Gasquet, *The Black Death of 1348 and 1349* (London: George Bell and Sons, 1908), p. 83.

189 "offer a fuller picture": Philip S. Ziegler, *The Black Death* (New York: Harper & Row, 1969), p. 125.

190 plague's initial assault: Ibid., p. 137.

191 Dorset had to fill: Gasquet, *The Black Death of 1348 and 1349,* pp. 90, 91.

191 pointing out the Baiter: Ibid., p. 92.

191 Bridport doubled its normal complement of bailiffs: Ibid.

192 "Cruel death": Knighton, "Chronicon Henrici Knighton," in Horrox, *The Black Death,* p. 77.

192 "The plague raged": Reverend Samuel Seyer, *Memoirs Historical and Topographical of Bristol and Its Neighbourhood; from the Earliest Period Down to the Present Time* (Bristol: 1823, printed for the author by J. M. Gutch, 1821–23 [1825]), p. 143.

192 municipal death rate: Ziegler, *The Black Death,* p. 135.

192 Christmas of 1348: Gasquet, *The Black Death of 1348 and 1349,* p. 195.

193 in Gloucester: Geoffrey le Baker, "Chronicon Galfridi le Baker," in Horrox, *The Black Death,* p. 81.

193 "Wishing, as is our duty": Ralph of Shrewsbury, "Register of Bishop Ralph of Shrewsbury," in Horrox, *The Black Death,* pp. 112–13.

194 half its normal complement of priests: Ziegler, *The Black Death,* p. 128.

194 holy relics: W. M. Ormrod, "The English Government and the Black Death of 1348–9," in *England in the Fourteenth Century: Proceedings of the 1985 Harlaxton Symposium,* ed. by Boydell and Brewer (Woodbridge, Suffolk: Boydell Press, 1986), p. 176.

194 "Certain sons of perdition": Ralph of Shrewsbury, *The Register of Ralph of Shrewsbury, Bishop of Bath and Wells, 1329–1363,* ed. by Thomas Scott Holmes (Somerset Record Society, 1896), p. 596.

195 "go around the parish church": Ibid., p. 598.

195 Tilgarsley: Ziegler, *The Black Death,* p. 139.

195 Woodeaton: Ibid., p. 140.

195 "university is ruined": Anna M. Campbell, *The Black Death and Men of Learning* (New York: Columbia University Press, 1931), p. 162.
195 "ye University of Oxenford": Richard Fitzralph, in F. D. Shrewsbury, *Bubonic Plague in the British Isles* (Cambridge: Cambridge University Press, 1970), p. 81.
195 "houses in the country retired": Anthony Wood, *History and Antiquities of the University of Oxford,* vol. 1 (Oxford: Printed for the editor, 1792–96), p. 449.
196 "In the same year": Knighton, "Chronicon Henrici Knighton," in Horrox, *The Black Death,* p. 77.
196 In Uzbekistan: Samuel K. Cohn, Jr., *The Black Death Transformed: Disease and Culture in Early Renaissance Europe* (London: Arnold, 2002), p. 132.
196 enjoy better health: Prestwich, *The Three Edwards,* p. 137.
197 According to one estimate: Dyer, *Making a Living in the Middle Ages,* p. 272.
197 "Scepter and Crown": Ziegler, *The Black Death,* p. 132.
197 Joan Plantagenet: Norman F. Cantor, *In the Wake of the Plague: The Black Death and the World It Made* (New York: Free Press, 2001), pp. 32–39.
198 On September 2, Princess Joan: Ibid., p. 44.
198 Bishop of Carlisle: Ibid., p. 48.
198 "No fellow human being": "Letter of Edward III," in Horrox, *The Black Death,* p. 250.
199 Southampton: Ziegler, *The Black Death,* p. 138.
199 ecclesiastical death in Southampton: Gasquet, *The Black Death of 1348 and 1349,* p. 131.
199 "A voice has been heard in Rama": Bishop William Edendon, "Vox in Rama," in Horrox, *The Black Death,* pp. 116, 117.
200 James de Grundwell: Gasquet, *The Black Death of 1348 and 1349,* pp. 189, 190.
200 "Supreme pontiff": Ibid., p. 127.
200 attacked a monk: W. L. Woodland, *The Story of Winchester* (London: J. M. Dent & Sons, 1932), p. 114.
201 fined forty pounds: Richard Britnell, "The Black Death in English Towns," *Urban History* 21, part 2 (Oct. 1994): 204.
201 clerical mortality rates: George Gordon Coulton, *Medieval Panorama: The English Scene from Conquest to Reformation* (Cambridge: Cambridge University Press, 1938–39), p. 496.
201 town of Winchester: Josiah Cox Russell, *British Medieval Population* (Albuquerque: University of New Mexico Press, 1948), p. 285.
201 four thousand: Ziegler, *The Black Death,* p. 146.
201 Crawley: Norman Scott Brien Gras and Ethel Culbert Gras, *The Economic and Social History of an English Village (Crawley, Hampshire) A.D. 909–1928* (Cambridge, Mass.: Harvard University Press, 1930), p. 153.
201 contemporary records indicate: Shrewsbury, *Bubonic Plague in the British Isles,* p. 91.
201 "Since the greater part": Edward III, in Ziegler, *The Black Death,* p. 146.

202 "death without sorrow": John of Reading, "Chronica Johannis de Reading," in Horrox, *The Black Death,* p. 74.
202 "With his friends": Ziegler, *The Black Death,* p. 133.
203 crop yields: Dyer, *Making a Living in the Middle Ages,* p. 238.
204 manor rolls for 1348: E. Robo, "The Black Death in the Hundred of Farnham," *English Historical Review* 44, no. 176 (Oct. 1929): 560–72.
204 740 people died: Ibid., p. 562.
204 more than a third: Ibid., p. 571.
205 "A man could have": Knighton, "Chronicon Henrici Knighton," in Horrox, *The Black Death,* p. 78.
205 amounted to 305 pounds: Robo, "The Black Death in the Hundred of Farnham," p. 565.
205 Forty times: Ibid., p. 566.
206 "No workman or laborer": "Chronicle of Cathedral Priory of Rochester," in Horrox, *The Black Death,* p. 78.
206 dairymaid's output: Robo, "The Black Death in the Hundred of Farnham," p. 567.
206 twenty-two shillings: Ibid., p. 568.

Chapter Nine: Heads to the West, Feet to the East

210 "among the noble cities": William FitzStephen, in R. A. Pelham, "The Fourteenth Century," in *Historical Geography of England Before 1800,* ed. by H. C. Darby (Cambridge: Cambridge University Press, 1951), p. 222.
210 Cheapside, London's premier commercial district: A. R. Myers, *London in the Age of Chaucer* (Norman: University of Oklahoma Press, 1972), pp. 17–23. See also: Christopher Dyer, *Making a Living in the Middle Ages: The People of Britain, 850–1520* (New Haven, Conn.: Yale University Press, 2002), pp. 119, 217.
210 "citizens of London": FitzStephen, in Pelham, "The Fourteenth Century," p. 222.
210 John Rykener: David Lorenzo Boyd and Ruth Mazo Karras, "'Ut cum muliere': A Male Transvestite Prostitute in Fourteenth-Century London," in *Premodern Sexualties,* ed. by Louise Fradenburg and Carl Freccero (London: Routledge, 1996), pp. 99–116.
211 "They drive me to death": Myers, *London in the Age of Chaucer,* p. 23.
211 Richard the Raker: Philip S. Ziegler, *The Black Death* (New York: Harper & Row, 1969), p. 154.
211 "city is very much corrupted": B. Lambert, *The History and Survey of London and Its Environs from the Earliest Period to the Present Time*, vol. 1 (London: Printed for T. Hughes and M. Jones by Dewick and Clarke, 1806), p. 241.
211 "To this city": FitzStephen, in Pelham, p. 222.
212 Southwark, a squalid little suburb: Myers, *London in the Age of Chaucer,* pp. 110–11.

212 Edward's initial response: W. M. Ormrod, "The English Government and the Black Death of 1348–1349," in *England in the Fourteenth Century: Proceedings of the 1985 Harlaxton Symposium*, eds. Boydell and Brewer (Woodbridge, Suffolk: Boydell Press, 1986), pp. 175, 176.

213 epidemic spreads eastward: le Baker, "Chronicon Galfridi le Baker," in *The Black Death: Manchester Medieval Sources,* trans. and ed. by Rosemary Horrox (Manchester: Manchester University Press, 1994), p. 81.

213 infected before the surrounding countryside: Ziegler, *The Black Death,* p. 156. See also Horrox, *The Black Death,* p. 10.

213 Descriptions of pestilential London: Thomas Vincent and Daniel Defoe, quoted in "A Curse on All Our Houses," *BBC History Magazine* 5, no. 10 (October 2004): 36.

214 mixture of caskets: Britnell, "Black Death in English Towns," p. 204. See also: Duncan Hawkins, "The Black Death and the New London Cemeteries of 1348," *Antiquity* 54 (1990): 640.

214 Sir Walter Manny: Hawkins, "The Black Death and the New London Cemeteries," p. 637.

214 "grew so powerful": Robert of Avesbury, "Robertus de Avesbury de Gestis Mirabilibus Regis Edwards Text," in Horrox, *The Black Death,* pp. 63–64.

215 "A great plague raging": John Stow, "A Survey of London," in Horrox, *The Black Death,* pp. 266–67.

215 London's overall mortality: Ziegler, *The Black Death,* p. 157–58.

215 egalitarian abandon: Ibid., p. 159.

216 population of the postplague capital: Josiah Russell, *British Medieval Population* (Albuquerque: University of New Mexico Press, 1948), p. 285. See also: Pelham, "The Fourteenth Century," p. 233.

216 "forgetful of their profession": John of Reading, "Chronica Johannis de Reading," in Horrox, *The Black Death,* p. 75.

216 "wasted their goods": Knighton, "Chronicon Henrici Knighton," in Horrox, *The Black Death,* p. 130.

216 "both in the East and West": Ibn Khaldun, in Robert S. Gottfried, *The Black Death: Natural and Human Disaster in Medieval Europe* (New York: Free Press, 1983), p. 41.

216 "Objective studies indicate": Herman Kahn, *On Thermonuclear War* (Princeton: Princeton University Press, 1961), p. 21.

217 "sights that haunt": in Lord Dufferin, "Black Death of Bergen," *Letters from High Latitudes* (London: Oxford University Press, 1910), p. 38.

217 "superficial yet fevered gaiety": James Westfall Thompson, "The Aftermath of the Black Death and the Aftermath of the Great War," *American Journal of Sociology* 26 (1920–21): 23.

217 unpromising environment blossom: Dyer, *Making a Living in the Middle Ages*, p. 167.

218 Norwich: Ibid., p. 190.

219 "year ending 1350": Reverend Augustus Jessop, "The Black Death in East Anglia," in *The Coming of the Friars, and Other Historic Essays* (London: T. Fisher Unwin, 1894), pp. 206–07.
219 "was in the most flourishing state": Ziegler, *The Black Death,* p. 170.
219 "Most . . . of the dwelling places": Francis Aidan Gasquet, *The Black Death of 1348 and 1349* (London: George Bell and Sons, 1908), p. 152.
219 peasants of Conrad Pava: Jessop, "The Black Death in East Anglia," p. 200.
220 twenty-one families: Ibid., p. 201.
220 Emma Goscelin's life: Ibid., p. 202.
220 "threading [through] the filthy alleys": Ibid., p. 219.
220 "Tumbrels discharging": Ibid., p. 211.
221 "times of plague": G. B. Niebuhr, in Ziegler, *The Black Death,* p. 259.
221 One-Day Priest: Jessop, "The Black Death in East Anglia," p. 231.
221 Alice Bakeman: Ibid., p. 232.
221 Catherine Bugsey: Ibid., p. 234.
221 "Whether by chance": Gasquet, *The Black Death of 1348 and 1349,* p. 251.
222 In the city: Hamilton Thompson, "The Registers of John Gynwell, Bishop of Lincoln for the years 1347–1350," *Archeological Journal* 68 (1911): 326.
222 "Almighty God": Ralph of Shrewsbury, *The Register of Ralph of Shrewsbury, Bishop of Bath and Wells, 1329–1363*, ed. Thomas Scott Holmes (Somerset Record Society, 1896), p. 596.
222 "O ye of little faith": Johannes Nohl, *The Black Death: A Chronicle of the Plague Compiled from Contemporary Sources,* trans. by C. H. Clarke (London: 1926), p. 231.
223 "Let us look": Thomas Brinton, Bishop of Rochester, "The Sermons of Thomas Brinton," in Horrox, *The Black Death,* p. 141.
223 "And a good job, too!": Knighton, "Chronicon Henrici Knighton," in Horrox, *The Black Death,* p. 76,
223 "And no wonder": John of Reading, "Chronica Johannis de Reading," in Horrox, *The Black Death,* p. 133.
223 "before the pestilence": Knighton, "Chronicon Henrici Knighton," in Horrox, *The Black Death,* pp. 78, 79.
223 Norwich, where sixty clerks: Gasquet, *The Black Death of 1348 and 1349,* p. 238.
224 "the monastic orders": Ibid., p. 251.
224 "The picture one forms": Ziegler, *The Black Death,* p. 261.
224 population of almost eleven thousand: Russell, *British Medieval Population,* p. 142.
225 "Almighty God": William Zouche, Archbishop of York, "Historical Letters and Papers from the Northern Registers," in Horrox, *The Black Death,* p. 111.
225 Clerical losses: Ziegler, *The Black Death,* p. 181.
225 Abbey of Meaux: Friar Thomas Burton, "Chronica Monasterii de Melsa," in Horrox, *The Black Death,* p. 68.
225 Siamese twins: Ibid., p. 70.

225 seven tax collectors: Ormrod, "English Government and the Black Death," p. 178.
225 "Considering the waste": Ziegler, *The Black Death,* p. 183.
226 ten local parishes: Ibid., p. 184.
226 Wakebridge family's brush with annihilation: Horrox, *The Black Death,* pp. 250, 251.
226 William of Liverpool: Ziegler, *The Black Death,* p. 184.
226 refusing to pay fines: Ibid., p. 185.
227 mad, lone peasant: Ibid., p. 186.
227 "Laughing at their enemies": Knighton, "Chronicon Henrici Knighton," in Horrox, *The Black Death,* p. 78.
227 "Within a short space of time": Ibid.
227 A third of Scotland: Ziegler, *The Black Death,* pp. 190, 191.
227 "We see death": Jeuan Gethin, "The Black Death in England and Wales as Exhibited in Manorial Documents," ed. by W. Rees, *Proceedings of the Royal Society of Medicine* 16, part 2 (1920): 27.
228 Madoc Ap Ririd: Ziegler, *The Black Death,* p. 192.
228 "cut down the English": le Baker, "Chronicon Galfridi le Baker," in Horrox, *The Black Death,* p. 82.
228 Irish historian: Maria Kelly, *A History of the Black Death in Ireland* (Stroud, Glouchestershire: Tempus, 2001), p. 38.
228 plague seems to have landed: Ibid., pp. 21–42.
228 death rate: Ibid., p. 41.
229 "And Here it seems": in Horrox, *The Black Death,* p. 82.

Chapter Ten: God's First Love

232 "anyone suffering the effects": Confession of Barber Surgeon Balavigny, "Strassburg Urkundenbuch," in *The Black Death: Manchester Medieval Sources,* trans. and ed. by Rosemary Horrox (Manchester: Manchester University Press, 1994), p. 214.
232 "on pain of excommunication": Ibid.
232 surgeon's transcript: "Strassburg Urkundenbuch," Ibid., pp. 212–14.
233 "it has been brought": Pope Clement VI, "Bull: Sicut Judeis," in Horrox, *The Black Death,* p. 221.
233 "No human condition": Primo Levi, *Survival in Auschwitz: The Nazi Assault on Humanity,* trans. Stuart Woolf (New York: Collier Books, 1961), p. 23.
234 "this is the spring": Balavigny Confession, "Strassburg Urkundenbuch," in Horrox, *The Black Death,* p. 214.
234 "We Go": Friedrich Heer, *God's First Love: Christians and Jews over Two Thousand Years,* trans. by Geoffrey Skelton (New York: Weybright and Talley, 1967), p. 5.
234 Jewish casualty figures in uprising: Paul Johnson, *A History of the Jews* (New York: Harper & Row, 1987).

234 Dio Cassius and Tacitus: Ibid., p. 148.
234 "a sad people": Ibid., p. 143.
235 Benjamin of Tudela: *The Itinerary of Benjamin of Tudela,* ed. by A. Adler (London: 1840).
235 from eight million: Johnson, *A History of the Jews,* p. 171.
235 Ashkenazi community: *German-Jewish History in Modern Times,* ed. by Michael A. Meyer and Michael Brenner (New York: Columbia University Press, 1996), vol. 1: *Tradition and Enlightenment 1600–1780,* by Mordechai Breuer and Michael Graetz, p. 17.
235 skilled, nonnational elite: Amy Chua, *World on Fire: How Exporting Free Market Democracy Breeds Ethnic Hatred and Global Instability* (New York: Doubleday, 2003).
236 "By virtue of their experience": Breuer and Graetz, *Tradition and Enlightenment,* p. 13.
236 "Jews and other merchants": Ibid.
236 "greatest misfortune": *Letters of Medieval Jewish Traders,* trans. by S. D. Goitein (Princeton: Princeton University Press, 1973), p. 207.
236 "the term 'dark early'": Breuer and Graetz, *Tradition and Enlightenment,* p. 13.
236 Aaron of Lincoln: James Parkes, *The Jew in the Medieval Community* (New York: Hermon Press, 1976).
237 Abraham of Bristol: Heer, *God's First Love,* p. 85.
237 *Adversus Judaeos*: Dan Cohn-Sherbok, *Anti-Semitism: A History* (Stroud, Gloucestershire: Sutton Publishing, 2002), pp. 36–39.
238 Mention of Jews in gospel: Elaine Pagels, *The Origin of Satan* (New York: Vintage, 1995), pp. 99–105.
238 "May the *minim*": Cohn-Sherbok, *Anti-Semitism,* p. 45.
238 orthodox establishment also disparaged Christ: Heer, *God's First Love,* p. 33.
238 "I know that many people": John Chrysostom in James Carroll, *Constantine's Sword* (Boston: Houghton Mifflin, 2001), p. 213.
238 Bodo, father-confessor: Heer, *God's First Love,* p. 60.
239 "Now he lives in": Ibid., p. 60.
239 expelled and their goods confiscated: Johnson, *A History of the Jews,* p. 213.
239 policy of expulsion: Max L. Margolis and Alexander Marx, *A History of the Jewish People* (Philadelphia: Jewish Publication Society, 1927), pp. 399–400.
239 "wild with grief": Augustine of Hippo, *The Confessions,* book 5, ch. 8, trans. by R. S. Pine-Coffin (New York: Penguin, 1961), p. 100.
240 "lovely brainwave": Mendelssohn in Carroll, *Constantine's Sword,* p. 219.
240 "Judaism endured": Neusner in Carroll, *Constantine's Sword,* p. 218.
240 Ashkenazi settlements: Breuer and Graetz, *Tradition and Enlightenment,* p. 15.
241 "We depart to wage war": Ibid., p. 21.
241 "In a great voice": "Chronicle of Solemon bar Simson," *The Jews and the Crusaders: The Hebrew Chronicles of the First and Second Crusades,* trans. and

ed. by Shlomo Eidelberg (Madison: University of Wisconsin Press, 1977), pp. 30–31.

241 "Hear, O Israel": Ibid., pp. 23–24.

241 pretty young Jewess taunted: Carroll, *Constantine's Sword,* p. 247.

242 *Civitas Dei*: Jeremy Cohen, *The Friars and the Jews: The Evolution of Medieval Anti-Judaism* (Ithaca: Cornell University Press, 1982), p. 241.

242 "hateful to Christ": Geoffrey Chaucer, in Philip S. Ziegler, *The Black Death* (New York: Harper & Row, 1969), p. 99.

242 blood libel: Johnson, *A History of the Jews,* p. 209.

243 hemorrhoidal suffering: Ibid., p. 210.

243 "Wherefore the leaders and rabbis": Marc Saperstein, *Moments of Crisis in Jewish-Christian Relations* (Philadelphia: Trinity Press International, 1989), p. 22.

243 "Jews and Saracens": Cohen, *The Friars and the Jews,* p. 266.

243 English Jews: Cohn-Sherbok, *Anti-Semitism,* p. 56.

243 petty degradations: Johnson, *A History of the Jews,* p. 214.

244 "That the Jews were victims": Norman Cantor, *In the Wake of the Plague* (New York: Free Press, 2001), pp. 151–52.

244 "most of our people": Rabbi Solomon, in Cohen, *The Friars and the Jews,* p. 54.

245 "not every Louis": Ibid., p. 70.

245 "have partnerships with Christians": Johnson, *A History of the Jews,* p. 218.

246 "One would be accusing God": Heer, *God's First Love,* p. 68.

246 *The Treasure and the Law*: Rudyard Kipling, in James Parkes, *The Jew in the Medieval Community*, p. 345.

247 lender could charge: Ibid., p. 373.

247 Guillaume, Lord of Drace: Ibid, p. 355.

247 "One should not": Johnson, *A History of the Jews,* p. 174.

247 "Jewish moneylenders": Norman Rufus Colin Cohn, *Pursuit of the Millennium* (Fair Lawn, N.J.: Essential Books, 1957), p. 124.

248 Jewish moneylenders: Johnson, *A History of the Jews,* p. 174.

248 Attacks on Minna and Jacob Tam: Margolis and Marx, *A History of the Jewish People,* p. 366.

248 major outbreaks: Breuer and Graetz, *Tradition and Enlightenment,* p. 24.

249 lepers of France were exterminated: Carlo Ginzburg, *Ecstasies: Deciphering the Witches' Sabbath,* trans. by Raymond Rosenthal (New York: Pantheon, 1991), p. 33.

249 "You see how the healthy Christians": Ibid., p. 41.

249 Pregnant lepers: Ibid., pp. 33, 34.

249 "Beware the friendship": Ibid., p. 38.

250 secret covenant: Ibid., p. 45.

251 Around Narbonne: Ibid., p. 46.

251 Pedro the Ceremonious: David Nirenberg, *Communities of Violence: Persecution*

of Minorities in the Middle Ages (Princeton: Princeton University Press, 1996), p. 237.

252 "some wretched men": Heyligen, "Breve Chronicon Clerici Anonymi," in Horrox, *The Black Death,* p. 45.

252 In Cervera: Nirenberg, *Communities of Violence,* p. 238.

252 "Without any reason": Ibid., pp. 245, 240.

252 "rivers and fountains": Barbara W. Tuchman, *A Distant Mirror: The Calamitous 14th Century* (New York: Ballantine Books, 1978), p. 109.

252 tried for contaminating local wells: Ginzburg, *Ecstasies,* p. 65.

253 "it cannot be true": Pope Clement VI, "Bull: Sicut Judeis," in Horrox, *The Black Death,* p. 222.

253 Marrying the Jews to the well poisonings: Ginzburg, *Ecstasies,* p. 65.

253 "poison which killed the Jews": Margolis and Marx, *A History of the Jewish People,* p. 406.

253 Belieta's interrogation: Belieta, "Strassburg Urkundenbuch," in Horrox, *The Black Death,* p. 215.

254 Aquetus told his interrogators: Aquetus, in Horrox, *The Black Death,* p. 216.

254 turning point in the pogroms: Cantor, *In the Wake of the Plague,* p. 154.

254 "forbids his wife and family": Balavigny Confession, "Strassburg Urkundenbuch," in Horrox, *The Black Death,* p. 213.

254 another plotter: Ibid., pp. 214–19.

255 notion of conspiracy: Ibid., pp. 215–17.

255 "Within the revolution of one year": Heinrich Truchess, "Fontes Rerum Germanicarum," in Horrox, *The Black Death,* p. 208.

256 "The people of Speyer": Cantor, *In the Wake of the Plague,* p. 156.

256 pogroms reached Strassburg: Alfred Haverkamp, "Zur Geschichte Der Juden Im Deutschland Des Spaten Mittelalters Und Der Fruhen Newzeils," in *Judenverflogugen zur Zeil des Schwarzen Todes im Gesellschaftsgefuge deutscher Stadte* (Stuttgart: Hiersmann, 1981), pp. 62–64.

256 half of Strassburg's Jewish population: Cantor, *In the Wake of the Plague,* p. 157.

257 A few weeks later: Johannes Nohl, *The Black Death: A Chronicle of the Plague Compiled from Contemporary Sources,* trans. C. H. Clarke (London: 1926), pp. 184–94.

257 "They were burnt": Truchess, "Fontes Rerum Germanicarum," in Horrox, *The Black Death,* p. 209.

257 "Not a single one": Jizchak Katzenelson, in Heer, *God's First Love,* p. 12.

Chapter Eleven: "O Ye of Little Faith"

259 arrive in force in Central Europe: Jean-Noël Biraben, *Les hommes et la peste en France* (Paris: Mouton, 1975), pp. 75–76.

260 Balkans infected: Ole J. Benedictow, *The Black Death, 1346–1353: Complete History* (Woodbridge, Suffolk: Boydell Press, 2004), p. 72. See Also: Francis

Aidan Gasquet, *The Black Death of 1348–49* (London: George Bell & Sons, 1908), pp. 68–69.

260 condolence to survivors: Gasquet, *The Black Death of 1348–49,* p. 68.

260 Muhldorf misdating: Benedictow, *The Black Death, 1346–1353,* p. 189.

260 Description of plague encirclement of Germany: Ibid., pp. 186–200.

261 Mortality in German cities: Philip S. Ziegler, *The Black Death* (New York: Harper & Row, 1969), pp. 84–86.

262 "small children": Johannes Nohl, *The Black Death: A Chronicle of the Plague Compiled from Contemporary Sources,* trans. C. H. Clarke (London: 1926), pp. 34, 35.

262 "race without a head": Henrici de Hervordia, "Chronicon Henrici de Hervordia," in *The Black Death: Manchester Medieval Sources,* trans. and ed. by Rosemary Horrox (Manchester: Manchester University Press, 1994), p. 150.

262 "naked and covered in blood": Norman Cohn, *Pursuit of the Millennium* (New York: Oxford University Press, 1970), p. 124.

263 "Your hands above": Nohl, *The Black Death,* p. 229.

263 walked in a circle: Ibid., p. 230.

263 position of the particular sin: Ziegler, *The Black Death,* p. 89.

264 "iron spikes": de Hervordia, in Horrox, *The Black Death,* p. 150.

264 "Come here for penance": Nohl, *The Black Death,* p. 228.

264 Heavenly Letter: Ibid., p. 231.

265 enthusiastic self-floggers: Ziegler, *The Black Death,* p. 87.

265 appease divine wrath: Ibid.

265 origins of Flagellants: Cohn, *Pursuit of the Millennium,* pp. 124–30.

265 German wing: Ibid., pp. 127, 129, 137.

266 "gigantic women from Hungary": Nohl, *The Black Death,* p. 227.

266 spiritual home in Germany: Cohn, *Pursuit of the Millennium,* pp. 124–27.

266 "now laughing now weeping": Thomas Walsingham, *Historia Anglicana 1272–1422,* in Horrox, *The Black Death,* p. 154.

266 rules of Flagellant behavior: Cohn, *Pursuit of the Millennium,* p. 132.

267 changing demographic: Ziegler, *The Black Death,* pp. 94–95.

267 in Frankfurt: Ibid., p. 93.

267 In 1349 Strassburg: Cohn, *Pursuit of the Millennium,* p. 130.

267 less enthusiastic: Ibid., pp. 139–40.

268 "Pope took part": Heyligen, "Breve Chronicon Clerici Anonymi," in Horrox, *The Black Death,* p. 44.

268 "Already the Flagellants": Nohl, *The Black Death,* p. 239.

268 King Casimir: Norman F. Cantor, *In the Wake of the Plague* (New York: Free Press, 2001), p. 163.

268 the plague in Spain: Benedictow, *The Black Death, 1346–1353,* pp. 77–82.

269 King Alfonso: Ziegler, *The Black Death,* p. 114.

269 stopped at a little country inn: Li Muisis, "Recueil de Chroniques de Flandres," in Horrox, *The Black Death,* p. 47.

270 the plague in Poland: Benedictow, *The Black Death, 1346–1353*, pp. 218–21.

270 the plague in Bohemia: Ibid., pp. 221, 224.
271 regional data from the Netherlands: W. P. Blockmans, "The Social and Economic Effects of the Plague in the Low Countries, 1349–1500," *Revue Belge de Philologie et d'Histoire* 58, 833–63. See also Benedictow, *The Black Death, 1346–1353,* pp. 203–6.
271 Flanders, which had: David Nicholas, *Medieval Flanders* (London: Longman, 1992), p. 226.

Chapter Twelve: "Only the End of the Beginning"

273–74 Norway infected: Ole J. Benedictow, *The Black Death, 1346–1353: The Complete Story* (Woodbridge, Suffolk: Boydell Press, 2004), pp. 153, 154.
274 "People did not live": Ole J. Benedictow, "Plague in the Late Medieval Nordic Countries," *Epidemiological Studies* (1992): p. 44.
274 point to bubonic plague: Ibid.
274 two recent outbreaks: Wendy Orent, *Plague: the Mysterious Past and Terrifying Future of the World's Most Dangerous Disease* (New York: Free Press, 2004), p. 57.
275 sole surviving cleric in the diocese of Drontheim: Aidan Gasquet, *The Black Death of 1348 and 1349* (London: George Bell & Sons, 1908), p. 77.
275 Rype: Philip S. Ziegler, *The Black Death* (New York: Harper & Row, 1969), p. 112.
275 "struck the world": Gasquet, *The Black Death of 1348 and 1349,* p. 78.
276 reentered Russia: David Herlihy, *Black Death and the Transformation of the West* (Cambridge, Mass.: Harvard University Press, 1997), p. 25.
276 Survivors drank intoxicatingly: Matteo Villani, in Herlihy, *Black Death and the Transformation of the West,* pp. 46–47.
276 "There are three things": Robert S. Gottfried, *The Black Death: Natural and Human Disaster in Medieval Europe* (New York: Free Press, 1983), p. 81.
276 "It was thought": Matteo Villani, in Herlihy, *Black Death and the Transformation of the West,* p. 65.
277 "No one could": Agnolo di Tura, *Cronaca senese*, ed. Alessandro Lisini and F. Iacometti (Bologna, 1931–1937), p. 566.
277 "The life we lead": Petrarch, "Letter from Parma," in *The Black Death: Manchester Medieval Sources,* trans. and ed. by Rosemary Horrox (Manchester: Manchester University Press, 1994), p. 249.
277 "In 1361, a grave pestilence": John of Reading, "Chronica Johannes de Reading," in Horrox, *The Black Death,* p. 86.
278 second pestilence: Gottfried, *The Black Death,* p. 130.
278 "Children's Plague": Knighton, "Chronicon Henrici Knighton," in Horrox, *The Black Death,* p. 85.
278 "a multitude of boys": John Hatcher, *Plague, Population and the English Economy, 1348–1530* (London: Macmillan, 1977), p. 30.

278 Modern scientific opinion: Ibid., p. 59.
278 In the Netherlands: Gottfried, *The Black Death,* p. 133.
278 differed from its predecessor: Orent, *Plague,* pp. 144–45.
279 "The Black Death became": Ibid., p. 138.
279 "There is no doubt": Ibid., p. 138.
279 plague foci: Dr. Ken Gage, Chief Plague Division, U.S. Centers for Disease Control, personal communication.
280 *Piers Plowman*: Hatcher, *Plague, Population and the English Economy,* p. 40.
280 outbreak of smallpox: Christiane Klapisch-Zuber, "Plague and Family Life," in *The New Cambridge Medieval History,* vol. 6 (Cambridge: Cambridge University Press, 2001), p. 13.
280 Statistics on infectious disease: "Plagues," in *The Dictionary of the Middle Ages,* ed. Joseph Strayer (New York: Scribner, 1982), p. 680.
281 "Many a man": Ibid., p. 683.
281 from 30 to 40 percent: M. Levi-Bacci, *A Concise History of World Population* (Oxford: Oxford University Press, 1997), pp. 31, 53.
281 as high as 60 to 75 percent: "Plagues," in *The Dictionary of the Middle Ages,* p. 681.
281 Florence shrank: Anne G. Carmichael, *Plague and the Poor in Renaissance Florence* (Cambridge: Cambridge University Press, 1986), pp. 60, 66.
281 England by perhaps as much: Hatcher, *Plague, Population and the English Economy,* p. 38.
281 Eastern Normandy: David Herlihy, *The Black Death and the Transformation of the West,* ed. Samuel K. Cohn, Jr. (Cambridge, Mass.: Harvard University Press, 1997), p. 35.
281 "women conceived": Jean de Venette, "The Chronicle of Jean de Venette," in Horrox, *The Black Death,* p. 57.
282 would have replaced: Hatcher, *Plague, Population and the English Economy,* p. 40.
282 killed their caregivers: Carmichael, *Plague and the Poor,* pp. 93–94.
282 "birth dearth": Klapisch-Zuber, "Plague and Family Life," pp. 138–42.
282 "life expectancies were": Herlihy, *Black Death and the Transformation of the West,* p. 43.
282 twelve-year-old peasant boy: Christopher Dyer, *Making a Living in the Middle Ages: The People of Britain, 850–1520* (New Haven: Yale University Press, 2002), p. 275.
283 English peer: Klapisch-Zuber, "Plague and Family Life," p. 136.
283 median age on the continent: *Economist,* Nov. 25, 2003, p. 28.
283 Florence had the same percentage: Herlihy, *Black Death and the Transformation of the West,* p. 43.
283 convent of Longchamp: Klapisch-Zuber, "Plague and Family Life," p. 137.
284 "The roof of an old house": Dyer, *Making a Living in the Middle Ages,* p. 265.
284 "serving girls . . . want": Matteo Villani, *Chronica di Matteo Villani,* ed. by I. Moutier, book 1, ch. 5 (Florence: Magheri, 1825), p. 11.

284 "all essentials": Knighton, "Chronicon Henrici Knighton," in Horrox, *The Black Death,* p. 80.
284 "For many have certainly": Herlihy, *Black Death and the Transformation of the West,* p. 41.
284 "No man now alive": Dyer, *Making a Living in the Middle Ages,* p. 266.
285 winners and losers: Ibid., p. 278.
285 "The common people": Matteo Villani, *Chronica di Matteo Villani,* book 1, ch. 4, p. 10.
285 "workers of the land": Gottfried, *The Black Death,* p. 148.
285 often prosperous enough: Ibid., p. 140.
286 "Take this job and shove it": Dyer, *Making a Living in the Middle Ages,* p. 279.
286 Joan Edwaker: Ibid., p. 267.
286 "The world": Ibid., p. 210.
287 "the ruling groups": Ibid., p. 28.
287 there were insurrections: Ziegler, *The Black Death,* p. 275.
287 industrial Europe: Gottfried, *The Black Death,* p. 140.
287 technological innovation: Herlihy, *The Black Death and the Transformation of the West,* pp. 50–51; Gottfried, *The Black Death,* pp. 142–43.
288 innovations in the medical profession: Gottfried, *The Black Death,* pp. 117–223.
289 hospital also began to move: Ibid., p. 121.
289 birth of public health: Ibid., pp. 123–24.
289 theory of contagion: Herlihy, *The Black Death and the Transformation of the West,* p. 72.
289 medieval higher education: Ibid., p. 70.
290 "privatization of Christianity": Norman F. Cantor, *In the Wake of the Plague: The Black Death and the World It Made* (New York: Free Press, 2001), p. 205.
290 "About what can you preach": Henry Charles Lea, *A History of the Inquisition of the Middle Ages,* vol. 1 (New York: Harper & Brothers, 1882–88), p. 290.
291 "Parsons and parish priests": William Langland in Ziegler, *The Black Death,* p. 264.
291 Lollards: Cantor, *In the Wake of the Plague,* p. 207.
291 "no other epoch": Johan Huizinga, *The Waning of the Middle Ages: A Study of the Forms of Life, Thought and Art in France and the Netherlands in the Fourteenth and Fifteenth Centuries* (Mineola, N.Y.: Dover Publications, 1999), p. 12.
292 tomb of Cardinal Jean de Lagrange: John Aberth, *From the Brink of the Apocalypse: Confronting Famine, War, Plague, and Death in the Later Middle Ages* (New York: Routledge, 2001), pp. 230–31.
292 *The Three Living and the Three Dead:* Ibid., pp. 196–205.
293 *Dance of Death*: Ibid., pp. 205–15.
293 "A more diversified economy": Herlihy, *The Black Death and the Transformation of the West,* pp. 50–51.

Afterword: The Plague Deniers

295 Plague Deniers: Cohn, Samuel K., *The Black Death Transformed: Disease and Culture in Early Renaissance Europe* (New York: Oxford University Press, 2003); Graham Twigg, *The Black Death: A Biological Reappraisal* (New York: Schocken, 1985); Susan Scott and Christopher J. Duncan, *The Biology of Plagues: Evidence from Historical Populations* (New York: Cambridge University Press, 2001).
296 Anthrax: Twigg, *The Black Death,* p. 200.
296 Disease X: Cohn, *The Black Death Transformed,* p. 247.
296 hemorrhagic plague: Scott and Duncan, *The Biology of Plagues,* pp. 107–8, 385, 388.
297 Gasquet's symptom list: Francis Aidan Gasquet, *The Black Death of 1348 and 1349* (London: George Bell and Sons, 1908), pp. 8–9.
297 "Breath spread the infection": da Piazza, "Bibliotheca Scriptorum," in *The Black Death: Manchester Medieval Sources,* trans. and ed. by Rosemary Horrox (Manchester: Manchester University Press, 1994), p. 36.
297 "The disease is threefold": Heyligen, "Breve Chronicon Clerici Anonymi," in Horrox, *The Black Death,* pp. 42–43.
298 list of symptoms: Cohn, *The Black Death Transformed,* pp. 41–54; Scott and Duncan, *The Biology of Plagues,* pp. 107–9; Twigg, *The Black Death,* pp. 202–10.
298 describe the bubo differently: Cohn, *The Black Death Transformed,* pp. 64–65.
298 Rat die-offs: Philip Ziegler, *The Black Death* (New York: Harper & Row, 1969), p. 27; Norman F. Cantor, *In the Wake of the Plague: The Black Death and the World It Made* (New York: Free Press, 2001), p. 172.
299 In Vietnam: Wendy Orent, *Plague: The Mysterious Past and Terrifying Future of the World's Most Dangerous Disease* (New York: Free Press, 2004), p. 57.
299 during the Third Pandemic: Jeremy Cohen, *The Friars and the Jews* (Ithaca, N.Y.: Cornell University Press, 1982), p. 176.
299 immune to climatic effects: Scott and Duncan, *The Biology of Plagues,* p. 364.
300 finding DNA: Raoult et al., "Molecular Identification by Suicide 'PCR' of Yersinia pestis," PNAS 97: 12800–12803.
300 marmot version of the cough: Orent, *Plague,* pp. 56–57.
301 the human flea: Ibid., p. 138.
301 evolutionary terms: Dr. Robert R. Brubaker, Professor of Microbiology at Michigan State University, personal communication.
301 "By the late nineteenth century": Ibid.
301 "Indian Plague Commissioners": Anne G. Carmichael, "Plagues and More Plagues," *Early Science and Medicine* 8, no. 3 (2003): 7.
302 nutrition and decent nursing care: Ibid.
302 "fatigue, destitution": Ibid.
303 human flea as a plague vector: Dr. Ken Gage, Chief Plague, U.S. Centers for Disease Control, personal communication.
303 south of the Alps: Cohn, *The Black Death Transformed.*
303 "There may have been": Anne G. Carmichael, personal communication.

ACKNOWLEDGMENTS

I WOULD LIKE TO THANK WILLIAM H. MCNEILL, PROFESSOR emeritus of history at the University of Chicago, and Ann G. Carmichael, associate professor of history at the University of Indiana, for reading the manuscript and offering suggestions. For answering my questions about Marseille, the medieval period, and the Mongol Golden Horde, I would like to thank, respectively, historians Daniel Lord Smail at Fordham University, Robert Lerner at Northwestern University, and Uli Schamiloglu at the University of Wisconsin. For patiently answering my questions about the biology of plague, I owe a debt of gratitude to microbiologists Robert Brubaker at Michigan State University, Robert Perry at the University of Kentucky, Stanley Falkow at Stanford University, Arturo Casadevall at Albert Einstein Medical College, Christopher Wills at the University of California at Davis, and Ken Gage at the U.S. Centers for Disease Control. For information about ecological change in the fourteenth century, M. G. L. Baillie at Queens College, Belfast, Ireland, has been an indispensable source, and for information about the medieval climate, Brian Fagan at the University of California at Santa Barbara was most helpful. I would also like to thank archivist and historian Guy Fringer for first igniting my interest in the Black Death.

This book could not have been written without the help of my research assistants: the incomparable Laurie Sarney, who can find any document or reference, no matter how obscure; my team of graduate students at Columbia University: Ed Reno, R. R. Rozos, and George Fiske, who helped shepherd me through the mysteries of medieval Latin; and Jennifer Jue-Steuck, who turned several cabinets full of ill-filed papers and books into a crisp accounting of more than 800 footnotes. I would also like to thank the staffs at Columbia University's Butler Library and the New York Academy of Medicine for their help and assistance.

This book also could not have been written without the personal support of several individuals, among them Loren Fishman, whose help was vital during the difficult early months of composition, and my cousin Timothy Malloy and his wife, Maureen, and Elizabeth Weller, who provided critical assistance at several points in the project.

There are three people I would particularly like to thank: my agent, Ellen Levine, who believed in this project from the start; my editor at HarperCollins, Marjorie Braman, whose editorial judgment, unflagging support, good humor, and inexhaustible patience made this book possible; and my wife, Sheila Weller Kelly, who suffered through all the agonies of authorship with me and whose constant rereadings of the manuscript and shrewd editorial judgment and suggestions immensely improved the quality of the pages.

All errors and mistakes herein are mine alone.

—John Kelly, August 20, 2004

INDEX

Insights,
Interviews
& More . . .

About the author

About the book

Read on

About the author

Meet John Kelly

© Laura Pedrick

JOHN KELLY grew up in Boston, the only child of a salesman and an administrator for Filene's department store. He passed much of his youth reading military history and drawing comic-book characters. This last obsession would, years later, give him a "deep understanding" of the antihero in the 2003 film *American Splendor.*

"He passed much of his youth reading military history and drawing comic-book characters."

He graduated from Boston University and earned a master's degree from New York University. "Being a young father," he says, "meant giving up grad school pre-Ph.D. That thesis on Andre Gide and the Communist Ethic never got finished."

He pursued a career writing about science and medicine. His articles appeared in a series of medical and then mainstream magazines. Still, though, he carried a torch for his collegiate love, European history, to which he devoted his leisure hours.

He is the author of nine books, including

Three on the Edge: The Stories of Ordinary American Families in Search of a Medical Miracle (Bantam, 1999). Writing *Three on the Edge,* he says, awakened him to "the greater satisfaction of narrative nonfiction." *Publishers Weekly* called the book "compelling, touching . . . rendered without sentiment by an expert storyteller."

The Great Mortality marked his winning combination of profession and pastime—of writing about science and medicine and reading about history. His next book, *A Visitation of Providence,* will tackle the Irish Potato Famine. He is drawn to catastrophe, which, as he puts it, "brings out the extremes of human nature: the greatest cruelty and inhumanity, as well as astonishing acts of selflessness, love, and courage; moments when the social fabric is in tatters, and others when, against all odds, it miraculously holds." He adds, "Writing about such epochal events has spoiled me for other subjects."

He has been an enthusiastic runner for thirty years. He relaxes, it comes as no surprise, by reading history; he lately enjoyed, and would recommend, *Armageddon: The Battle for Germany, 1944–1945,* by Max Hastings (Knopf, 2004).

He lives in New York City and in Berkshire County, Massachusetts, with his wife, writer Sheila Weller. He has a son and daughter and two grandchildren.

"He is drawn to catastrophe, which, as he puts it, 'brings out the extremes of human nature.'"

John Kelly on Writing *The Great Mortality*

AN AUTHOR AND SUBJECT can't spend three or four years together without developing a relationship. Sometimes the two become fast friends; other times, they become like a long-married couple who have come to know each other's tricks and idiosyncrasies all too well. There are books that grow out of a deep love affair between author and subject, and others where the subject is so emotionally draining that, at the book's completion, the author feels battered and depleted. I was in such a state the September morning I arrived at Columbia University's Butler Library to begin research on what would become *The Great Mortality.* Several months earlier, Romy Hochman, an eighteen-year-old cancer patient who had played a major role in my previous book on experimental medicine, *Three on the Edge,* had died. Now here I was on a perfect late-summer day surrounded by students Romy's age, about to enter the realm of death again. Except this time—I thought, walking up the library steps under a lovely September sky—I would be dealing with not death in the singular, but death in the millions: everywhere and at all times; death piling up at the front door, coming through the roof and windows and across the fields. I wasn't sure I had much stomach left for that.

My feelings that morning were further complicated by a strain of ambivalence regarding the period I was about to study. I had taken several courses on the Middle Ages in graduate school and had emerged from those classes feeling no particular affinity for the era. The thirteenth and fourteenth centuries had seemed distant and unresonant to me, and medieval people—with their God-haunted mentality and their almost animal-like docility—seemed a species of alien being. Such "otherness" might be fine for a historian accustomed to moving across the centuries, but I was a writer of science and medicine who was drawn to the Black Death for a very modern reason. In an age of the avian flu, Ebola, and AIDS, I wanted to take an anticipatory glance backward at the greatest pandemic in human history.

This complex of feelings made the first few months of research difficult. If death in the singular is vivid and heartbreaking, then death en mass, I quickly discovered, is numbing and depressing. At first I also found it difficult to relate to the medieval mentality. It seemed claustrophobic, static, hierarchical, and full of either/ors. For medieval man and woman, there was good or bad, God or the Devil, Heaven or Hell, eternal salvation or eternal ▶

"Medieval people—with their God-haunted mentality and their almost animal-like docility—seemed a species of alien being."

John Kelly on Writing *The Great Mortality*
(continued)

damnation, and almost nothing in between except purgatory, where the sinful could expect to spend a few thousand years hanging by their tongues from trees of fire. Coming from a culture where the concept of human progress was part of the fabric of life, it was also strange to enter a world where that concept did not exist—at least, not progress as we understand the term today. For medieval people, ancient Rome represented the pinnacle of human achievement. Thus, to the extent that medieval man and woman had a notion of progress, it involved recapitulating the past, not capturing the future.

However, as the months passed, a subtle alteration in my perspective occurred. The first time I noticed the change was the morning I opened a new collection of source material from the University of Manchester. By chance, my eye happened to fall upon a fragment of chronicle dated November 1347. Previously, looking at a nearly seven-hundred-year-old date had made me feel like an astronomer viewing a distant planet in a far-away galaxy. On this particular morning, I still felt like an astronomer, but now, suddenly, the planet seemed close enough to belong to my own galaxy. This process of telescoping continued until, by the time I was midway through *The Great Mortality,* I was

“I was beginning to confuse past with present. One day when my wife and I were discussing a European trip we had taken, I said, ‘Yes, that was in 1380 wasn’t it?’”

beginning to confuse past with present. One day when my wife and I were discussing a European trip we had taken, I said, "Yes, that was in 1380 wasn't it?" I meant to say "1980."

No doubt, simple acculturation played a role in my mistake. Any historical period will come to life if you spend enough time thinking and writing about it—and I spent nearly five years researching and writing *The Great Mortality.* But even more than acculturation, I think what ultimately made me feel at home in the Middle Ages was the people I met there. A lot has changed since 1347, but not human nature. The fourteenth century had as gaudy a collection of schemers and dreamers, fools and knaves, heroes and villains as the twenty-first.

One of my favorite schemers was a tough, larcenous old Marseille peasant named Jacme de Podio. Jacme lost a son, granddaughter, and daughter-in-law to the plague in January and February of 1348, but as March dawned in Marseille, Jacme was consumed not with grief, but with greed. As his daughter-in-law's sole surviving heir, the old man was determined to get his hands on her estate, but to do that he had to overcome a large hurdle. The court would not release the estate until it received evidence that the daughter-in-law and her two ▸

John Kelly on Writing *The Great Mortality*
(continued)

immediate heirs—Jacme's son and granddaughter—were dead. So in March of 1348, the worst month of the plague in Marseille, the old man spent weeks walking the streets of his deceased son and daughter-in-law's neighborhood looking for witnesses who would testify to their deaths. And such was Jacme's persistence that he managed to find two.

Another schemer who left a deep impression on me was Queen Joanna of Naples and Sicily. One of the most glamorous women of the Middle Ages, the beautiful young Joanna could have given old Jacme a lesson or two in duplicity. One night, a few years before the Black Death struck Europe, Joanna's husband, Prince Andreas of Hungary, was found dangling from a hangman's rope in the Neapolitan moonlight. Thereafter, Joanna had a series of adventures that would seem improbable for the heroine of a Harlequin Romance. These included the invasion of her domains by angry Hungarian in-laws who believed Joanna was complicit in Andreas's death; a midnight escape to Provence, where the young queen reunited with her lover, the glamorous and handsome Luigi of Taratino; and then, in the midst of the plague, Joanna's trial for murder, an event which featured a mix of sex and celebrity every bit as heady as the O. J. Simpson trial and

"[Queen Joanna's trial for murder] featured a mix of sex and celebrity every bit as heady as the O. J. Simpson trial."

ended in what some would call an equally improbable verdict—innocent.

However, it was an Everyman from Siena named Agnolo di Tura who made me feel most at home in the Middle Ages. A devout family man who worked two and three jobs to support his wife and five children, Agnolo felt like someone who could have come from my old neighborhood in Boston. I particularly liked the pride Agnolo took in his enormous bulk—he often signed his correspondence "Agnolo di Tura del Grasso" (Agnolo the Fat). I also admired his drive. In a society where class and gender pigeonholed people from the moment of birth, Agnolo was that exceptional thing—a striver, and as far as can be determined, a successful striver. In the years before the Black Death, Agnolo, who seems to have been born into the medieval equivalent of a working- or lower-middle-class family, managed to achieve a middle-, and, perhaps even an upper-middle-class life.

What I most admired about Agnolo, however, was his Everyman's eloquence. When the Black Death destroyed everything he cared about, this plainspokenness would make Agnolo the author of what, in my opinion, is the single most haunting sentence in all the literature of the Black Death: "And I, Agnolo di Tura, called the fat, buried my wife and five children with my own hands." ▸

> "The Black Death caused death and destruction on an unimaginable scale, yet nowhere did it succeed in shattering the fabric of civilization."

John Kelly on Writing *The Great Mortality*
(continued)

All the way to the end of the book, writing about mass death continued to depress me, but as I followed the plague across Europe, I found myself deriving solace from humanity's seemingly inexhaustible capacity for resilience. The Black Death caused death and destruction on an unimaginable scale, yet nowhere did it succeed in shattering the fabric of civilization. There were, it is true, countless instances of plague-driven panic and abandonment, of cowardice and greed, but in every afflicted town and village a cadre of men and women always stepped forward to begin the necessary business of recovery: to bury the dead, to nurture the sick, and—equally critical at a time when half the population of a city or region could vanish in a matter of months—to insure a reasonably orderly transfer of property from the dead to the living.

If the Black Death was a great disaster for humanity, it was a great triumph for the human spirit.

Strange Parallels

The Black Death—the Indian Ocean Tsunami

HISTORY, we are told, does not repeat itself, but watching the news reports from South Asia over the holidays of December 2004, I was struck by the many parallels between the tsunami which ravaged the region the day after Christmas and the Black Death. Both disasters arrived on a morning tide, struck without warning under a lovely tropical sky, and lead to grand improvisations of rescue, mourning, and reckoning that tested human character and civil polity. Both catastrophes also produced mortality rates far in excess of anything contemporaries had experienced, indeed, had dreamed possible.

However, in modern South Asia as in medieval Europe, wealth and privilege played a role in deciding who lived and who died. In both Phuket, Thailand, and Ahangama, Sri Lanka, affluent Western tourists received immediate care in clean, well-equipped medical facilities while tens of thousands of peasants were left untended for days in open-air relief stations where the smell of putrefying flesh made the air unbreathable and a burning tropical sun made cooking utensils and surgical ▸

“Both disasters arrived on a morning tide [and] struck without warning under a lovely tropical sky.”

instruments hot to the touch. Though even great wealth could not buy a medieval European competent medical care, it did provide them with something as valuable—the option of flight to a plague-free refuge. "Dear ladies," says Pampinea, one of the young aristocrats in the *Decameron,* Giovanni Boccaccio's fable of Black Death Florence, "Here we linger for no purpose [other] than to count the number of corpses. If this be so and we plainly perceive that it is [why don't we instead] go and stay together in one of our various country estates." Still, in itself, even privilege was no guarantee of survival in either disaster. The jet-skiing grandson of Thailand's King Bhumibol was engulfed by one of the killer waves that came ashore at Phuket; the beautiful teenage daughter of England's King Edward III was felled by the Black Death en route to Spain for her wedding. "No fellow human being could be surprised if we were inwardly desolated by the sting of grief," wrote a mournful Edward, "for we are human, too."

> "The jet-skiing grandson of Thailand's King Bhumibol was engulfed by one of the killer waves; the beautiful teenage daughter of England's King Edward III was felled by the Black Death."

Echoes of the tsunami's instant ravagement—the survivors' tales of turning and suddenly finding a child, a spouse, a parent swallowed by a turbulent, debris-filled sea—can also be heard in the literature of the Black Death, most notably in a letter by the poet Petrarch. "Where are our dear friends now?" he wrote to a friend toward the end of the Black Death. "What lightning bolt devoured them? . . . What earthquake toppled them? What tempest drowned them? . . . There was a crowd of us; now we are almost alone."

Another link between the two tragedies is higher female mortality rates. And in both cases, they seemed to have been higher for the same reason—women were more likely to be at home when disaster struck. In Black Death Europe, a woman's housebound existence increased her exposure to plague-bearing rodent fleas; in contemporary coastal South Asia, it meant being near the sea when the tsunami struck rather than in inland hills where the men went to tend their farms.

Though science has long known what causes tsunamis—undersea earthquakes—modern South Asians were as quick as medieval Europeans to interpret the disaster that struck them as a sign of divine displeasure with a wicked humanity. A notable medieval proponent of the "wicked humanity" school was the English monk Henry Knighton, who was sure that God had killed a third or more of Europe because many of England's well-born young women had become tournament groupies. "Whenever and wherever tournaments were held," Knighton wrote a few decades after the Black Death, "a troupe of ladies would turn up . . . mounted on charges, ▸

"Modern South Asians were as quick as medieval Europeans to interpret the disaster that struck them as a sign of divine displeasure with a wicked humanity."

Strange Parallels ***(continued)***

[wearing] thick belts studded with gold and silver slung across their hips . . . deaf to the demands of modesty." Knighton's modern counterpart was the imam of tsunami-devastated Meulaboh, Indonesia, who interpreted the disaster as God's punishment on townsfolk. "Some Muslim people," he declared, "celebrated Christmas, they drank alcohol, and they danced on the seashore in violation of the Muslim way. This was a big mistake."

Both catastrophes also left behind images "that haunt the soul forever / Poisoning life till life is done" ("The Black Death of Bergen," by Lord Dufferin). After a tour of mass graves at Meulaboh, a *Washington Post* reporter wrote, "Even by the standards of carnage inflicted elsewhere by the catastrophe, what happened here will evoke horror and amazement for generations to come." A Black Death chronicler named Marchione di Coppo Stefani left a similarly memorable account of his encounter with mass death. Describing how people were buried in Florence's municipal plague pits, Stefani wrote, the dead are laid out "layer upon layer just like one puts layers of cheese on lasagna." Accounts of animals feasting on human remains on Indonesian beaches and bodies decaying under an intense Thai sun also could have been plucked whole from almost any page of the Black Death literature. Still, both then and now there was a touchingly stubborn insistence on decent burial: the patient, hopeful teams of DNA decoders in Phuket harken back to medieval London's mandate that the plague dead be sorted by age and gender and buried in neat rows, with heads to the west and feet to the east.

Finally, both tragedies featured luminous moments of selflessness and heroism. The aftermath of the tsunami disaster saw a mother and her son, Julie and Casey Sobolewski of California, rescue survivors in dramatic fashion. Sailing off the Thai coast, the Sobolewskis pulled aboard survivors trapped in the undertow; Casey, moreover, jumped into his lifeboat to rescue floundering children. Another hero, Dr. Ruvan Samarasinghe of Sri Lanka, steadfastly finished a Cesarean delivery while waves crashed against the hospital walls; after the procedure, he led the mother and her newborn to safety. These are modern counterparts of Antonio de Benito, Giudotto de Bracelli, and Domenico Tarrighi, three Genoese notaries who unwaveringly made out wills and legal documents rather than abandon their crucial task. Equally gallant was a woman named Simonia who, ignoring danger to herself, nursed her friend Aminigina through the final days of a bitter plague death, changing her ▸

> "Both tragedies featured luminous moments of selflessness and heroism."

Strange Parallels *(continued)*

soiled nightshirts, wiping her mouth, and holding her hand when she cried.

In a catastrophic tragedy, the barriers of time and culture fade, revealing the fundamental character of humanity.